Advances in Experimental Medicine and Biology

Volume 1037

More information about this series at http://www.springer.com/series/5584

Hiroyoshi Ariga • Sanae M. M. Iguchi-Ariga
Editors

DJ-1/PARK7 Protein

Parkinson's Disease, Cancer and Oxidative Stress-Induced Diseases

Springer

Editors
Hiroyoshi Ariga
Faculty of Pharmaceutical Sciences
Hokkaido University
Sapporo, Japan

Sanae M. M. Iguchi-Ariga
Faculty of Agriculture
Hokkaido University
Sapporo, Japan

ISSN 0065-2598 ISSN 2214-8019 (electronic)
Advances in Experimental Medicine and Biology
ISBN 978-981-10-6582-8 ISBN 978-981-10-6583-5 (eBook)
https://doi.org/10.1007/978-981-10-6583-5

Library of Congress Control Number: 2017958141

Printed on acid-free paper

This Springer imprint is published by Springer Nature
The registered company is Springer Nature Singapore Pte Ltd.
The registered company address is: 152 Beach Road, #21-01/04 Gateway East, Singapore 189721, Singapore

Preface

As society ages, there is a subsequent increase in the number and onsets of age-related diseases, including cancer and neurodegenerative conditions such as Alzheimer's disease and Parkinson's disease (PD), as well as of lifestyle-related diseases such as obesity and diabetes, especially in developed countries. Their increase is giving rise to economic pressure by government insurance regimes against the expanding cost of medical care. It is therefore necessary to elucidate mechanisms of the onset and pathogenesis of these diseases and to develop methods for their diagnosis and cure. Because the onsets of cancer and neurodegenerative disease occur by abnormal cell growth and neuronal cell death, respectively, it has long been considered that both diseases are in different classes. Although both are sporadic, about 10% of them are genetic, and analyses of such familial forms of gene products have contributed to an understanding of molecular mechanisms underlying their onsets and pathogenesis.

The *DJ-1* gene was first found to be a novel oncogene in 1997 and later, in 2003, also found to be a causative gene for a familial form of PD, park7. The *DJ-1* gene is therefore the first gene discovered that is known to cause cancer and neurodegenerative diseases, including PD. After identification of *DJ-1* as *park7*, DJ-1 research has developed extensively, and approximately 1300 papers regarding DJ-1 have been published. DJ-1 protein plays multiple roles and explores an anti-oxidative stress function by using its various functions, as loss or reduction of those functions leads to onsets of oxidative stress-related diseases, including cancer, neurodegenerative diseases, type 2 diabetes, and male infertility (Introduction in Chap. 1).

This book comprises topics covering the structural biology of the DJ-1 superfamily (Chap. 2), the clinical significance of DJ-1 in neurodegenerative diseases (Chap. 3), cancer (Chap. 4), fertilization (Chap. 5) and diabetes mellitus (Chap. 11), basic functions of DJ-1 concerning anti-oxidative reaction (Chap. 6), transcriptional regulation (Chap. 7), regulation of signal transduction (Chap. 8) and protein repair reaction (Chap. 9), a biomarker of Parkinson's disease (Chap. 12), and therapeutic application for PD (Chap. 12) and cancer (Chap. 13). The reader will discover that DJ-1 is a promising protein both for basic cell biology and for the mechanism and therapy for oxidative stress-related diseases.

We hope that this work will provide timely reviews of DJ-1 that will stimulate further progress in this field and contribute to medical treatment of oxidative stress-related diseases.

We express our great thanks to all of the scientists who contributed to this book, and we also wish to thank Dr. Yasutaka Okazaki and Ms. Momoko Asawa of Springer Japan for their patient help and proficient editing.

Sapporo, Japan Hiroyoshi Ariga
August 2017 Sanae M. M. Iguchi-Ariga

Contents

Chapter 1
Introduction/Overview

Hiroyoshi Ariga and Sanae M. M. Iguchi-Ariga

Abstract The *DJ-1* gene is an oncogene and also causative gene for a familial form of Parkinson disease. Although exits of cancer and neurodegenerative diseases, including Parkinson disease, are completely opposite, there are some common points of view between both diseases, including growth and death signaling pathways, and oxidative stresses affect the onset and pathogenesis of both cancer and neurodegenerative diseases. DJ-1 has versatile functions and plays a role in protection against oxidative stress. Inactivation and/or excess activation of DJ-1 functions, therefore, leads to onsets of oxidative stress-related diseases such as type 2 diabetes and male infertility in addition to cancer and neurodegenerative diseases, and studies about DJ-1 will give rise to the common mechanism among these diseases. Furthermore, secreted DJ-1 levels in serum and DJ-1-binding compounds will be a diagnostic biomarker and therapeutic drug for neurodegenerative diseases, respectively.

Keywords DJ-1 • Parkinson disease • Cancer • Neurodegenerative diseases • Oxidative stress • Mitochondria • Signal transduction • Type 2 diabetes • Oncogene • Male infertility • Biomarker • Therapeutic drug

The *DJ-1* gene has been identified as a novel oncogene that transformed cells in corporation with the activated *ras* gene in 1997 (Nagakubo et al. 1997), and overexpression of DJ-1 has been reported in many kinds of cancer cells and tissues, especially in cancers with a high grade of malignancy. The *DJ-1* gene has also been found to be a causative gene for the familial form of Parkinson disease (PD), *park7*,

H. Ariga (✉)
Faculty of Pharmaceutical Sciences, Hokkaido University,
Kita 12, Nishi 6, Kita-ku, Sapporo 060-0812, Japan
e-mail: hiro@pharm.hokudai.ac.jp

S.M.M. Iguchi-Ariga
Faculty of Agriculture, Hokkaido University,
Kita 9, Nishi 9, Kita-ku, Sapporo 060-8589, Japan
e-mail: myan@res.agr.hokudai.ac.jp

H. Ariga, S.M.M. Iguchi-Ariga (eds.), *DJ-1/PARK7 Protein*,
Advances in Experimental Medicine and Biology 1037,
https://doi.org/10.1007/978-981-10-6583-5_1

in 2003 (Bonifati et al. 2003), and mutations of the *DJ-1* gene, including several deletion and substitution mutations, have been found in PD patients.

Structure and functions of DJ-1 have extensively been studied for this decade and showed that DJ-1 has versatile functions (see review and references therein; Bandyopadhyay and Cookson 2004; Kahle et al. 2009; Wilson 2011; Ariga et al. 2013; Pantcheva et al. 2014; Saito 2014; Ariga 2015; Cao et al. 2015). These include transcriptional regulation, quenching reaction oxygen species (ROS), regulation of signal transduction pathways, and roles as chaperone and enzymes, all of which lead to anti-oxidative stress reaction and mitochondrial regulation of DJ-1. Human DJ-1 is comprised of 189 amino acids with three cysteine residues at amino acid numbers 46, 53, and 106 (C43, C54, and C106, respectively) (Nagakubo et al. 1997). Of three cysteine residues, C106 is the most sensitive against oxidative stress and easily oxidized to forms with SOH, SO_2H, and SO_3H (Kinumi et al. 2004; Taira et al. 2004; Canet-Avilés et al. 2004). It is thought that DJ-1 at C106 with reduced SH to SO_2H and at C106 with SO_3H is active and inactive forms of DJ-1, respectively, (Taira et al. 2004; Canet-Avilés et al. 2004; Martinat et al. 2004), and the presence of different proportions between reduced and oxidized C106 DJ-1 was found in patients with cancer and PD (see reviews cited above). These studies indicate that functions of DJ-1 are regulated or determined by the oxidative status of C106.

In addition to participation of DJ-1 in onsets of familial PD and cancer, a number of experimental results suggest that DJ-1 is also related to onset or physiology of several oxidative stress-induced diseases, including stroke (Aleyasin et al. 2007; Yanagisawa et al. 2008), type 2 diabetes mellitus (Jain et al. 2012), male infertility (Okada et al. 2002; An et al. 2011), and various neurodegenerative diseases such as sporadic PD (Bandopadhyay et al. 2004), Alzheimer disease (Choi et al. 2006), Huntington disease (Sajjad et al. 2014), and amyotrophic lateral sclerosis (Lev et al. 2015).

How does DJ-1 contribute to physiology of several oxidative stress-induced diseases? We think roles of DJ-1 in cancer and PD, for instance. ROS are an important player in terms of oxidative stress, and a majority of ROS is produced in mitochondria. ROS have two functions: The low level of ROS works as a second messenger to stimulate the cell cycle movement. The high level of ROS is, however, harmful, attacking nucleic acids, proteins, and lipids, thereby inducing cell death. Since DJ-1 works as an anti-oxidative stress protein using its various functions as described above, DJ-1 is activated by oxidization of its C106 after cells receive oxidative stresses and reduces the ROS level, contributing to maintenance of cell homeostasis. Since the substantia nigra in the middle brain is more susceptible to oxidative stresses due to lesser anti-oxidative stress machineries than other tissues, ROS is accumulated, inducing excessive oxidation of DJ-1, being inactivation of DJ-1, thereby leading to PD. In cancer cells, on the other hand, DJ-1 is overexpressed, increasing the level of active forms of DJ-1, reducing ROS levels, attenuating cell cell-killing activity of ROS, thereby leading to cancer (Fig. 1.1). Further studies of DJ-1 will clarify these models at the molecular level.

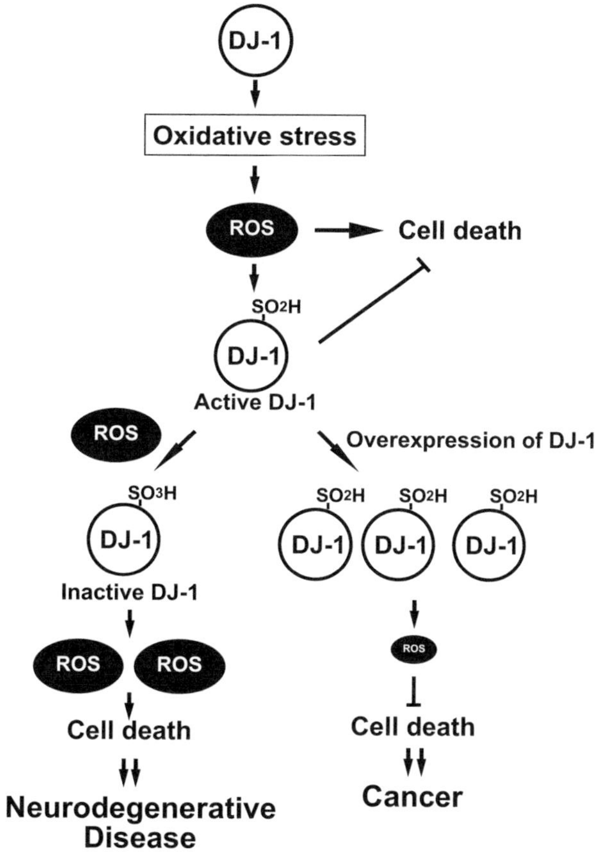

Fig. 1.1 Schematic model of onsets of Parkinson's disease and cancer by DJ-1

In this book, front-runners working in various specific fields described roles of DJ-1 in their targets. DJ-1 is a promising protein both for basic cell biology and for mechanism and therapy toward oxidative stress-related diseases. These articles will contribute to giving ideas for clarification mechanism of cell homeostasis and for development therapeutic drugs against diseases.

References

Aleyasin H, Rousseaux MW, Phillips M, Kim RH, Bland RJ, Callaghan S, Slack RS, During MJ, Mak TW, Park DS (2007) The Parkinson's disease gene DJ-1 is also a key regulator of stroke-induced damage. Proc Natl Acad Sci U S A 104:18748–18753

An CN, Jiang H, Wang Q, Yuan RP, Liu JM, Shi WL, Zhang ZY, XP P (2011) Down-regulation of DJ-1 protein in the ejaculated spermatozoa from Chinese asthenozoospermia patients. Fertil Steril 96:19–23

Ariga H (2015) Common mechanisms of onset of cancer and neurodegenerative diseases. Biol Pharm Bull 38:795–808

Ariga H, Takahashi-Niki K, Kato I, Maita H, Niki T, Iguchi-Ariga SMM (2013) Neuroprotective function of DJ-1 in Parkinson's disease. Oxidative Med Cell Longev 2013:683920

Bandopadhyay R, Kingsbury A, Cookson M, Reid A, Evans I, Hope A, Pittman A, Lashley T, Canet-Aviles R, Miller D, Mc Lendon C, Strand C, Leonard A, Ariga H, Wood N, de Silva R, Hardy J, Holton J, Lees A, Revesz T (2004) The expression of DJ-1 (PARK7) in normal human CNS and idiopathic Parkinson's disease. Brain 127:420–430

Bandyopadhyay S, Cookson MR (2004) Evolutionary and functional relationships within the DJ1 superfamily. BMC Evol Biol 4:6

Bonifati V, Rizzu P, van Baren MJ, Schaap O, Breedveld GJ, Krieger E, Dekker MC, Squitieri F, Ibanez P, Joosse M, van Dongen JW, Vanacore N, van Swieten JC, Brice A, Meco G, van Duijn CM, Oostra BA, Heutink (2003) Mutations in the DJ-1 gene associated with autosomal recessive early-onset parkinsonism. Science 299:256–259

Canet-Avilés RM, Wilson MA, Miller DW, Ahmad R, McLendon C, Bandyopadhyay S, Baptista MJ, Ringe D, Petsko GA, Cookson MR (2004) The Parkinson's disease protein DJ-1 is neuroprotective due to cysteine-sulfinic acid-driven mitochondrial localization. Proc Natl Acad Sci U S A 101:9103–9108

Cao J, Lou S, Ying M, Yang B (2015) DJ-1 as a human oncogene and potential therapeutic target. Biochem Pharmacol 93:241–250

Choi J, Sullards MC, Olzmann JA, Rees HD, Weintraub ST, Bostwick DE, Gearing M, Levey AI, Chin LS, Li L (2006) Oxidative damage of DJ-1 is linked to sporadic Parkinson and Alzheimer diseases. J Biol Chem 281:10816–10824

Jain D, Jain R, Eberhard D, Eglinger J, Bugliani M, Piemonti L, Marchetti P, Lammert E (2012) Age- and diet-dependent requirement of DJ-1 for glucose homeostasis in mice with implications for human type 2 diabetes. J Mol Cell Biol 4:221–230

Kahle PJ, Waak J, Gasser T (2009) DJ-1 and prevention of oxidative stress in Parkinson's disease and other age-related disorders. Free Radic Biol Med 47:1354–1361

Kinumi T, Kimata J, Taira T, Ariga H, Niki E (2004) Cysteine-106 of DJ-1 is the most sensitive cysteine residue to hydrogen peroxide-mediated oxidation in vivo in human umbilical vein endothelial cells. Biochem Biophys Res Commun 317:722–728

Lev N, Barhum Y, Lotan I, Steiner I, Offen D (2015) DJ-1 knockout augments disease severity and shortens survival in a mouse model of ALS. PLoS One 10:e0117190

Martinat C, Shendelman S, Jonason A, Leete T, Beal MF, Yang L, Floss T, Abeliovich A (2004) Sensitivity to oxidative stress in DJ-1-deficient dopamine neurons: an ES- derived cell model of primary parkinsonism. PLoS Biol 2:e327

Nagakubo D, Taira T, Kitaura H, Ikeda M, Tamai K, Iguchi-Ariga SMM, Ariga H (1997) DJ-1, a novel oncogene which transforms mouse NIH3T3 cells in cooperation with ras. Biochem Biophys Res Commun 231:509–513

Okada M, Matsumoto K, Niki T, Taira T, Iguchi-Ariga SMM, Ariga H (2002) DJ-1, a target protein for an endocrine disrupter, participates in the fertilization in mice. Biol Pharm Bull 25:853–856

Pantcheva P, Elias M, Duncan K, Borlongan CV, Tajiri N, Kaneko Y (2014) The role of DJ-1 in the oxidative stress cell death cascade after stroke. Neural Regen Res 9:1430–1433

Saito Y (2014) Oxidized DJ-1 as a possible biomarker of Parkinson's disease. J Clin Biochem Nutr 54:138–144

Sajjad MU, Green EW, Miller-Fleming L, Hands S, Herrera F, Campesan S, Khoshnan A, Outeiro TF, Giorgini F, Wyttenbach A (2014) DJ-1 modulates aggregation and pathogenesis in models of Huntington's disease. Hum Mol Genet 23:755–766

Taira T, Saito Y, Niki T, Iguchi-Ariga SM, Takahashi K, Ariga H (2004) DJ-1 has a role in antioxidative stress to prevent cell death. EMBO Rep 5:213–218

Wilson MA (2011) The role of cysteine oxidation in DJ-1 function and dysfunction. Antioxid Redox Signal 15:111–122

Yanagisawa D, Kitamura Y, Inden M, Takata K, Taniguchi T, Morikawa S, Morita M, Inubushi T, Tooyama I, Taira T, Iguchi-Ariga SM, Akaike A, Ariga H (2008) DJ-1 protects against neurodegeneration caused by focal cerebral ischemia and reperfusion in rats. J Cereb Blood Flow Metab 28:563–578

Chapter 2
Structural Biology of the DJ-1 Superfamily

Nathan Smith and Mark A. Wilson

Abstract The DJ-1 (also called the DJ-1/PfpI, ThiJ/PfpI, or DJ-1/ThiJ/PfpI) superfamily is a structural and functional diverse group of proteins that are present in most organisms. Many of these proteins remain poorly characterized at the biochemical level, but include some known chaperones, proteases, and various stress response proteins that remain mechanistically mysterious. This chapter outlines what is known from a structural perspective about the cellular and biochemical functions of many of these proteins from distinct clades of the superfamily in several organisms. In humans, DJ-1 appears to function primarily as a redox-responsive protein that may act as a sensor for imbalances in cellular redox state. Because mutations in human DJ-1 cause certain types of heritable Parkinson's disease, the role of oxidative posttranslational modifications and pathogenic mutations in human DJ-1 is emphasized in the latter sections of this chapter.

Keywords DJ-1 • X-ray crystallography • Structural biology • Cysteine oxidation • Sulfinic acid • Sulfonic acid • Oxidative stress • Chaperone • Protease • Glyoxalase

2.1 Introduction

Human DJ-1 (PARK7) is a small (20 kDa) protein that belongs to the large, multi-clade DJ-1 (or sometimes DJ-1/PfpI) superfamily. The DJ-1 superfamily is present in most organisms from bacteria to humans, and often several different members of the superfamily can be found in a single species. Human DJ-1 is the most extensively studied member of this superfamily, due to its role in multiple diseases including cancer (Nagakubo et al. 1997), ischemia-reperfusion injury (Wang et al. 2017), and, as indicated by its gene name PARK7, autosomal recessive, early-onset Parkinson's disease (Bonifati et al. 2003). Many studies spanning more than a decade have shown that human DJ-1 and several of its animal homologues lower cellular reactive oxygen species (ROS) levels and help maintain proper

N. Smith • M.A. Wilson (✉)
Department of Biochemistry and the Redox Biology Center, University of Nebraska, Lincoln, NE 68588, USA
e-mail: mwilson13@unl.edu

H. Ariga, S.M.M. Iguchi-Ariga (eds.), *DJ-1/PARK7 Protein*, Advances in Experimental Medicine and Biology 1037,
https://doi.org/10.1007/978-981-10-6583-5_2

mitochondrial function. However, despite extensive study of this protein, the precise biochemical mechanism by which animal DJ-1 accomplishes this is unclear. Human DJ-1 has been proposed to possess several activities: cellular redox sensor, chaperone, RNA-binding protein, glutathione-independent glyoxalase, protease, redox-regulated esterase, and others. Some of these are discussed in other chapters in this volume. Of these, the molecular function with the clearest connection to observed DJ-1 phenotypes is redox sensing (Canet-Aviles et al. 2004; Meulener et al. 2005, 2006; Pardo et al. 2006; Taira et al. 2004; Takahashi-Niki et al. 2004; Blackinton et al. 2009); however, this does not exclude other activities as potentially relevant.

The small size, high solubility, NMR tractability, and proclivity for forming well-diffracting crystals have made human DJ-1 a popular target for structural and biophysical characterization. More broadly, structural studies of the DJ-1 superfamily have helped rationalize the surprisingly divergent functions of several member proteins. Three-dimensional structural information has allowed researchers to elucidate the functions of myriad proteins over the past 60 years. However, an underappreciated facet of high-resolution structural information is that it allows not only the post hoc rationalization of functional data but also for the formulation of new hypotheses concerning molecular function. DJ-1 itself is a prime example of how structural studies can be used to formulate and test new hypotheses about protein function and regulation. Furthermore, the DJ-1 superfamily provides a remarkably rich set of examples that illustrate the ways in which evolution can profoundly alter protein function by relatively parsimonious changes in a conserved protein architecture, which we discuss throughout this chapter.

As of May 2017, 84 structures from the DJ-1 superfamily have been deposited in the Protein Data Bank (PDB), although not all are distinct members (e.g., there are many structures of human DJ-1 with various mutations, etc.). The combination of this wealth of structural information with phylogenetic approaches has identified key similarities and differences between members of the DJ-1 superfamily, thus facilitating their classification into functionally distinct clades (Fig. 2.1). Furthermore, these structural approaches have provided valuable information about the cladistic structure and likely evolution of function in this superfamily (Bandyopadhyay and Cookson 2004; Lucas and Marin 2007; Wei et al. 2007). For example, human DJ-1 is more structurally similar to the prokaryotic YajL-like proteins than it is to other eukaryotic superfamily members, such as the Hsp31 proteins found in microbes or the human protein ES1 (KNP-I/C21orf33). This close structural kinship between DJ-1 and YajL suggests that the study of bacterial YajL-like homologues could provide a window into animal DJ-1 function, as the manipulation of unicellular organisms is typically easier than vertebrates (Bandyopadhyay and Cookson 2004; Lucas and Marin 2007). While YajL has been studied to a limited extent, the potential functional connection between YajL and DJ-1 has been underexplored, and structural data suggests that it is likely to be informative.

In addition, the combined analysis of three-dimensional structures and sequence-based conservation has helped locate features of DJ-1 superfamily proteins that appear to be essential for their function. For example, the nucleophile elbow region between an α-helix and a β-strand contains a highly conserved cysteine residue

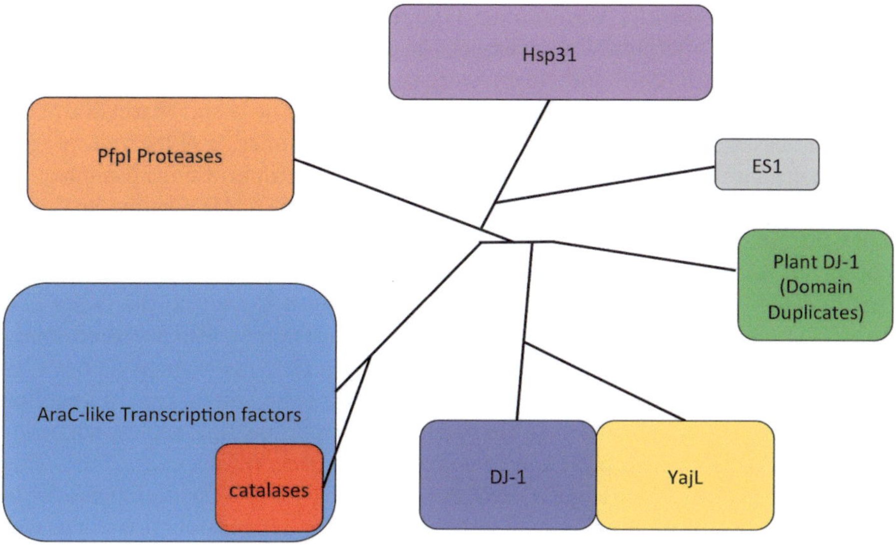

Fig. 2.1 Schematic cladogram of the DJ-1 superfamily. Distinct clades are shown as colored ovals with arbitrary branch length (Modified from Bandyopadhyay and Cookson 2004)

(Cys106 in human DJ-1). This residue is reactive in most superfamily members and readily oxidized in human DJ-1, which appears to be necessary for DJ-1's activity as a redox sensor (Honbou et al. 2003; Huai et al. 2003; Lee et al. 2003; Wilson et al. 2003; Tao and Tong 2003; Lin et al. 2012). The three-dimensional structures of reduced and oxidized DJ-1 have clarified the detailed interactions between the oxidized cysteine and the surrounding conserved residues that could not have been understood solely through studying the sequence. Furthermore, structural studies have elucidated how certain pathogenic mutations associated with parkinsonism cause loss of function through destabilization of DJ-1. Although destabilization seems to be the most common mechanism by which parkinsonian mutations diminish DJ-1 protective function, the molecular basis of this destabilization is intriguingly diverse among these mutations. Considered in total, the current understanding of the DJ-1 superfamily has been informed to a considerable degree by the abundance of structural information now available, which has been used both to form and to test hypotheses about the function of these proteins (Wilson 2011).

2.2 Close Bacterial DJ-1 Homologues

Human DJ-1's role in multiple diseases, coupled with lingering uncertainty about its specific function, has led to the search for homologues in tractable lower organisms for in-depth functional characterization. Of the several clades of bacterial DJ-1 homologues, the family exemplified by YajL from *Escherichia coli* shares the closest sequence identity with human DJ-1 (40%), while other bacterial homologues from the superfamily are members of distinct clades, such as ElbB, YhbO, and Hsp31 (all in *E. coli*) and have lower sequence identity (Bandyopadhyay and Cookson 2004; Quigley et al. 2003; Wilson et al. 2005; Lucas and Marin 2007).

Previously there had been confusion surrounding the function of the YajL protein due to a misannotation that has stubbornly persisted. The roots of this misannotation are in a report that a mutation of YajL (called at that time ThiJ) resulted in *E. coli* that were auxotrophic for thiamine. Additional work showed that the thiamine auxotrophy of this strain was due to a disruption in the neighboring ThiI locus and not in YajL/ThiJ (Mueller et al. 1998). The resulting confusion led to the belief that YajL/ThiJ was an important protein for thiamine production, which would imply that it is not be likely to share a function with human DJ-1, as humans do not synthesize thiamine. The clarification that the YajL/ThiJ gene had no role in thiamine biosynthesis led to the reversion to the systematic gene name YajL and the acknowledgment that the encoded protein is of unknown function (Wilson et al. 2005). This clarification about the YajL gene product has invited study of and comparison to human DJ-1, as these proteins may have more closely related functions than was previously suspected (Wilson et al. 2005).

The crystal structure of YajL was solved in 2005, 2 years after that of the human DJ-1. This structure revealed that YajL is highly similar in structure to human DJ-1, with both proteins containing eight α-helices and eight β-strands in a flavodoxin-like fold, which is common to all members of the DJ-1 superfamily. Remarkably, YajL and human DJ-1 superimpose with a C_α root mean squared deviation (RMSD) of 0.97 Å, indicating that they are nearly identical at the backbone level (Fig. 2.2). Like human DJ-1 and most of the other superfamily members, YajL forms a homodimer with a large (~2500 Å^2) buried interface comprising mainly hydrophobic residues. As the exposure of hydrophobic residues to water would destabilize the protein, it is probable that YajL is only stable in its dimeric form. A minor distinction

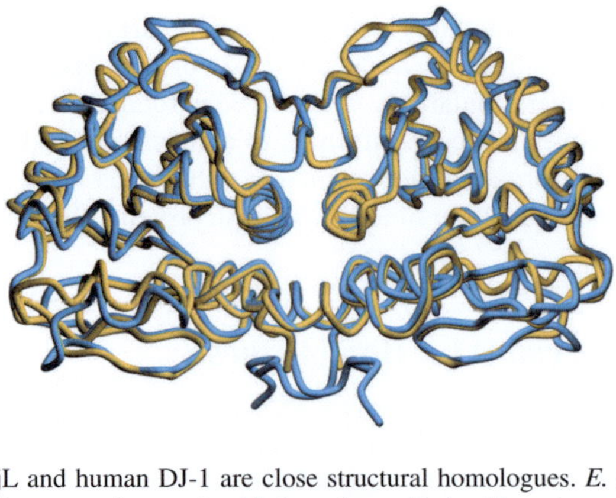

Fig. 2.2 *E. coli* YajL and human DJ-1 are close structural homologues. *E. coli* YajL (blue) and human DJ-1 (yellow) are superimposed, with the polypeptide backbone trace represented as a coil. The excellent superposition of these proteins (0.9 Å RMSD) is evident, emphasizing their relatedness (Reproduced with permission from Wilson et al. 2005)

between YajL and DJ-1 is the Tyr-rich C-terminal extension of YajL, which is found in some other prokaryotes but no known animal DJ-1 homologues. The role of this extension is unknown. This high degree of structural similarity between YajL and DJ-1 suggests the likelihood of some corresponding functional similarity.

All members of the DJ-1 superfamily which have been structurally characterized contain a characteristic "nucleophile elbow": a strand-turn-helix motif which contains a cysteine residue in most members (this residue is Cys106 in human DJ-1). This cysteine falls into the unfavorable region in the Ramachandran plot in nearly all members of the superfamily. Furthermore, this cysteine residue is easily oxidized to the sulfenic (Cys-SOH), sulfinate (Cys-SO_2^-), and sulfonate (Cys-SO_3^-) forms. This high propensity for oxidization had led to DJ-1's proposed role as a redox/oxidative stress sensor, as loss of this Cys oxidation results in a loss of the oxidative stress-resistant phenotype in almost all cases (see below for a discussion of the C106DD/EE mutations). In YajL, the corresponding cysteine residue is oxidized to both the sulfenate and sulfinate in the two molecules in the asymmetric unit of the crystal. The crystal structure of YajL also permits the identification of likely hydrogen bonding partners for stabilization of oxidized Cys106. One of these, Glu17, is important for both cysteine pK_a depression and stabilization of the oxidized Cys residue in human DJ-1 (where it is Glu18) and presumably has similar roles in YajL. Additionally, the backbone amide moieties of Gly75 and Ala107 also stabilize the –SOH and –SO_2^- modifications (Wilson et al. 2005). Separate from its role in stabilizing Cys106 oxidation, Glu17 has also been proposed as a possible partner for the creation of a "catalytic dyad" with Cys106, which may be involved in a reported glutathione-independent glyoxalase activity for YajL and DJ-1 (Abdallah et al. 2016), although this activity is weak and its physiological relevance unclear (Zhao et al. 2014; Hasim et al. 2014).

2.3 Other Prokaryotic DJ-1 Superfamily Members

E. coli contains three other DJ-1 superfamily members, all of which are structurally characterized and one of which has an unambiguously defined function. YhbO, Hsp31 (YedU), and sigma cross-reacting protein 27a (SCRP27a/ElbB) all contain the α/β flavodoxin-like fold, but a variety of insertions into the core fold provide distinct oligomerization interfaces and thus different dimeric quaternary structures (Fig. 2.3). In addition to these four canonical DJ-1 superfamily members in *E. coli*, HPII catalase also contains a domain that possesses the flavodoxin-like glutamine amidotransferase fold and is in DJ-1 superfamily, although the function of the DJ-1-like domain in HPII is unknown (Horvath and Grishin 2001), and we do not further discuss it here. Hsp31 has been shown to have both holdase chaperone and glutathione-independent methylglyoxalase activities (Subedi et al. 2011; Sastry et al. 2002). YhbO was initially thought to be a cysteine protease based on similarity to the archaeal PfpI-like proteases but appears to lack protease activity. It may have a role in the stress response, although the mechanism of its action is unclear

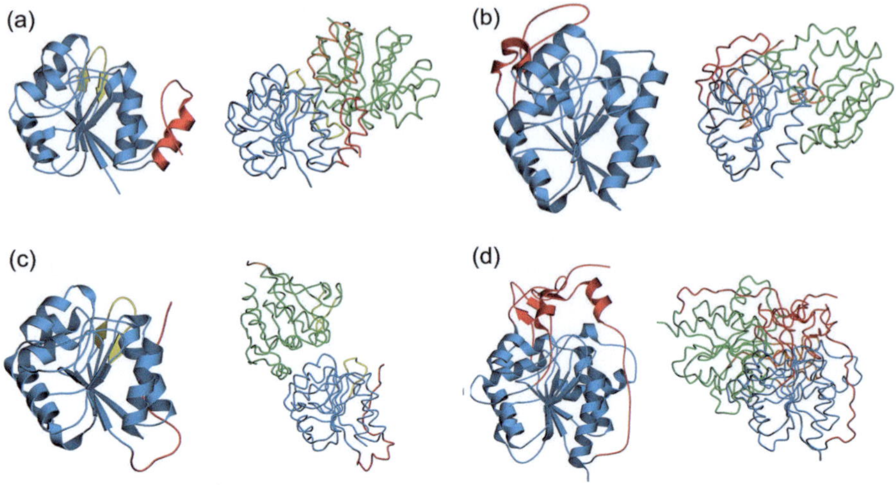

Fig. 2.3 *E. coli* members of the DJ-1 superfamily are structurally diverse at the tertiary and quaternary levels. For each protein, the monomer structure is shown as a ribbon diagram, and the dimer structure is represented as a coil. The conserved elements of the core fold are shown in blue, while the protein-specific insertions are shown in red. For the dimer structures, one monomer is blue (insertions are red), and the other monomer is green (insertions are orange). In all cases, the blue monomer is in the same orientation as in the ribbon diagram. (a) YajL, (b) ElbB/SCRP 27a (PDB code 1VHQ), (c) YhbO (PDB code 1OI4), (d) Hsp31 (PDB code 1 N57). In YajL (a) and YhbO (c), the β-hairpin involved in dimerization is shown in yellow. While not an insertion, the residues in this β-hairpin are in different conformations in other homologues (Reproduced with permission from Wilson et al. 2005)

(Abdallah et al. 2007). SCRP27a/ElbB was identified as a protein that enhances lycopene biosynthesis in engineered *E. coli* and is highly conserved in vertebrates, although its detailed function is also unclear. In vertebrates, the homologous protein (ES1/KNP-I/C21orf33) is a mitochondrial protein that facilitates the formation of mega-mitochondria in cone cells of zebrafish retina (Masuda et al. 2016).

Interestingly, each of the four *E. coli* DJ-1 superfamily members forms dimers, but each uses a different dimeric interface despite having similar core folds (Fig. 2.3). Hsp31 has an inserted helix that allows for the creation of a Cys-His-Glu catalytic triad in its active site as well as an N-terminal extension that participates in many of the dimer interface contacts (Lee et al. 2003; Quigley et al. 2003). Despite their clear differences in structure, study of Hsp31 has been important for advancing the DJ-1 field. Comparison of Hsp31 and human DJ-1 motivated the search for a specific chaperone activity for human DJ-1, although the existence and relevance of this activity in human DJ-1 is uncertain. Hsp31 also has a weak peptidase activity, although the physiological relevance of this activity is unclear (Lee et al. 2003; Malki et al. 2005).

In addition, Hsp31 was identified as the protein responsible for a previously detected but unassigned glutathione-independent glyoxalase (GlxIII or GloIII) activity in *E. coli* (Subedi et al. 2011). Although this activity is detectable for many

DJ-1 homologues (including human DJ-1), it is weak in metazoan DJ-1 homologues, and its physiological relevance for those proteins is uncertain (Zhao et al. 2014; Hasim et al. 2014). In contrast, the methylglyoxalase activity of the Hsp31 clade proteins has been better established and appears to serve as a "backup" glyoxalase that can protect various yeasts (and possibly bacteria) in methylglyoxal-rich conditions (Bankapalli et al. 2015; Hasim et al. 2014; Subedi et al. 2011). There has been confusion about this issue in the literature, as some researchers have conflated the more active Hsp31 proteins with their less active DJ-1 homologues from a different clade of the superfamily, which are distinct proteins with different active sites, structures, and functions. This confusion has been compounded by the identification of several DJ-1 superfamily proteins as deglycases that act on the early adducts of glyoxals with amino acids (Abdallah et al. 2016), which is controversial and may be artifactual (Richarme and Dairou 2017; Pfaff et al. 2017).

YhbO is an *E. coli* representative of the clade that includes PH1704 and PfpI, archaeal proteases that were characterized in the late 1990s to early 2000s (Du et al. 2000). Both YhbO and PH1704 lack specific insertions present in the other DJ-1 superfamily members and are thus closest to a reduced "core" DJ-1 superfamily protein fold (Wilson et al. 2005). However, the quaternary structures of PH1704 and YhbO are drastically different from one another, with PH1704 forming a hexameric ring while YhbO forming only a dimer (Wei et al. 2007). The PH1704 hexamer comprises a trimer of dimers in which the active site containing the active site nucleophile Cys100 is located between two monomers. Although PH1704 exhibits some weak proteolytic activity in lower oligomeric states, it is highly active as a hexamer (Du et al. 2000). The distinct oligomerization states of YhbO and PH1704 were unexpected and highlight the importance of structure elucidation for homologous proteins. These differing quaternary structures point to divergent functions, and YhbO does not possess the robust degradative proteolytic activity of its archaeal homologues. Incidentally, both YhbO and SCRP27a/ElbB are examples of proteins whose structures were solved by structural genomic initiatives but not published (PDB OI41 and 1VHQ, respectively).

2.4 The Yeast Hsp31 Proteins

Yeasts are popular model organisms because of their rapid growth, ease of culture, and single-celled eukaryotic nature. The identification of DJ-1 superfamily members in yeast was therefore important in the study of the DJ-1 superfamily. The budding yeast *Saccharomyces cerevisiae* contains four Hsp31-clade proteins, the exemplar of which is YDR533C/Hsp31. YDR533C/Hsp31 is a close relative of *E. coli* Hsp31, as confirmed by elucidation of its structure (Wilson et al. 2004; Graille et al. 2004). The three additional homologues of Hsp31 in *S. cerevisiae* are YPL280W/Hsp32, YOR391C/Hsp33, and YMR322C/Hsp34. These are 99% identical to each other and ~70% identical to YDR533C. This 99% similarity likely occurred via gene duplication followed by recombination into a sub-telomeric region. Interestingly, these homologues respond somewhat differently to stress than does YDR533C/Hsp31. YDR533C is upregulated in response to heat stress, accumulation of misfolded proteins, and oxidative stress, while the other three are more responsive to nitrogen starvation (Miller-Fleming et al. 2014; Trotter et al. 2002). Recent work has suggested that these proteins are involved in autophagy, survival in stationary phase, protein quality control, and defense against reactive α-ketoaldehydes and other glycolytic metabolites (Hasim et al. 2014; Miller-Fleming et al. 2014; Natkanska et al. 2017).

YDR533C/Hsp31 contains a catalytic triad that is reminiscent of that found in the proteases PfpI and PH1704. A conserved His that immediately follows the catalytic Cys is a distinguishing feature of the catalytic triad-containing members of the DJ-1 superfamily populating the Hsp31 and PfpI clades, but is absent in other clades of the superfamily. Similar to *E. coli* Hsp31, YDR533C contains an additional domain (P domain) that is also characteristic of the Hsp31 clade. The P domain caps the core flavodoxin-like domain and contributes the acidic amino acid (Glu170 in *S. cerevisiae* Hsp31) that completes the triad with Cys138 and His139 (Wilson et al. 2004). A recently characterized glutathione-independent glyoxalase activity that converts methylglyoxal to D-lactate has been measured for Hsp31 proteins from *S. cerevisiae*, *S. pombe*, and *C. albicans*. As discussed above for *E. coli* Hsp31 (*see Other prokaryotic DJ-1 superfamily members*), this activity appears to function as a backup for the more active glyoxalase I system in these organisms.

YDR533C is a dimer both in solution and crystal; however, this dimer has a different interface than other members of the DJ-1 superfamily, including the bacterial Hsp31s, which was surprising. YDR533C/Hsp31 has a modest dimerization interface of ~1100A^2 total buried surface area. The principal difference between the yeast Hsp31 and bacterial Hsp31 dimerization interfaces is the involvement of an N-terminal extension in monomer-monomer interaction that is absent in YDR533C and other yeast Hsp31s. The YDR533C/Hsp31 interface only buries approximately half of the total surface area of the *E. coli* Hsp31 dimer, suggesting the yeast protein forms a less stable dimer (Wilson et al. 2004). Additionally, YDR533C lacks the two His residues required for the Zn^{2+} binding observed in *E. coli* Hsp31, although the functional role of metal binding by *E. coli* Hsp31 is not known.

Despite having differing oligomeric interfaces, structural comparison of YDR533C/Hsp31 and *E. coli* Hsp31 reveals that they share negatively charged clefts on either side of the dimer (Lee et al. 2003). Furthermore, the Cys-138-His139-Glu-170 catalytic triad in YDR533C/Hsp31 is similar to the Cys185-His186-Asp214 triad present in *E. coli* Hsp31. In *S. cerevisiae* YDR533C/Hsp31, these residues form a hydrogen bonding network that also includes Glu30, which donates a hydrogen bond to the S_γ atom of Cys138. Although absent in the more distantly related glutamine aminotransferases, this highly conserved Glu residue is present in DJ-1, Hsp31, PfpI, and most other members of the DJ-1 superfamily (Wilson et al. 2004). The high degree of conservation suggests that this Glu (residue 18 in humans) may be important for the functions of diverse superfamily members. The details of Glu18's peculiar protonation state and contribution to cysteine reactivity were mentioned above and will be discussed in more detail below for human DJ-1, where it has been thoroughly characterized.

Contrary to budding yeast *S. cerevisiae,* which lacks a close homologue of animal DJ-1, the fission yeast *Schizosaccharomyces pombe* contains a DJ-1 superfamily member (SPAC22E12.03c/SpDJ-1) that is highly homologous and structurally similar to human DJ-1 (Section 6) (Madzelan et al. 2012). This homologue is present in addition to the Hsp31 homologues that are also present in *S. pombe*, suggesting that these proteins do not have identical functions. Like human DJ-1, the reactive cysteine in SpDJ-1 is prone to oxidation and was observed exclusively as the Cys-SO_2^- (sulfinate) in the crystal (Madzelan et al. 2012). The presence of a close homologue of animal DJ-1 in *S. pombe* makes this lesser-studied model yeast a potentially powerful system for exploring the evolutionarily conserved functions of DJ-1-like proteins in Eukarya (Lucas and Marin 2007).

One surprising difference between SpDJ-1 and human DJ-1 is the substitution of Pro114 in human DJ-1 with a Thr in the *S. pombe* protein. Structural and biochemical studies show that the hydroxyl group of this Thr donates a hydrogen bond to the backbone carbonyl oxygen atom of Cys 111, depressing the pK_a of the Cys thiol via a hydrogen bond polarization mechanism (Madzelan et al. 2012). This study and others illustrate the value of the DJ-1 superfamily in exploring the structural determinants of cysteine chemistry and reactivity in the context of a conserved core fold.

2.5 DJ-1 Homologues in Animals

Due to the high level of evolutionary conservation of the DJ-1 superfamily, the presence of homologues of human DJ-1 in other animals allows for the study of this protein across multiple metazoan model organisms. Although both vertebrate and invertebrate models have been used, including mouse (Lopert and Patel 2014), rat (Dave et al. 2014), roundworm (Harrington et al. 2010), and fruit flies (*Drosophila melanogaster*) (Meulener et al. 2005), only *Drosophila* DJ-1 homologues have had their three-dimensional structure determined to date.

Interestingly, like many insects, *Drosophila melanogaster* contains two DJ-1 homologues (DJ-1α and DJ-1β). These two proteins have 67% sequence identity but have distinct tissue expression patterns and phenotypes when ablated (Meulener et al. 2005; Park et al. 2005). DJ-1α is found mainly in the testis, while DJ-1β is expressed in many tissues and provides greater protection against oxidative stress in several studies (Park et al. 2005; Meulener et al. 2006). Structurally DJ-1β is very similar to human DJ-1, with most differences confined to solvent-exposed regions that are intrinsically mobile in the human protein (e.g., residues 58–64 and 125–134). Similar to most other members of the DJ-1 superfamily, DJ-1β contains a Cys residue (Cys104) at the nucleophile elbow that is in the unfavorable region of the Ramachandran plot and is readily oxidized (Lin et al. 2012). Both *Drosophila* and human DJ-1 are markedly stabilized by the formation of the Cys-SO_2^- modification (ΔT_m ~ +12 °C), likely due in part to hydrogen bonding to the nearby Glu residue

(Glu17 in DJ-1β). This oxidative stabilization may play a role in regulating DJ-1 function in oxidative stress conditions, particularly in light of the probable destabilization of DJ-1 by further oxidation to the $-SO_3^-$ form (Lin et al. 2012).

The crystal structure of *Drosophila* DJ-1β also highlighted the relatively poor conservation of a proposed Cys-His catalytic dyad in close DJ-1 homologues. The Cys106 and His126 residues in human DJ-1 were suggested to form a catalytically important dyad (Chen et al. 2010; Huai et al. 2003), despite the imidazole side-chain rotamer in His126. In *Drosophila* DJ-1β, His126 is replaced by a Tyr (Tyr124). The structure of *Drosophila* DJ-1β showed that the phenolic oxygen atom of the Tyr124 is positioned away from the Cys104 residue and therefore unable to hydrogen bond with the thiol(ate) or participate in acid-base catalysis. Like His126 in human DJ-1, the primary importance of the residue at this position appears to be making hydrogen bonds that span the dimer interface and stabilize the active dimer (Lin et al. 2012).

An additional and informative difference between human DJ-1 and *Drosophila* DJ-1β is the substitution of a Leu residue at the position of Met26 in the human protein. The M26I mutation in human DJ-1 results in Parkinson's disease due to an enhanced sampling of unstable conformations at physiological temperature (Milkovic et al. 2015). The presence of Leu at this position in DJ-1β, as well as corresponding studies of the artificial M26L and M26V mutants of human DJ-1, shows that the presence of a β-branched amino acid at this position is disruptive to the DJ-1 fold, while ϒ-branched ones are tolerated. Despite these differences between human and *Drosophila* DJ-1β, many other important residues associated with disease development are conserved, such as Leu10, Ala104, Asp149, Glu163, and Leu166 (Lin et al. 2012).

2.6 DJ-1 Superfamily Members in Plants

Arabidopsis thaliana contains six DJ-1 superfamily homologues: three that are close homologues of human DJ-1 (AtDJ1A-C) and three that are homologues of *E. coli* YhbO (AtDJ1D-F). All are tandem duplications of the core flavodoxin-like domain. The tandem duplication results in a pseudo-dimer in which both subunits are contained within a single polypeptide chain. Interestingly, and unlike all previously studied DJ-1 superfamily members, AtDJ1C is essential for viability in *Arabidopsis* (Lin et al. 2011). It was determined that this protein is required for proper chloroplast development and inactivation of the AtDJ1C gene results in albino seedlings that do not mature. An additional curiosity of AtDJ1C is that it lacks the conserved Cys residue in the nucleophile elbow that is present in almost all other members of the DJ-1 superfamily. Although AtDJ1C does not appear to have the same function as human DJ-1, it illustrates that the DJ-1 superfamily permits a surprising diversity of function in the context of a conserved core fold. The other two close homologues of DJ-1 in *Arabidopsis* (AtDJ1A and AtDJ1B) both contain the conserved cysteine in both DJ-1 domains. AtDJ1A appears to be more closely related to human DJ-1, both owing to the presence of the conserved Cys residues and the phenotype of accelerated cell death in older plants when its gene is disrupted (Xu et al. 2010). It has been suggested that this protein enhances the oxidative stress response through increased activation of cytosolic superoxide dismutase 1 (CSD) with a proposed role as a copper chaperone to CSD (Girotto et al. 2014), although that requires additional substantiation. Structures are available for AtDJ1D and indicate that these proteins form a trimeric ring that is structurally similar to the hexamer formed by the PfpI homologue PH1704 (Du et al. 2000; Choi et al. 2014). The AtDJ1D-F proteins, which are members of the PfpI-like clade of the superfamily, have also been reported to be active glutathione-independent glyoxalases (Ghosh et al. 2016; Choi et al. 2014).

2.7 Human DJ-1

Human DJ-1 is a 189-residue dimeric protein that is implicated in autosomal recessive, early-onset Parkinson's disease as well as cancer and other diseases (Bonifati et al. 2003; Nagakubo et al. 1997). The link to neurodegeneration has made DJ-1 a popular subject of research for many years. Consistent with its recessive pattern of inheritance, loss of human DJ-1 function causes dopaminergic neurodegeneration and consequent parkinsonism. Although a considerable body of evidence points to human DJ-1 serving as a redox sensor and being involved in mitochondrial maintenance, the biochemical basis of its activity is still incompletely understood. Because human DJ-1 is easily expressed, purified, and crystallized, it has been an ideal target for structure-based hypothesis generation. Consistent with its ease of handling, the structure of human DJ-1 was solved almost simultaneously by five independent groups (Wilson et al. 2003; Tao and Tong 2003; Honbou et al. 2003; Huai et al. 2003; Lee et al. 2003).

DJ-1 contains 11 β-strands (b1–11) and 8 α-helices (a1–8) the α/β flavodoxin-like fold found throughout the superfamily. The DJ-1 structure features a central parallel β-sheet flanked by α-helices as well as a β3–4 hairpin on one end and an additional antiparallel three-stranded β-sheet on the other end of the molecule (Honbou et al. 2003; Huai et al. 2003; Lee et al. 2003; Wilson et al. 2003; Tao and Tong 2003). The C-terminal helix (α8) distinguishes DJ-1 clade proteins from other members of the superfamily that share its core fold and is important for dimerization (Tao and Tong 2003).

At the time that the crystal structure of human DJ-1 was solved, the only other members of the superfamily that were structurally characterized were *E. coli* Hsp31 and PH1704 from *Pyrococcus horikoshii* (see above). The latter protein in particular invited comparisons between DJ-1 and the PfpI clade of cysteine proteases, as they shared similar monomeric structures (Du et al. 2000). Despite similar core structures, the two proteins have differing oligomerization states and active site architectures, making it unlikely that they share a molecular function. In particular, DJ-1 lacks the Cys-His-Glu catalytic triad found in PfpI and homologues and presumed to be required for the proteolytic activity of the PfpI enzymes. Despite these divergent active sites, there was some speculation that DJ-1 could form a caspase-like catalytic dyad between Cys106 and His126 (Tao and Tong 2003; Huai et al. 2003); however, most studies show only weak or no proteolytic activity for DJ-1. As discussed above, His126 is not conserved in some other animal homologues such as *Drosophila* DJ-1β, which is functionally similar to human DJ-1 but cannot support the same proposed proteolytic activity due to its divergent active site. Therefore, a physiologically relevant proteolytic activity of DJ-1 that would involve His126 is unlikely from a structural perspective.

DJ-1 has also been reported to possess chaperone activity. The structure of DJ-1 features hydrophobic patches between helices G and H near the dimer interface that have been proposed to be potential substrate binding sites for non-native proteins (Lee et al. 2003). Chaperone activity has been detected for DJ-1 by some groups (Shendelman et al. 2004; Zhou et al. 2006) but not by others using common in vitro substrates such as citrate synthase (Olzmann et al. 2004). However, several groups have measured an oxidation-dependent ability of DJ-1 to suppress the fibrillation of α-synuclein, a poorly structured protein also involved in Parkinson's disease. Although the oxidation state of Cys106 appears to be an important regulator of this chaperone function, the structural basis of this activity remains unclear (Zhou et al. 2006; Lee et al. 2003; Shendelman et al. 2004).

The best-characterized function of DJ-1 is protection against oxidative stress and mitochondrial damage. Cys106 has been identified as an important residue whose oxidation state regulates these responses. As mentioned above, Cys106 is located in the nucleophile elbow motif and is in an unfavorable region of the Ramachandran plot in nearly all structures. This is commonly observed for residues in nucleophile elbows and contributes to the elbow-like configuration of the backbone. In several DJ-1 structures, Cys106 is spontaneously oxidized to the sulfinate ($-SO_2^-$) (Wilson et al. 2003; Lin et al. 2012; Lee et al. 2003; Madzelan et al. 2012; Canet-Aviles et al. 2004), which suggests a role for Cys106 in allowing DJ-1 to sense redox hemostasis or to participate in cytoprotective signaling. Mutation of Cys106 results in a loss of protective ability for DJ-1, possibly due to loss of an oxidizable residue at this location. However, mutation of Cys106 also eliminates any other possible function of a thiol at this location, thereby confounding the interpretation of these mutant phenotypes from the perspective of cysteine oxidation. Testing the importance of Cys106 oxidation therefore presents challenges that are not easily addressed by direct mutation of Cys106.

Structural information was used to design mutations at the nearby Glu18 residue that alter Cys106 oxidation in order to provide a more direct test of the importance of Cys106 oxidation for DJ-1 cytoprotection. The highly conserved Glu18 residue near Cys106 is protonated at physiological pH and donates a hydrogen bond to Cys106, depressing the thiol pK_a value (Witt et al. 2008). In addition, the interaction of Glu18 with Cys106-SO2- stabilizes oxidized DJ-1 (Prahlad et al. 2014). Conservative mutation of Glu18 was therefore used to modulate Cys106 oxidation while retaining the thiol. The E18N and E18Q mutations in DJ-1 permit or enhance

Cys106 oxidation under physiological conditions, while the E18D mutation suppresses both Cys106 oxidation and DJ-1's ability to protect cells against rotenone, a mitochondrial complex I inhibitor that causes reactive oxygen species generation and mitochondrial dysfunction (Blackinton et al. 2009). Subsequent studies have confirmed that E18D is not cytoprotective against oxidative challenge but E18Q is (Eltoweissy et al. 2016; Cao et al. 2014). A separate study of Cys106's role in DJ-1 function showed that while C106D DJ-1 was not protective, the C106DD mutant (a combined missense and insertion mutation) was, indicating that Cys106 modification is important. In addition, the C106DD mutant removes a thiolate entirely at residue 106, raising questions about the necessity of a nucleophile at these locations for the oxidative stress-responsive aspects of DJ-1 function (Waak et al. 2009).

The exact mechanism by which Cys106 oxidation modifies DJ-1 function is still unclear; however, Cys106 can be oxidized either to the sulfinate ($-SO_2^-$) or sulfonate ($-SO_3^-$), each of which has distinct effects on protein stability. In contrast to Cys106-SO_2^-, which is more stable than unmodified Cys106, the third oxygen atom of Cys106-SO_3^- is predicted to clash with nearby residues and may lead to changes in DJ-1 structure. These structural changes and attendant destabilization may modulate DJ-1 function, either leading to loss of protection or more subtle changes that might be functionally important for DJ-1 cytoprotection (Zhou et al. 2006; Hulleman et al. 2007). The biphasic response of DJ-1 stability to oxidation may also allow DJ-1 to discriminate between mild and extreme oxidative stress and to commit the cell to different responses as a consequence of varying oxidative burden. In addition, the Cys106 pocket is prone to interaction with other molecules (Landon et al. 2009), and oxidation of Cys106 may therefore modulate the interaction of DJ-1 with its binding partners, thereby transducing the oxidant signal received by Cys106 into a downstream cellular outcome.

2.8 Structural Insights into the Pathogenicity of Disease-Causing Mutations in Human DJ-1

Parkinson's disease (PD) is a common neurological disorder characterized by the loss of dopaminergic neurons in the substantia nigra. PD presents with a spectrum of symptoms related to insufficient striatal dopamine including tremor, bradykinesia, gait abnormality, and postural instability. There are several treatments for PD, most aimed at restoring dopamine levels, but no cure. Most PD cases are classified as idiopathic and have no clear single cause. However, a small percentage has been found to be heritable and can be traced to specific mutations in a collection of genes associated with PD. DJ-1, encoded by the PARK7 gene, is one example and an infrequent cause of recessively inheritable PD.

The missense mutations of DJ-1 that are known or strongly suspected of being parkinsonian include L166P, L10P, ΔP158, A104T E163K, M26I, D149A, and E64D. DJ-1 structure is severely affected by the L166P, L10P, and ΔP158 mutations, which all cause a profound loss of protein stability and disrupted dimerization. The best studied of these is L166P, which is unable to form dimers and tends to aggregate, causing it to be degraded in the cell (Miller et al. 2003; Gorner et al. 2004). L10P has a similar effect; it cannot form homodimers, but is able to dimerize to a limited extent when presented with WT DJ-1 (Repici et al. 2013; Ramsey and Giasson 2010). The L10P mutation, like L166P, results in a protein that is unstable and rapidly degraded in the cell (Ramsey and Giasson 2010). ΔP158 DJ-1 also appears incapable of forming homodimers, but it does retain the ability to form heterodimers with wild-type DJ-1 (Repici et al. 2013). Tellingly, each of these mutations involves proline (Pro): either deleted in ΔP158 or pathologically substituted in L10P and L166P. Because Pro is a conformationally restricted amino acid, it is not well tolerated in the middle of secondary structural elements (such as α-helices) where its preferred backbone torsion angles are incompatible with those required by the secondary structural context. In contrast, Pro is favored at helical termini, in loop regions, or in intrinsically disordered portions of proteins. Therefore, the loss of Pro at the beginning of α-helix a7 in ΔP158 may compromise the stability of that helix, whereas the L10P and L166P missense mutations both introduce Pro into the middle of α-helices near the dimer interface, which acts to destabilize both the affected α-helix and the dimeric interface, leading to poor folding of DJ-1.

Among the class of less structurally disruptive parkinsonian mutations in DJ-1 are M26I and A104T (Malgieri and Eliezer 2008; Repici et al. 2013; Milkovic et al. 2015; Lakshminarasimhan et al. 2008). M26 is located in α-helix 1 in the core of the protein (Fig. 2.4); however, its mutation to Ile does not cause loss of dimerization or severe structural defects. It is only mildly destabilizing, creating a minor (0.7 Å) steric conflict with the neighboring Ile31 and reducing the thermal melting

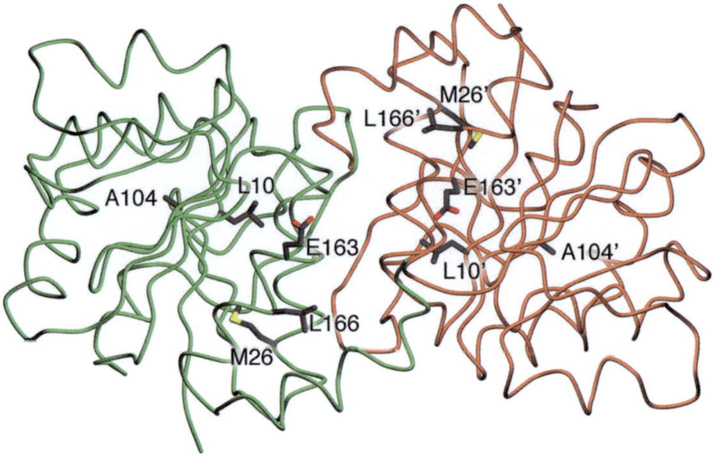

Fig. 2.4 Human DJ-1 dimer with the locations of some parkinsonian mutants indicated. The protomers of the DJ-1 dimer are colored green and orange. A selection of the residues mutated in certain forms of DJ-1-linked parkinsonism are labeled in each protomer and distinguished by primes

temperature by ~3 °C, thus raising questions about why the M26I mutant was reliably found to be unstable in cells (Malgieri and Eliezer 2008; Milkovic et al. 2015). Recent NMR, spectroscopic, and cross-linking studies showed that M26I DJ-1 samples aberrant, aggregation-prone conformations at physiological temperatures but not below, resulting in temperature-sensitive destabilization of the protein in a cellular environment (Milkovic et al. 2015). Like M26I, A104T DJ-1 is also able to homodimerize, although it is markedly (~8 °C) less stable than wild-type protein in thermal denaturation (Malgieri and Eliezer 2008; Lakshminarasimhan et al. 2008). A104T is located on the C-terminal side of β-strand 7 and has been shown by X-ray crystallography to cause local environmental changes that introduce a partially occupied water molecule into the core of the protein (Lakshminarasimhan et al. 2008). This mutation may have a similar effect to M26I in transiently exposing hydrophobic residues to the solvent, whose recognition by the cell leads to the premature destruction of A104T DJ-1.

Although the L10P, L166P, M26I, and A104T mutations are all located in buried regions of DJ-1, several disease-associated mutations occur in solvent-exposed regions. The E163K mutation disrupts a salt bridge interaction with Arg145 that spans the dimer interface (Fig. 2.4), and this mutation substantially destabilizes DJ-1 by ~11 °C as measured using differential scanning calorimetry (Lakshminarasimhan et al. 2008). In cells, E163K DJ-1 loses its cytoprotective ability against oxidative stress while maintaining mitochondrial protection, which is unexpected and mechanistically intriguing (Ramsey and Giasson 2008). Also surface exposed, the D149A mutation forms a stable homodimer whose stability is diminished by ~8 °C compared to wild-type DJ-1. D149A appears to lose some secondary structure compared to the wild-type protein, although its NMR chemical shifts indicate that the protein is well folded and highly similar to wild-type DJ-1 (Malgieri and Eliezer 2008). The most structurally conservative characterized parkinsonian DJ-1 mutant is E64D, which shows almost no structural change, homodimerizes well, and is comparably stable to the wild-type protein in vitro. However, there is some evidence of a slightly increased turnover rate in the cell (Repici et al. 2013), although not all studies agree on this point. Located at the end of α-helix 2, E64D is solvent exposed, and thus the impact of a conservative Glu to Asp mutation is expected to be very minor. It is possible that this missense mutation alters DJ-1's ability to interact with other proteins, although this is speculative and untested.

2.9 Summary

The structural study of DJ-1 has been useful in rationalizing some aspects of its function, in clarifying the impact of disease-associated mutations, and in formulating new hypotheses about its regulation in protecting cells from oxidative stress and in redox sensing. Additionally, the large number and diversity of members of the DJ-1 superfamily has facilitated the identification of homologues in model organisms that have significantly advanced our understanding of DJ-1 biology. However, the breadth of the DJ-1 superfamily and the functional diversity of its various clades have also caused difficulties, including the misannotation of many proteins as thiamine anabolic genes and lingering confusion about the degree of functional relatedness of DJ-1 superfamily proteins from different clades. Structural analysis has been critical for the proper classification of distinct classes of proteins in the DJ-1 superfamily and for identifying their essential differences in the vicinity of the highly conserved and functionally important active site cysteine residue. There remain several unanswered questions about the functions of DJ-1 superfamily proteins, and based on prior work, there is every reason to believe that further structural study will play a major role in addressing those questions.

References

Abdallah J, Caldas T, Kthiri F, Kern R, Richarme G (2007) YhbO protects cells against multiple stresses. J Bacteriol 189(24):9140–9144. https://doi.org/10.1128/JB.01208-07

Abdallah J, Mihoub M, Gautier V, Richarme G (2016) The DJ-1 superfamily members YhbO and YajL from Escherichia coli repair proteins from glycation by methylglyoxal and glyoxal. Biochem Biophys Res Commun 470(2):282–286. https://doi.org/10.1016/j.bbrc.2016.01.068

Bandyopadhyay S, Cookson MR (2004) Evolutionary and functional relationships within the DJ1 superfamily. BMC Evol Biol 4:6. https://doi.org/10.1186/1471-2148-4-6

Bankapalli K, Saladi S, Awadia SS, Goswami AV, Samaddar M, D'Silva P (2015) Robust glyoxalase activity of Hsp31, a ThiJ/DJ-1/PfpI family member protein, is critical for oxidative stress resistance in Saccharomyces cerevisiae. J Biol Chem 290(44):26491–26507. https://doi.org/10.1074/jbc.M115.673624

Blackinton J, Lakshminarasimhan M, Thomas KJ, Ahmad R, Greggio E, Raza AS, Cookson MR, Wilson MA (2009) Formation of a stabilized cysteine sulfinic acid is critical for the mitochondrial function of the parkinsonism protein DJ-1. J Biol Chem 284(10):6476–6485. https://doi.org/10.1074/jbc.M806599200

Bonifati V, Rizzu P, van Baren MJ, Schaap O, Breedveld GJ, Krieger E, Dekker MC, Squitieri F, Ibanez P, Joosse M, van Dongen JW, Vanacore N, van Swieten JC, Brice A, Meco G, van Duijn CM, Oostra BA, Heutink P (2003) Mutations in the DJ-1 gene associated with autosomal recessive early-onset parkinsonism. Science 299(5604):256–259. https://doi.org/10.1126/science.1077209

Canet-Aviles RM, Wilson MA, Miller DW, Ahmad R, McLendon C, Bandyopadhyay S, Baptista MJ, Ringe D, Petsko GA, Cookson MR (2004) The Parkinson's disease protein DJ-1 is neuroprotective due to cysteine-sulfinic acid-driven mitochondrial localization. Proc Natl Acad Sci U S A 101(24):9103–9108. https://doi.org/10.1073/pnas.0402959101

Cao J, Ying M, Xie N, Lin G, Dong R, Zhang J, Yan H, Yang X, He Q, Yang B (2014) The oxidation states of DJ-1 dictate the cell fate in response to oxidative stress triggered by 4-hpr: autophagy or apoptosis? Antioxid Redox Signal 21(10):1443–1459. https://doi.org/10.1089/ars.2013.5446

Chen J, Li L, Chin LS (2010) Parkinson disease protein DJ-1 converts from a zymogen to a protease by carboxyl-terminal cleavage. Hum Mol Genet 19(12):2395–2408. https://doi.org/10.1093/hmg/ddq113

Choi D, Kim J, Ha S, Kwon K, Kim EH, Lee HY, Ryu KS, Park C (2014) Stereospecific mechanism of DJ-1 glyoxalases inferred from their hemithioacetal-containing crystal structures. FEBS J 281(24):5447–5462. https://doi.org/10.1111/febs.13085

Dave KD, De Silva S, Sheth NP, Ramboz S, Beck MJ, Quang C, Switzer RC, 3rd, Ahmad SO, Sunkin SM, Walker D, Cui X, Fisher DA, McCoy AM, Gamber K, Ding X, Goldberg MS,

Benkovic SA, Haupt M, Baptista MA, Fiske BK, Sherer TB, Frasier MA (2014) Phenotypic characterization of recessive gene knockout rat models of Parkinson's disease. Neurobiol Dis 70:190–203. doi:https://doi.org/10.1016/j.nbd.2014.06.009

Du X, Choi IG, Kim R, Wang W, Jancarik J, Yokota H, Kim SH (2000) Crystal structure of an intracellular protease from Pyrococcus horikoshii at 2-A resolution. Proc Natl Acad Sci U S A 97(26):14079–14084. https://doi.org/10.1073/pnas.260503597

Eltoweissy M, Dihazi GH, Muller GA, Asif AR, Dihazi H (2016) Protein DJ-1 and its anti-oxidative stress function play an important role in renal cell mediated response to profibrotic agents. Mol BioSyst 12(6):1842–1859. https://doi.org/10.1039/c5mb00887e

Ghosh A, Kushwaha HR, Hasan MR, Pareek A, Sopory SK, Singla-Pareek SL (2016) Presence of unique glyoxalase III proteins in plants indicates the existence of shorter route for methylglyoxal detoxification. Sci Rep 6:18358. https://doi.org/10.1038/srep18358

Girotto S, Cendron L, Bisaglia M, Tessari I, Mammi S, Zanotti G, Bubacco L (2014) DJ-1 is a copper chaperone acting on SOD1 activation. J Biol Chem 289(15):10887–10899. https://doi.org/10.1074/jbc.M113.535112

Gorner K, Holtorf E, Odoy S, Nuscher B, Yamamoto A, Regula JT, Beyer K, Haass C, Kahle PJ (2004) Differential effects of Parkinson's disease-associated mutations on stability and folding of DJ-1. J Biol Chem 279(8):6943–6951. https://doi.org/10.1074/jbc.M309204200

Graille M, Quevillon-Cheruel S, Leulliot N, Zhou CZ, Li de la Sierra Gallay I, Jacquamet L, Ferrer JL, Liger D, Poupon A, Janin J, van Tilbeurgh H (2004) Crystal structure of the YDR533c S cerevisiae protein, a class II member of the Hsp31 family. Structure 12(5):839–847. https://doi.org/10.1016/j.str.2004.02.030

Harrington AJ, Hamamichi S, Caldwell GA, Caldwell KA (2010) C. elegans as a model organism to investigate molecular pathways involved with Parkinson's disease. Dev Dyn 239(5):1282–1295. https://doi.org/10.1002/dvdy.22231

Hasim S, Hussin NA, Alomar F, Bidasee KR, Nickerson KW, Wilson MA (2014) A glutathione-independent glyoxalase of the DJ-1 superfamily plays an important role in managing metabolically generated methylglyoxal in Candida albicans. J Biol Chem 289(3):1662–1674. https://doi.org/10.1074/jbc.M113.505784

Honbou K, Suzuki NN, Horiuchi M, Niki T, Taira T, Ariga H, Inagaki F (2003) The crystal structure of DJ-1, a protein related to male fertility and Parkinson's disease. J Biol Chem 278(33):31380–31384. https://doi.org/10.1074/jbc.M305878200

Horvath MM, Grishin NV (2001) The C-terminal domain of HPII catalase is a member of the type I glutamine amidotransferase superfamily. Proteins 42(2):230–236

Huai Q, Sun Y, Wang H, Chin LS, Li L, Robinson H, Ke H (2003) Crystal structure of DJ-1/RS and implication on familial Parkinson's disease. FEBS Lett 549(1-3):171–175

Hulleman JD, Mirzaei H, Guigard E, Taylor KL, Ray SS, Kay CM, Regnier FE, Rochet JC (2007) Destabilization of DJ-1 by familial substitution and oxidative modifications: implications for Parkinson's disease. Biochemistry 46(19):5776–5789. https://doi.org/10.1021/bi7001778

Lakshminarasimhan M, Maldonado MT, Zhou W, Fink AL, Wilson MA (2008) Structural impact of three Parkinsonism-associated missense mutations on human DJ-1. Biochemistry 47(5):1381–1392. https://doi.org/10.1021/bi701189c

Landon MR, Lieberman RL, Hoang QQ, Ju S, Caaveiro JM, Orwig SD, Kozakov D, Brenke R, Chuang GY, Beglov D, Vajda S, Petsko GA, Ringe D (2009) Detection of ligand binding hot spots on protein surfaces via fragment-based methods: application to DJ-1 and glucocerebrosidase. J Comput Aided Mol Des 23(8):491–500. https://doi.org/10.1007/s10822-009-9283-2

Lee SJ, Kim SJ, Kim IK, Ko J, Jeong CS, Kim GH, Park C, Kang SO, Suh PG, Lee HS, Cha SS (2003) Crystal structures of human DJ-1 and Escherichia coli Hsp31, which share an evolutionarily conserved domain. J Biol Chem 278(45):44552–44559. https://doi.org/10.1074/jbc.M304517200

Lin J, Nazarenus TJ, Frey JL, Liang X, Wilson MA, Stone JM (2011) A plant DJ-1 homolog is essential for Arabidopsis thaliana chloroplast development. PLoS One 6(8):e23731. https://doi.org/10.1371/journal.pone.0023731

Lin J, Prahlad J, Wilson MA (2012) Conservation of oxidative protein stabilization in an insect homologue of parkinsonism-associated protein DJ-1. Biochemistry 51(18):3799–3807. https://doi.org/10.1021/bi3003296

Lopert P, Patel M (2014) Brain mitochondria from DJ-1 knockout mice show increased respiration-dependent hydrogen peroxide consumption. Redox Biol 2:667–672. https://doi.org/10.1016/j.redox.2014.04.010

Lucas JI, Marin I (2007) A new evolutionary paradigm for the Parkinson disease gene DJ-1. Mol Biol Evol 24(2):551–561. https://doi.org/10.1093/molbev/msl186

Madzelan P, Labunska T, Wilson MA (2012) Influence of peptide dipoles and hydrogen bonds on reactive cysteine pKa values in fission yeast DJ-1. FEBS J 279(22):4111–4120. https://doi.org/10.1111/febs.12004

Malgieri G, Eliezer D (2008) Structural effects of Parkinson's disease linked DJ-1 mutations. Protein Sci 17(5):855–868. https://doi.org/10.1110/ps.073411608

Malki A, Caldas T, Abdallah J, Kern R, Eckey V, Kim SJ, Cha SS, Mori H, Richarme G (2005) Peptidase activity of the Escherichia Coli Hsp31 chaperone. J Biol Chem 280(15):14420–14426. https://doi.org/10.1074/jbc.M408296200

Masuda T, Wada Y, Kawamura S (2016) ES1 Is a mitochondrial enlarging factor contributing to form mega-mitochondria in zebrafish cones. Sci Rep 6:22360. https://doi.org/10.1038/srep22360

Meulener M, Whitworth AJ, Armstrong-Gold CE, Rizzu P, Heutink P, Wes PD, Pallanck LJ, Bonini NM (2005) Drosophila DJ-1 mutants are selectively sensitive to environmental toxins associated with Parkinson's disease. Curr Biol 15(17):1572–1577. https://doi.org/10.1016/j.cub.2005.07.064

Meulener MC, Xu K, Thomson L, Ischiropoulos H, Bonini NM (2006) Mutational analysis of DJ-1 in Drosophila implicates functional inactivation by oxidative damage and aging. Proc Natl Acad Sci U S A 103(33):12517–12522. https://doi.org/10.1073/pnas.0601891103

Milkovic NM, Catazaro J, Lin J, Halouska S, Kizziah JL, Basiaga S, Cerny RL, Powers R, Wilson MA (2015) Transient sampling of aggregation-prone conformations causes pathogenic instability of a Parkinsonian mutant of DJ-1 at physiological temperature. Protein Sci 24(10):1671–1685. https://doi.org/10.1002/pro.2762

Miller DW, Ahmad R, Hague S, Baptista MJ, Canet-Aviles R, McLendon C, Carter DM, Zhu PP, Stadler J, Chandran J, Klinefelter GR, Blackstone C, Cookson MR (2003) L166P mutant DJ-1, causative for recessive Parkinson's disease, is degraded through the ubiquitin-proteasome system. J Biol Chem 278(38):36588–36595. https://doi.org/10.1074/jbc.M304272200

Miller-Fleming L, Antas P, Pais TF, Smalley JL, Giorgini F, Outeiro TF (2014) Yeast DJ-1 superfamily members are required for diauxic-shift reprogramming and cell survival in stationary phase. Proc Natl Acad Sci U S A 111(19):7012–7017. https://doi.org/10.1073/pnas.1319221111

Mueller EG, Buck CJ, Palenchar PM, Barnhart LE, Paulson JL (1998) Identification of a gene involved in the generation of 4-thiouridine in tRNA. Nucleic Acids Res 26(11):2606–2610

Nagakubo D, Taira T, Kitaura H, Ikeda M, Tamai K, Iguchi-Ariga SM, Ariga H (1997) DJ-1, a novel oncogene which transforms mouse NIH3T3 cells in cooperation with ras. Biochem Biophys Res Commun 231(2):509–513. https://doi.org/10.1006/bbrc.1997.6132

Natkanska U, Skoneczna A, Sienko M, Skoneczny M (2017) The budding yeast orthologue of Parkinson's disease-associated DJ-1 is a multi-stress response protein protecting cells against toxic glycolytic products. Biochim Biophys Acta 1864(1):39–50. https://doi.org/10.1016/j.bbamcr.2016.10.016

Olzmann JA, Brown K, Wilkinson KD, Rees HD, Huai Q, Ke H, Levey AI, Li L, Chin LS (2004) Familial Parkinson's disease-associated L166P mutation disrupts DJ-1 protein folding and function. J Biol Chem 279(9):8506–8515. https://doi.org/10.1074/jbc.M311017200

Pardo M, Garcia A, Thomas B, Pineiro A, Akoulitchev A, Dwek RA, Zitzmann N (2006) The characterization of the invasion phenotype of uveal melanoma tumour cells shows the presence of MUC18 and HMG-1 metastasis markers and leads to the identification of DJ-1 as a potential serum biomarker. Int J Cancer 119(5):1014–1022. https://doi.org/10.1002/ijc.21942

Park J, Kim SY, Cha GH, Lee SB, Kim S, Chung J (2005) Drosophila DJ-1 mutants show oxidative stress-sensitive locomotive dysfunction. Gene 361:133–139. https://doi.org/10.1016/j.gene.2005.06.040

Pfaff DH, Fleming T, Nawroth P, Teleman AA (2017) Evidence against a role for the parkinsonism-associated protein DJ-1 in methylglyoxal detoxification. J Biol Chem 292(2):685–690. https://doi.org/10.1074/jbc.M116.743823

Prahlad J, Hauser DN, Milkovic NM, Cookson MR, Wilson MA (2014) Use of cysteine-reactive cross-linkers to probe conformational flexibility of human DJ-1 demonstrates that Glu18 mutations are dimers. J Neurochem 130(6):839–853. https://doi.org/10.1111/jnc.12763

Quigley PM, Korotkov K, Baneyx F, Hol WG (2003) The 1.6-A crystal structure of the class of chaperones represented by Escherichia coli Hsp31 reveals a putative catalytic triad. Proc Natl Acad Sci U S A 100(6):3137–3142. https://doi.org/10.1073/pnas.0530312100

Ramsey CP, Giasson BI (2008) The E163K DJ-1 mutant shows specific antioxidant deficiency. Brain Res 1239:1–11. https://doi.org/10.1016/j.brainres.2008.09.009

Ramsey CP, Giasson BI (2010) L10p and P158DEL DJ-1 mutations cause protein instability, aggregation, and dimerization impairments. J Neurosci Res 88(14):3111–3124. https://doi.org/10.1002/jnr.22477

Repici M, Straatman KR, Balduccio N, Enguita FJ, Outeiro TF, Giorgini F (2013) Parkinson's disease-associated mutations in DJ-1 modulate its dimerization in living cells. J Mol Med (Berl) 91(5):599–611. https://doi.org/10.1007/s00109-012-0976-y

Richarme G, Dairou J (2017) Parkinsonism-associated protein DJ-1 is a bona fide deglycase. Biochem Biophys Res Commun 483(1):387–391. https://doi.org/10.1016/j.bbrc.2016.12.134

Sastry MS, Korotkov K, Brodsky Y, Baneyx F (2002) Hsp31, the Escherichia coli yedU gene product, is a molecular chaperone whose activity is inhibited by ATP at high temperatures. J Biol Chem 277(48):46026–46034. https://doi.org/10.1074/jbc.M205800200

Shendelman S, Jonason A, Martinat C, Leete T, Abeliovich A (2004) DJ-1 is a redox-dependent molecular chaperone that inhibits alpha-synuclein aggregate formation. PLoS Biol 2(11):e362. https://doi.org/10.1371/journal.pbio.0020362

Subedi KP, Choi D, Kim I, Min B, Park C (2011) Hsp31 of Escherichia coli K-12 is glyoxalase III. Mol Microbiol 81(4):926–936. https://doi.org/10.1111/j.1365-2958.2011.07736.x

Taira T, Iguchi-Ariga SM, Ariga H (2004) Co-localization with DJ-1 is essential for the androgen receptor to exert its transcription activity that has been impaired by androgen antagonists. Biol Pharm Bull 27(4):574–577

Takahashi-Niki K, Niki T, Taira T, Iguchi-Ariga SM, Ariga H (2004) Reduced anti-oxidative stress activities of DJ-1 mutants found in Parkinson's disease patients. Biochem Biophys Res Commun 320(2):389–397. https://doi.org/10.1016/j.bbrc.2004.05.187

Tao X, Tong L (2003) Crystal structure of human DJ-1, a protein associated with early onset Parkinson's disease. J Biol Chem 278(33):31372–31379. https://doi.org/10.1074/jbc.M304221200

Trotter EW, Kao CM, Berenfeld L, Botstein D, Petsko GA, Gray JV (2002) Misfolded proteins are competent to mediate a subset of the responses to heat shock in Saccharomyces cerevisiae. J Biol Chem 277(47):44817–44825. https://doi.org/10.1074/jbc.M204686200

Waak J, Weber SS, Gorner K, Schall C, Ichijo H, Stehle T, Kahle PJ (2009) Oxidizable residues mediating protein stability and cytoprotective interaction of DJ-1 with apoptosis signal-regulating kinase 1. J Biol Chem 284(21):14245–14257. https://doi.org/10.1074/jbc.M806902200

Wang H, Li YY, Qiu LY, Yan YF, Liao ZP, Chen HP (2017) Involvement of DJ1 in ischemic preconditioning induced delayed cardioprotection in vivo. Mol Med Rep 15(2):995–1001. https://doi.org/10.3892/mmr.2016.6091

Wei Y, Ringe D, Wilson MA, Ondrechen MJ (2007) Identification of functional subclasses in the DJ-1 superfamily proteins. PLoS Comput Biol 3(1):e10. https://doi.org/10.1371/journal.pcbi.0030010

Wilson MA (2011) The role of cysteine oxidation in DJ-1 function and dysfunction. Antioxid Redox Signal 15(1):111–122. https://doi.org/10.1089/ars.2010.3481

Wilson MA, Collins JL, Hod Y, Ringe D, Petsko GA (2003) The 1.1-A resolution crystal structure of DJ-1, the protein mutated in autosomal recessive early onset Parkinson's disease. Proc Natl Acad Sci U S A 100(16):9256–9261. https://doi.org/10.1073/pnas.1133288100

Wilson MA, St Amour CV, Collins JL, Ringe D, Petsko GA (2004) The 1.8-A resolution crystal structure of YDR533Cp from Saccharomyces cerevisiae: a member of the DJ-1/ThiJ/PfpI superfamily. Proc Natl Acad Sci U S A 101(6):1531–1536. https://doi.org/10.1073/pnas.0308089100

Wilson MA, Ringe D, Petsko GA (2005) The atomic resolution crystal structure of the YajL (ThiJ) protein from Escherichia coli: a close prokaryotic homologue of the Parkinsonism-associated protein DJ-1. J Mol Biol 353(3):678–691. https://doi.org/10.1016/j.jmb.2005.08.033

Witt AC, Lakshminarasimhan M, Remington BC, Hasim S, Pozharski E, Wilson MA (2008) Cysteine pKa depression by a protonated glutamic acid in human DJ-1. Biochemistry 47(28):7430–7440. https://doi.org/10.1021/bi800282d

Xu XM, Lin H, Maple J, Bjorkblom B, Alves G, Larsen JP, Moller SG (2010) The Arabidopsis DJ-1a protein confers stress protection through cytosolic SOD activation. J Cell Sci 123(10):1644–1651. https://doi.org/10.1242/jcs.063222

Zhao Q, Su Y, Wang Z, Chen C, Wu T, Huang Y (2014) Identification of glutathione (GSH)-independent glyoxalase III from Schizosaccharomyces pombe. BMC Evol Biol 14:86. https://doi.org/10.1186/1471-2148-14-86

Zhou W, Zhu M, Wilson MA, Petsko GA, Fink AL (2006) The oxidation state of DJ-1 regulates its chaperone activity toward alpha-synuclein. J Mol Biol 356(4):1036–1048. https://doi.org/10.1016/j.jmb.2005.12.030

Chapter 3
Expression of DJ-1 in Neurodegenerative Disorders

Daria Antipova and Rina Bandopadhyay

Abstract In 2003, autosomal recessive loss-of-function mutations were identified in *PARK7* gene that caused early-onset Parkinson's disease (PD). The *PARK7* gene encodes a conserved protein termed DJ-1. DJ-1 is a ubiquitous protein, and within the brain, it is present in the nucleus and cytoplasm of both neuronal and glial cells. DJ-1 is a multifunctional protein, and numerous studies have ascribed various roles, including antioxidative properties, chaperone function, protease activities, mitochondrial functions and regulation of transcription to the protein. The DJ-1 protein undergoes oxidation and post-translational modifications that are important for its function. Not only is DJ-1 linked to familial PD, but it is also associated with the pathogenic mechanisms of sporadic PD and other neurodegenerative disorders where oxidative stress is implicated. In this chapter we provide an overview on the expression of DJ-1 mRNA and protein in different neurodegenerative disorders and discuss some of its main functions together with DJ-1's potential for neuroprotection.

Keywords DJ-1 • DJ-1 oxidation • DJ-1 mRNA • Amyotrophic lateral sclerosis • Huntington's disease • Parkinson's disease • Sporadic inclusion body myositis • Synucleinopathies • Tauopathies

3.1 Introduction

Parkinson's disease (PD) is a common and an insidious neurological disorder usually affecting the elderly population. Numbers of patients are on the increase worldwide which inevitably causes a demand on social care. Current medications provide symptom control, but they do not halt or cure the incessant disease progression. Disease-modifying treatments or a permanent cure for PD therefore remains a

D. Antipova • R. Bandopadhyay (✉)
Reta Lila Weston Institute of Neurological Studies, UCL Institute of Neurology,
1, Wakefield Street, WC1N 1PJ London, UK
e-mail: rina.bandopadhyay@ucl.ac.uk

H. Ariga, S.M.M. Iguchi-Ariga (eds.), *DJ-1/PARK7 Protein*,
Advances in Experimental Medicine and Biology 1037,
https://doi.org/10.1007/978-981-10-6583-5_3

critical unmet need. The majority of PD cases are sporadic in origin with unknown aetiology, but specific genetic inheritance can account for around 10% of patients with familial PD (Quinn et al. 1987). Up to date, mutations in several genes have been found to be associated with familial PD (see review by Hernandez et al. 2016). The pathological hallmarks of PD comprise of the severe loss of dopaminergic neurons in the substantia nigra pars compacta region together with the accumulation of alpha-synuclein-positive Lewy bodies (LBs) (Jellinger 2012).

Recessively inherited loss of function mutations in *DJ-1/PARK7* gene was identified in 2003 by Bonifati and coworkers (Bonifati et al. 2003). DJ-1 protein has ubiquitous expression and is present in both neuronal and glial cells within the brain. DJ-1 is a conserved multifunctional protein and is implicated in a number of important homeostatic functions in cells (reviewed in Ariga et al. 2013). From the pathological point of view, a recent report has shown that DJ-1 mutation causes LB pathology (Taipa et al. 2016). In addition, DJ-1 is also implicated in the pathogenesis of sporadic PD and other neurodegenerative diseases. In this chapter we outline the cellular patterns of DJ-1 protein and mRNA expression in normal human brain and also in the pathology of PD and other neurodegenerative disorders. We also briefly discuss the implications of various DJ-1 functions in neurological diseases and the possibility of DJ-1 being neurotherapeutic.

3.2 Cellular Types Expressing DJ-1 in Neurodegenerative Conditions

3.2.1 DJ-1 in Alpha-Synucleinopathies

Alpha-synucleinopathies are a group of diverse neurodegenerative disorders associated with abnormal α-synuclein aggregation. There are three prominent members of this group of neurodegenerative conditions, among them Parkinson's disease (PD), multiple system atrophy (MSA) and dementia with Lewy bodies (DLB). The typical clinical features of these conditions are Parkinsonian syndrome associated with brady- and hypokinesia, rigidity and tremor (Jellinger 2001). Non-motor signs, such as autonomic dysfunction, hallucinations, dementia, depression, anxiety, etc., often occur in these patients as well (Garcia-Ruiz et al. 2014).

Post-mortem immunohistochemistry studies revealed DJ-1 expression in different disease-affected regions. Thus, strong DJ-1 immunoreactivity (IR) of reactive astrocytes was seen in the brainstem in PD and DLB cases and the cortical, cerebellar and brainstem regions of MSA patients (Neumann et al. 2004; Rizzu et al. 2004). Furthermore, Bandopadhyay and coworkers (2004, 2005) using the human frontal cortex and substantia nigra, two brain regions most commonly associated with LB PD pathology, demonstrated that DJ-1 was mostly expressed by astrocytic cell bodies and processes in both white and grey matter in idiopathic PD brains, PD subjects with DJ-1 R98Q polymorphisms and neurologically normal controls. DJ-1 staining was also seen in glial nuclei and cytoplasm. Different types of cells positive for

DJ-1 IR are shown in Fig. 3.1. Predominant astroglial staining is of particular interest for understanding functional properties of the DJ-1 protein, as two other Parkinson's disease genes, α-synuclein and parkin, are predominantly neuronal (Bandopadhyay et al. 2005). Interestingly, the hallmark lesions in PD, DLB and neurodegeneration with brain iron accumulation type 1 (NBIA-1, Hallervorden-Spatz disease) brains, i.e. LBs, occasionally showed DJ-1 immunostaining which localised to the peripheral rim (Bandopadhyay et al. 2004, 2005; Neumann et al. 2004; Jin et al. 2005). It is worth noting that neurofibrillary tau inclusions, such as neurofibrillary tangles (NFTs) and neuropil threads (NTs) in the substantia nigra (SN), caudate, putamen and temporal cortex of DLB cases, showed strong staining for DJ-1 (Rizzu et al. 2004). Neurons and the neuropil in diseased and control brains were weakly stained for DJ-1. Remarkably, the hallmark lesions in MSA, i.e. oligodendroglial cytoplasmic inclusions (GCIs), showed pronounced DJ-1 immunostaining co-localised with αSYN. A strong DJ-1 staining was also detected in neuroaxonal spheroids in NBIA-1 brain tissue (Neumann et al. 2004).

Although DJ-1 has been reported to be present primarily in glial cells in the cortex from cases with neurodegenerative conditions (Bandopadhyay et al. 2004; Meulener et al. 2005; Neumann et al. 2004; Rizzu et al. 2004), Olzmann et al. (2007) reported DJ-1 expression in neurons in subcortical regions of neurologically normal subjects. They identified weak DJ-1 IR in rare pyramidal neurons and evident DJ-1 immunostaining of cortical neuropil suggesting that DJ-1 regulates the structure and function of microtubules (Conde and Caceres 2009; Wang et al. 2011b; Sheng et al. 2013). DJ-1 labelling was also seen in neuromelanin-containing neurons in the SN pars compacta and reticulata, as well as in neuromelanin-negative neurons. These results are consistent with previously published papers indicating an important role of DJ-1 in the regulation of dopamine synthesis (Zhong et al. 2006; Ishikawa et al. 2009). Medium spiny neurons of the striatum showed moderate DJ-1 IR, and DJ-1 labelling was detected in the cytoplasm, processes and nuclei of glial cells in both grey and white matter across the cingulate cortex (Olzmann et al. 2007).

A number of two-dimensional polyacrylamide gel electrophoresis (2D-PAGE) studies have shown that DJ-1 undergoes modifications in response to oxidative stress. In particular, Kinumi et al. (2004) have demonstrated an acidic shift of the isoelectric point of DJ-1 due to post-translational modifications in cells treated with H_2O_2 (Canet-Avilés et al. 2004, Kinumi et al. 2004). Under oxidative stress conditions, cysteine residue in DJ-1 can directly add oxygen molecules with formation of cysteine-sulfenic acid (Cys-SOH), cysteine-sulfinic acid (Cys-SO_2H) and cysteine-sulfonic acid (Cys-SO_3H) (Saito, 2014). Cysteine residue 106 (Cys106) has appeared to be the most sensitive among three cysteine residues to H_2O_2 exposure suggesting that the protective function of DJ-1 is regulated by the oxidation of Cys106 (Kinumi et al. 2004; Wilson 2011). Notably, the oxidation of DJ-1 to Cys-SO_3H is associated with loss of its protective function (Zhou et al. 2006; Wilson 2011) which means that excessive oxidative stress can be associated with DJ-1 misfunctioning leading to neurodegeneration. Importantly in PD post-mortem brain tissue, different accumulation of over-oxidised DJ-1 is evident compared to controls

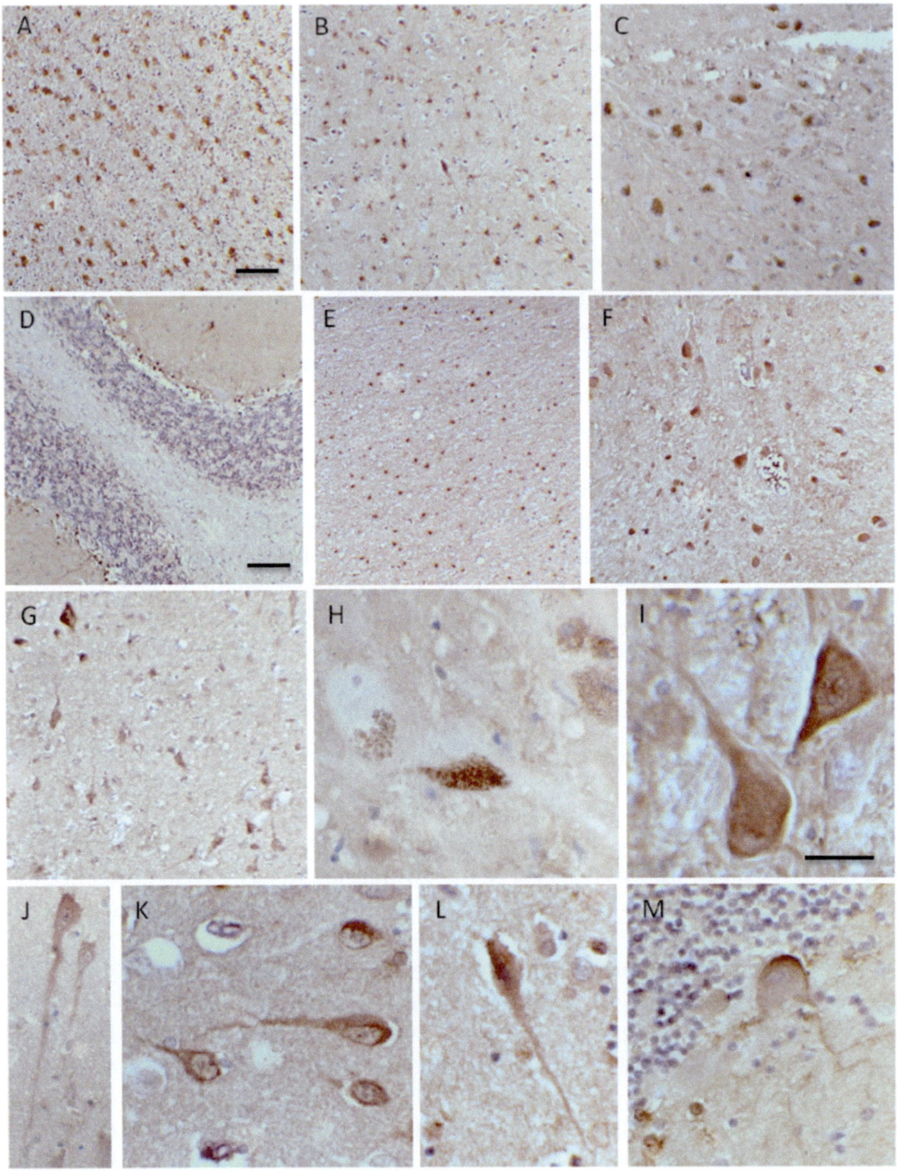

Fig. 3.1 Immunohistochemistry of DJ-1 in neurologically control and PD brains. Abundant astrocytic DJ-1 expression seen in control amygdala (**a**) and control frontal cortex (**b**). Reactive astrocytes in control substantia nigra (**c**). Immunoreactive Bergman glia positive for astrocytes in molecular layer of PD cerebellum (**d**) and PD cingulate gyrus (**e**). In PD, neurones of the inferior olives (**f**) and cingulate gyrus (**g**) show DJ-1 immunopositivity. Melanised dopaminergic neurons in a control case did not show any DJ-1 immunopositivity (**h**). Medullary neurons of a control case showed high DJ-1 immunopositivity (**i**). Neurones of a control case showing immunopositivity for DJ-1 in cingulate gyrus (**j**), amygdala (**k**) and frontal cortex (**l**). (**m**) A purkinje neuron in a control cerebellum showing DJ-1 immunopositivity (Scale bar in **a** is 20 μm for **a**, **b**, **c**, **f** and **g** and 40 μm for **d** and **e** and 10 μm in **h**–**m**)

in a number of brain anatomical regions (Bandopadhyay et al. 2004; Kumaran et al. 2009) (Fig. 3.2).

Since antibodies specifically recognising DJ-1 oxidised at Cys106 position (Cys106-oxDJ-1) were developed (Saito et al. 2009), the detailed distribution of Cys106-oxDJ-1 in human brain has been described only recently (Saito et al. 2014; Fig. 3.2). Unlike non-oxidised (total) DJ-1, Cys106-oxDJ-1 has been most prominently expressed by dopaminergic neuronal cell bodies and processes in the SN from PD patients. It has been shown to co-localise with LBs and astrocytic nuclei in the striatum, in neurons and glia in the red nucleus and in the inferior olivary nucleus. Interestingly, the most prominent Cys106-oxDJ-1 immunostaining has been detected in brain tissue from early PD patients meaning that DJ-1 oxidation might decrease dopamine synthesis leading to clinical manifestation (Saito et al. 2014). Decreased Cys106-oxDJ-1 IR detected in advanced PD cases might reflect other modifications of DJ-1, such as 4-hydroxy-2-nonenal-modified (Lin et al. 2012; Bandopadhyay et al. 2017), undetectable with antibodies used in the study performed by Saito et al. (2014). Elevated levels of oxDJ-1 in erythrocytes from unmedicated PD patients compared to medicated PD patients and neurologically healthy subjects have also been reported by Saito and colleagues (Saito et al. 2009). Oxidised DJ-1 co-localising with activated astrocytic marker Glial fibrillary acidic protein (GFAP) in human post-mortem tissue is shown in Fig. 3.3.

Only recently one case has been reported at autopsy for *DJ-1/PARK7* mutation (Taipa et al. 2016). This particular case harboured a novel c.515 T > A; p.L172Q mutation. The mutated protein is highly unstable and is degraded rapidly by the

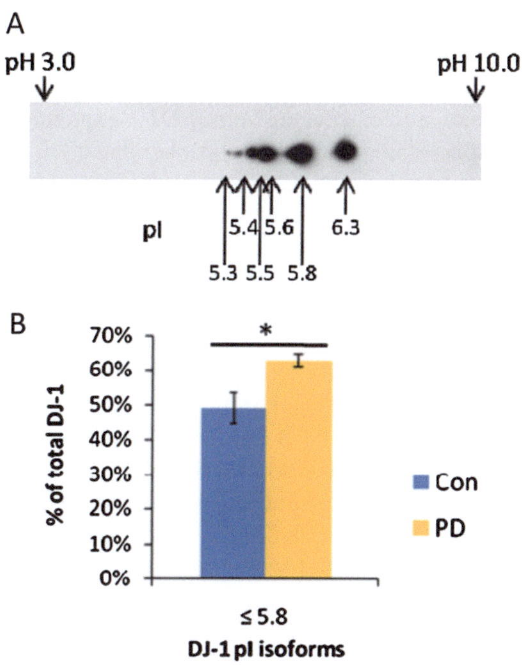

Fig. 3.2 (**a**) DJ-1 protein isoforms (pI) as seen by 2D-PAGE in control frontal cortex; (**b**) a significant accumulation of acidic pI isoforms is seen in PD patients compared to neurologically normal controls; $*P < 0.03$ using Student's *t*-test. Data represented as histograms are means ± standard error of means from an *n* of four cases from both categories

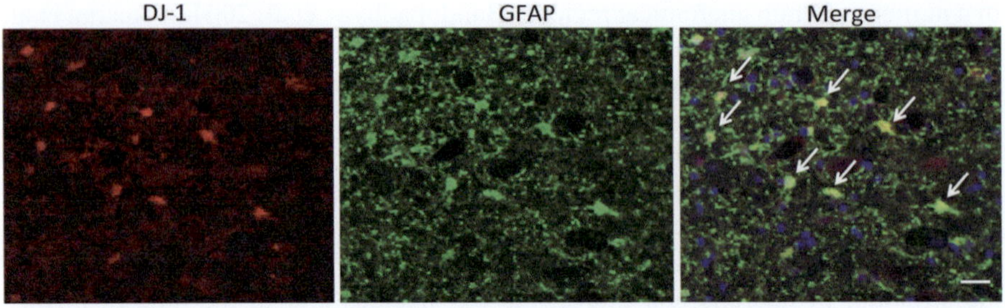

Fig. 3.3 Oxidised DJ-1 (red) and glial fibrillary acidic protein (green) overlap (yellow) in activated astrocytes in PD substantia nigra (Scale bar, 10 μm)

proteasome mimicking loss of function mutations seen with other DJ-1 mutations. The patient showed the classic pattern of dopaminergic cell loss in the substantia nigra and diffuse LB pathology at post-mortem. Earlier, using in vivo transcranial ultrasound scans, Schweitzer et al. (Schweitzer et al. 2007) demonstrated that PD-characteristic hyperechogenicity of the substantia nigra in *E64D* DJ-1 patient comparable to other monogenetically caused PD patients but less prominently than in sporadic PD patients. Positron emission tomography neuroimaging studies reported severe dopamine depletion in homozygous DJ-1 mutation carriers (Dekker et al. 2003, 2004; Hering et al. 2004). However, it needs to be seen if further DJ-1 mutation cases show LB pathology at autopsy.

3.2.2 DJ-1 in Taupathies

Several studies have examined DJ-1 expression in a number of neurodegenerative conditions in which hyperphosphorylated tau inclusions are known to be the major pathological signature (Fig. 3.3). Reactive astrocytes and neuroaxonal spheroids in disease-specific affected regions showed intense DJ-1 IR in Alzheimer's disease (AD), corticobasal degeneration (CBD), progressive supranuclear palsy (PSP), Pick's disease (PiD) and argyrophilic grains disease (AGD) (Neumann et al. 2004; Rizzu et al. 2004; Kumaran et al. 2007). The prominent and consistent labelling was observed in Pick bodies (Neumann et al. 2004; Rizzu et al. 2004; Kumaran et al. 2007), tau-positive glial inclusions (astrocytic plaques, coiled bodies and tufted astrocytes) in PSP, CBD and AGD, whereas neurofibrillary tangles and neuropil threads in AD, CBD, PSP, the LB variant of Alzheimer's disease (LBVAD), and frontotemporal dementia with parkinsonism linked to chromosome 17 (FTDP-17) showed DJ-1 IR only in a few cases studied (Neumann et al. 2004; Rizzu et al. 2004; Kumaran et al. 2007). Kumaran et al. (2007), for the first time, investigated the correlation between the presence of DJ-1 with two isoforms of tau, 3R and 4R. In cases of FTLD with MAPT mutations positive for 4R tau inclusions (i.e. N279 K and exon 10 + 16 mutations), DJ-1 IR was seen mostly in oligodendroglial coiled bodies, whereas in MAPT

R406W mutation cases positive for both 3R and 4R tau, DJ-1 IR was associated mainly with neurofibrillary tangles and neuropil threads (See Fig. 3.4). However, no DJ-1 IR was seen in FTLD with ubiquitin inclusions brains (Kumaran et al. 2007). DJ-1-positive neurites containing both 3R and 4R tau were seen in AD brains (Kumaran et al. 2007). Using double-label immunofluorescence, Rizzu et al. (2004) demonstrated the presence of DJ-1 in tau-positive inclusions within the hippocampus and dentate gyrus. The weak staining of ballooned neurons was seen in PiD, CBD and AGD brains, whereas argyrophilic grains in AGD appeared to be DJ-1 negative (Neumann et al. 2004). It is intriguing that DJ-1 is associated with neuronal and glial inclusions in tauopathies, but this could be in line with its antioxidant and/or chaperone role. This also emphasises that DJ-1 could be trapped in these inclusions in the disease process and therefore unable to perform its normal role.

3.2.3 DJ-1 in Ischaemia-Induced Neurodegeneration

Mullett and coworkers (Mullett et al. 2009) using brain tissue from human cortex demonstrated that ischaemia-induced neurodegeneration is also associated with enhanced DJ-1 IR. They reported that DJ-1 is predominantly expressed by reactive astrocytes within and surrounding cortical infarct region (Neumann et al. 2004; Rizzu et al. 2004) in both grey and white matter. Importantly, astrocytes present in subacute (weeks old) and chronic (months–years old) ischaemia regions are a lot more DJ-1 positive than those in acute infarct (days old) and non-infarct regions. Neurons inevitably expressed low levels of DJ-1 regardless of whether or not an infarct was present. Overall, these data support the idea that the major cellular DJ-1 response to stroke in the human brain is astrocytic (Mullett et al. 2009).

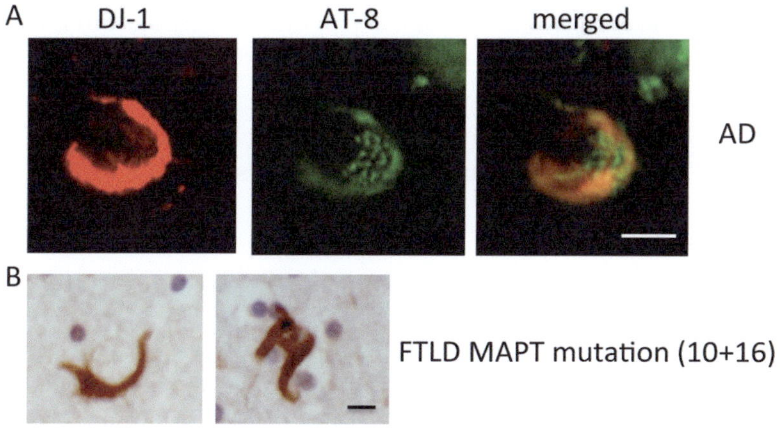

Fig. 3.4 (**a**) DJ-1 (red) is present in tau-positive (green) neurofibrillary tangle in an Alzheimer's disease case; (**b**) DJ-1-immunopositive coiled bodies from a FTLD case with MAPT mutation (10 + 16) (Scale bar is 10 μm in **a** and 5 μm in **b**)

3.2.4 DJ-1 in Sporadic Inclusion-Body Myositis (S-IBM)

S-IBM is known as the most common muscle disorder of the elderly population. It is a progressive condition associated with dramatic muscle weakness and wasting leading to severe disability (Engel and Askanas 2006) that currently cannot be successfully treated (Terracciano et al. 2008). Pathologically s-IBM is characterised by vacuolar degeneration and atrophy of muscle fibres with intra-muscle-fibre accumulations of misfolded, ubiquitinated, congophilic, multi-protein aggregates and lymphocytic inflammation (Askanas and Engel 2007; Dalakas 2007; Needham and Mastaglia 2007). Strikingly, the muscle fibre molecular phenotype shares the molecular pathology with both AD and PD, including accumulations of amyloid-β, phosphorylated tau, α-synuclein and parkin (Askanas and Engel 2007; Needham and Mastaglia 2007). There is also evidence of oxidative stress and mitochondrial abnormalities in s-IBM (Oldfors et al. 2006; Askanas and Engel 2007). Terracciano and coworkers (2008) showed that DJ-1 was highly oxidised in the muscle biopsies from s-IBM patients. DJ-1-immunoreactive cytoplasmic aggregates were identified only in a small number of s-IBM muscle fibres, whereas DJ-1 inclusions were not associated with the nuclei.

3.2.5 DJ-1 in Amyotrophic Lateral Sclerosis (ALS)

ALS is a rare adult-onset neurodegenerative condition with a selective degeneration of motor neurons in the motor cortex, brainstem and spinal cord (Milani et al. 2013). Immunostaining for DJ-1 showed a stronger signal in motor cortex sections from ALS patients compared to controls. DJ-1 IR was mostly detected in cortical neurons. A slightly increased DJ-1 signal was observed in human spinal cord sections of ALS patients in comparison to control tissue, whereas only reduced DJ-1 signal intensity was seen in human ALS muscle tissue (Knippenberg et al. 2013).

3.3 DJ-1 Protein Levels and Gene Expression

3.3.1 DJ-1 Protein Levels in Taupathies and α-Synucleinopathies

DJ-1 protein in human brain is known to migrate at molecular weight of approximately 20–24 kDa with the largest amount of the protein recovering in the buffer-soluble fraction (Neumann et al. 2004; Rizzu et al. 2004; Kumaran et al. 2009). Compared to controls, slightly elevated DJ-1 levels were detected in the insoluble fibrillar fraction from AD and PiD cortex (Neumann et al. 2004). DJ-1 in pons and cerebellum was also mostly buffer soluble. However, Neumann et al. failed to detect significant elevation of DJ-1 in the fibrillar fractions prepared from PSP pons. Furthermore, the presence of sarcosyl-insoluble modified DJ-1 in the fibrillar fractions from MSA, AD, PiD brains (Neumann et al. 2004; Rizzu et al. 2004; Kumaran et al. 2007) and FTLD with Pick bodies patients (Kumaran et al. 2007) was detected with biochemical analysis. Additionally, DJ-1 could also exist in high molecular weight complexes in normal and diseased tissue and may interact with α−synuclein (Meulener et al. 2005).

Compared to neurologically normal controls, a significant decrease in DJ-1 mRNA in PD brains was observed in putamen, frontal cortex, parietal cortex and cerebellum, whereas an up-regulation was detected in amygdala and entorhinal cortex (Kumaran et al. 2009). Analysis of DJ-1 mRNA levels indicated differential expression in diseased and control brains in different regions. Biochemical analysis revealed that soluble DJ-1 protein levels were lower in PD medulla, nigra, putamen, frontal cortex and cerebellum in comparison to age-matched controls. Similarly, the SDS-soluble DJ-1 levels were decreased in PD frontal cortex. Strikingly, the largest significant decrease in both DJ-1 mRNA and protein levels was detected in PD cerebellum, the brain region that is not typically associated with PD pathology (Kumaran et al. 2009).

Interestingly, several studies investigated the difference in the total DJ-1 levels between PD and AD cases and reported contradictory results. Thus, Moore et al. (2005) detected a fivefold increase in the DJ-1 protein level in the detergent-insoluble fraction of PD brain tissue compared to neurologically normal controls and AD brains. In contrast, DJ-1 levels significantly increased in the detergent-insoluble fraction of AD brains (Rizzu et al. 2004), whereas a significant decrease was noted in DJ-1 protein expression in PD compared to controls (Kumaran et al. 2007). Another report showed that total DJ-1 protein level is increased in PD and AD frontal cortex (Choi et al. 2006). The reason for these discrepancies may be due to the use of different anti-DJ-1 antibodies, individual differences between patient samples and/or extraction conditions (Choi et al. 2006).

A specific accumulation of DJ-1 acidic forms in brain tissue from PD (Bandopadhyay et al. 2004; Choi et al. 2006) and AD (Kumaran et al. 2007) patients was demonstrated in a number of studies. Indeed, up to 10 different DJ-1 isoforms, most importantly acidic forms (pI 5.5 and 5.7) with a molecular weight (MW) of 20 kDa, have been found in the PD and AD frontal cortex in comparison to age-matched controls (Choi et al. 2006). Interestingly, two mutually exclusive pools of DJ-1 were observed with a strong neural tissue specificity exhibited by acidic isoforms of DJ-1 monomer (Natale et al. 2010). Choi et al. (2006) have shown by using mass spectrometry analyses that DJ-1 is not only susceptible to cysteine oxidation but also to methionine oxidation to methionine sulfoxide at Met-133 and Met-134 in PD. Furthermore, Choi and coworkers revealed that basic isoforms of SDS-resistant DJ-1 dimers selectively accumulated in PD and AD are irreversibly oxidised by carbonylation. It has been recently reported that DJ-1 in idiopathic PD frontal cortex is S-nitrosylated (SNO-DJ-1) at the Cys106 site (Choi et al. 2014). Notably, a transnitrosylation reaction decreases the phosphatase activity of phosphatase and tensin homolog termed PTEN, and therefore SNO-DJ-1 plays a neuroprotective role by detoxifying nitric oxide. However, mutation or prior oxidation of cysteine residue 106 is likely to contribute to neurodegeneration by lacking the transnitrosylation activity of dysfunctional DJ-1 (Choi et al. 2014). Studying the fine-tuning of various modifications of the DJ-1 protein could hold a key in deciphering its dysfunction in neurodegenerative diseases.

3.3.2 DJ-1 Levels in HD

DJ-1 accumulation and its oxidised modification have been described in Huntington's disease (HD) brains, a neurodegenerative disorder associated with polyglutamine (polyQ) misfolding/aggregation (Sajjad et al. 2014). It has been demonstrated that total DJ-1 and DJ-1 isoforms oxidised at cysteine residue 106 are abnormally increased in HD human brains. Notably, the total level of DJ-1 protein was increased in the frontal cortex of HD patients, whereas the Cys-106 oxDJ-1 immunoreactivity was significantly enhanced in both the frontal cortex and cerebellum. In addition, Sajjad and coworkers demonstrated that DJ-1 can accelerate polyglutamine aggregation and toxicity in vitro in an oxidation-dependent manner (Sajjad et al. 2014).

3.3.3 DJ-1 Levels in ALS

Knippenberg and coworkers performed a comparative analysis of the DJ-1 mRNA and protein expression in post-mortem motor cortex and cervical and lumbar regions of spinal cord tissue and muscle biopsy samples from ALS patients and controls. They detected a non-significant increase in DJ-1 mRNA levels in human spinal cord and brain tissue from ALS patients. Reduction in DJ-1 gene expression was found in human ALS muscle tissue suggesting potential pathophysiological role for these proteins in sporadic ALS (Knippenberg et al. 2013). No difference in DJ-1 gene activity was observed in DA neurons of SN between IPD, schizophrenic patients and control subjects (Galter et al. 2007). Galter et al. (2007) also reported a predominantly neuronal expression of DJ-1 mRNA in human brain tissue. Moreover, CSF levels of DJ-1 were significantly higher in patients with sporadic ALS compared to controls (Yamashita et al. 2010).

3.3.4 DJ-1 Levels in S-IBM

Terracciano et al. (2008) demonstrated that the soluble and mitochondrial fraction DJ-1 23 kDa monomer and 46 kDa dimer, as well as DJ-1 mRNA, were significantly increased in s-IBM muscle fibres in comparison to dermatomyositis, polymyositis, morphologically nonspecific myopathy, amyotrophic lateral sclerosis, peripheral neuropathy and normal muscles. These findings might be suggestive of muscle fibres' self-defence against oxidative stress seen in s-IBM (Askanas et al. 1996; Broccolini et al. 1999, 2000; Nogalska et al. 2007; Tateyama et al. 2003; Yang et al. 1996, 1998) by up-regulation of DJ-1.

3.4 Functional Implications of DJ-1

The gene encoding multifunctional protein DJ-1 (*PARK7*) was first identified as an oncogene associated with gain of its function (Nagakubo et al. 1997). Recent evidence suggest that DJ-1 exhibits antioxidative function by altering the expression level of genes involved in antioxidative defence (Kahle et al. 2009; Im et al. 2012; Kato et al. 2013) and regulating various signalling pathways (Manning and Cantley 2007; Salmena et al. 2008; Kim et al. 2005, 2009) mediated by astrocytes (Mullett et al. 2013; Larsen et al. 2011). It increases cellular glutathione synthesis and, therefore, protects dopaminergic cells from oxidative stress (Zhou and Freed 2005). Several studies have demonstrated that DJ-1 eliminates oxidising agents, such as H_2O_2 and paraquat, and, therefore, prevents oxidative stress-induced cell death (Yokota et al. 2003; Taira et al. 2004). DJ-1 controls oxidative stress by activating superoxide dismutase 1, (Wang et al. 2011b; Girotto et al. 2014). Therefore, the important role of DJ-1 for mitochondrial integrity has been reported by numerous studies (Knippenberg et al. 2013). Taking into consideration the fact that muscle tissue contains a large amount of mitochondria and that DJ-1 regulates the muscle function, mitochondrial stress associated with DJ-1 dysfunction seen in ALS might lead to muscle degeneration (Knippenberg et al. 2013). Therefore, the role of DJ-1 in neurodegenerative muscle diseases should be explored further.

Another important role of DJ-1 is associated with the regulation of dopaminergic neurons homeostasis (Saito et al. 2014). In particular, DJ-1 activates dopamine synthesis up-regulating the transcription of tyrosine hydroxylase, an enzyme that converts L-tyrosine to L-3,4-dihydroxyphenylalanine (L-DOPA) (Zhong et al. 2006; Ishikawa et al. 2009). In addition, DJ-1 may regulate dopamine synthesis by virtue of oxidation under physiological conditions (Saito et al. 2014).

DJ-1 exhibits chaperone activity by preventing the aggregation of some proteins, including α-synuclein (Zondler et al. 2014) and microtubule-associated protein 1B (Wang et al. 2011a), "natively" unfolded proteins that tend to aggregate and fibrillate under certain conditions (Zhou et al. 2006). Also, DJ-1 inhibits aggregation and cytotoxicity caused by the mutant human α-synuclein by up-regulation of heat shock protein 70 (Zhou and Freed 2005). In addition, DJ-1 regulates the expression of neuronal β-tubulin III (Sheng et al. 2013) implying a role in the maintaining structure and function of microtubules (Conde and Caceres 2009). In addition a C-terminally truncated form of DJ-1 harbours active protease function that is cytoprotective against oxidative stress-induced apoptosis (Chen et al. 2010). Both the chaperone and protease functions of DJ-1 have important implications in our understanding of DJ-1 and neuroprotection.

DJ-1 has been reported to exhibit glyoxalase enzymatic activity (Thornalley et al. 1999; Bankpalli et al. 2015) associated with removing toxic metabolites of glucose oxidation, lipid peroxidation and DNA oxidation implicated in ageing and neurodegenerative disorders (Munch et al. 2000; Luth et al. 2005). More recently, studies showed that DJ-1 protects mouse embryonic fibroblasts and SH-SY5Y cells from glyoxals through protection of mitochondria (Lee et al. 2012). It is clear therefore that DJ-1 has the potential to be neuroprotective by virtue of its varied functions.

DJ-1 also localises, albeit to a lesser extent, at the outer membrane of the mitochondria to maintain a healthy mitochondrial environment (Miller et al. 2003; Kahle et al. 2009). It is of interest that PD affects DJ-1-associated mitochondrial targets and that loss of DJ-1 leads to impaired autophagy and accumulation of dysfunctional mitochondria (Krebiehl et al. 2010). A co-immunoprecipitation study suggested an interaction between Parkin, Pink1 and DJ-1, in vitro and in vivo situations and that they existed within the same complex (Xiong et al. 2009). The role of DJ-1 in this complex is to impart stability to PINK-1. Some heat shock proteins (HSP) such as HSPA4, HSP70 and HSP9 (mortalin) have been linked with Parkin, Pink1 and DJ-1 either through network connection studies or experimental evidence (van der Merwe et al. 2015; Li et al. 2005). These putative interactions could have important functional implications within the mitochondria; however, more work is warranted in this field in order to gain full understanding of the extent of the contribution of Parkin, PINK-1 and DJ-1 proteins in PD pathogenesis.

3.5 Therapeutic Potential

Currently available treatments for neurodegenerative diseases can only provide symptomatic relief without proven efficient cure. Taking into consideration the neuroprotective functions of DJ-1, it has been suggested as a novel therapeutic strategy in neurodegenerative disorders by reducing the formation of ROS-mediated neuronal cell loss. Thus, Yanagisawa et al. (2008) demonstrated a marked reduction of ischaemic size, behavioural dysfunction and nitrotyrosine formation after the intrastriatal injection of recombinant glutathione S-transferase-tagged human DJ-1 in rats with ischaemic neurodegeneration. They also showed a significant inhibition of H_2O_2-mediated ROS production in SH-SY5Y cell models. These results indicate that exogenous DJ-1 can reduce oxidative stress, and therefore it may be useful for the development of novel therapeutic targets in various neurodegenerative conditions, including cerebral ischaemia (Yanagisawa et al. 2008). These experimental data were corroborated by other studies using ischaemia-induced neurodegeneration models (Jeong et al. 2012b).

Lev et al. (2015) performed a study with the aim to examine the therapeutic potential of exogenous DJ-1 protein in the 6-hydroxydopamine and MPTP mouse models of PD and in cultured cells. They showed that DJ-1 increased cell survival, reduced ROS accumulation and protected against oxidative stress and neurotoxic insults using cultured cells. Moreover, DJ-1 administration restored dopamine content and improved behavioural abnormalities in PD mouse models. The improvement of mesencephalic dopaminergic neuron survival was also demonstrated by Toyoda et al. (2014) using D-lactate or glycolate, products of DJ-1 metabolism. The experimental data obtained by Lev et al. (2013) also pointed to a potential therapeutic use of DJ-1 as it has been shown to protect cells against dopamine toxicity, reduce oxidative stress and improve dopamine release in vivo using 6-hydroxydopamine hemiparkinsonian mouse model and transgenic DJ-1 knockout mice. The potential neuroprotective effect of DJ-1 has been reported in a number of other studies (Zhang et al. 2012; Jeong et al. 2012a).

DJ-1 modulating compounds, such as UCP0045037 and UCP0054278, have been reported to inhibit ROS production and oxidative stress-induced toxicity, prevent the complete oxidation of DJ-1 and, therefore, mimic the partially oxidised forms needed for normal DJ-1 functioning (Miyazaki et al. 2008; Yamane et al. 2009; Sajjad et al. 2014; Fig. 8). It is highly important to explore the therapeutic relevance of such compounds in polyQ-dependent pathology seen in HD (Sajjad et al. 2014).

3.6 Conclusion

It is evident that DJ-1 has ubiquitous expression in the brain and is present in both neurons and glia and plays a role in the pathogenesis of several neurodegenerative diseases. The role of DJ-1 is linked strongly with its antioxidative properties together with its putative role as a chaperone (Bandopadhyay and de Belleroche 2010) and/or protease activities and mitochondrial functions. These roles may not be mutually exclusive but could act in a cooperative manner. The DJ-1 levels are altered in several diseases, and therefore boosting the antioxidant role of DJ-1 could be a strategy to protect vulnerable neurons in all of the neurodegenerative disorders. DJ-1's interaction with PINK1 and Parkin could have implications in mitochondrial function. The recent link of DJ-1 and another kinase, LRRK2 (Heo et al. 2010), should also be explored in greater detail. A better understanding of DJ-1 functioning and misfunctioning in neurodegenerative conditions may offer the possibility to pharmacologically manipulate molecular pathways and antioxidant systems as promising neuroprotective therapies. The evidence of DJ-1 efficacy as a therapeutic tool is currently restricted to in vitro and in vivo animal studies. Establishing stem cell models using DJ-1 mutation carriers should pave the way for potential drug screening and ultimately for treatments in human patients in neurodegenerative and neuromuscular diseases.

Acknowledgements RB is funded by Reta Lila Weston Institute of Neurological Studies, UCL Institute of Neurology. Part of the work described here was funded by Parkinson's UK project grant award to RB. This work was supported in part by the Wellcome Trust/MRC joint call in Neurodegeneration award (WT089698) to the UK Parkinson's Disease Consortium (UKPDC) whose members are from the UCL Institute of Neurology, University of Sheffield and the MRC Protein Phosphorylation Unit Dundee. We also thank Dr. Ian Evans, Department of Medicine UCL Rayne Institute for his critical comments.

References

Ariga H, Takahashi-Niki K, Kato I, Maita H, Niki T, Iguchi-Ariga SMM (2013) Neuroprotective function of DJ-1 in Parkinson's disease. Oxidative Med Cell Longev 2013:1. https://doi.org/10.1155/2013/683920

Askanas V, Engel WK (2007) Inclusion-body myositis, a multifactorial muscle disease associated with aging: current concepts of pathogenesis. Curr Opin Rheumatol 19:550–559

Askanas V, Sarkozi E, Alvarez RB (1996) Superoxide Dismutase-1 gene and protein in vacuolated muscle fibers of sporadic inclusion body myositis, hereditary inclusion body myopathy, and cultured human muscle after β-amyloid precursor protein gene transfer. Neurology 46:A487

Bandopadhyay R, de Belleroche J (2010) Pathogenesis of Parkinson's disease: emerging role of molecular chaperones. Trends Mol Med 32(9):1199–1205

Bandopadhyay R, Kingsbury AE, Cookson MR, Reid AR, Evans IM, Hope AD, Pittman AM, Lashley T, Canet-Aviles R, Miller DW, McLendon C, Strand C, Leonard AJ, Abou-Sleiman PM, Healy DG, Ariga H, Wood NW, de Silva R, Revesz T, Hardy JA, Lees AJ (2004) The expression of DJ-1 (PARK7) in normal human CNS and idiopathic Parkinson's disease. Brain 127:420–430

Bandopadhyay R, Miller DW, Kingsbury AE, Jowett TP, Kaleem MM, Pittman AM, de Siva R, Cookson MR, Lees AJ (2005) Development, characterisation and epitope mapping of novel monoclonal antibodies for DJ-1 (PARK7) protein. Neurosci Lett 383:225–230

Bandopadhyay R, Antipova D, Strand C, Warner TT (2017) Investigation of the role of oxidised DJ-1 in Parkinson's disease progression. Neuropathol Appl Neurobiol 43(P09):36–36

Bankpalli K, Saladi S, Awadia SS, Goswami AV, Samaddar M, D'Silva P (2015) Robust glyoxalase activity of Hsp31, a ThiJ/DJ-1/PfpI family member protein, is critical for oxidative stress resistance in Saccharomyces cerevisiae. J Biol Chem 290(44):26491–26507

Bonifati V, Rizzu P, van Baren MJ, Schaap O, Breedvel GJ, Krieger E, Dekker MC, Squiteiri F, Ibanez P, Joosse M, van Dongen JW, Vanacore N, van Swieten JC, Brice A, Meco G, van Dujin CM, Oostra BA, Heutink P (2003) Mutations in the DJ-1 gene associated with autosomal recessive early-onset parkinsonism. Science 299(5604):256–259

Broccolini A, Mirault ME, Engel WK, Askanas V (1999) Abnormal accumulation of seleno-glutathione peroxidase-1 and catalase and their m-RNAs in sporadic inclusion body myositis. Neurology 52:A333

Broccolini A, Engel WK, Alvarez RB, Askanas V (2000) Redox factor-1 in muscle biopsies of patients with inclusion-body myositis. Neurosci Lett 287:1–4

Canet-Avilés RM, Wilson MA, Miller DW, Ahmad R, McLendon C, Bandyopadhyay S, Baptista MJ, Ringe D, Petsko GA, Cookson MR (2004) The Parkinson's disease protein DJ-1 is neuroprotective due to cysteine-sulfinic acid-driven mitochondrial localization. Proc Natl Acad Sci U S A 101:9103–9108

Chen J, Li L, Chin LS (2010) Parkinson disease protein DJ-1 converts from a zymogen to a protease by carboxyl-terminal cleavage. Hum Mol Genet 19(12):2395–2408

Choi J, Sullards MC, Olzmann JA, Rees HD, Weintraub ST, Bostwick DE, Gearing M, Levey AI, Chin LS, Li L (2006) Oxidative damage of DJ-1 is linked to sporadic Parkinson and Alzheimer diseases. J Biol Chem 281:10816–10824

Choi MS, Nakamura T, Cho SJ, Han X, Holland EA, Qu J, Petsko GA, Yates JR, Liddington RC, Lipton SA (2014) Transnitrosylation from DJ-1 to PTEN attenuates neuronal cell death in Parkinson's disease models. J Neurosci 34:15123–15131

Conde C, Caceres A (2009) Microtubule assembly, organization and dynamics in axons and dendrites. Nat Rev Neurosci 10:319Y32

Dalakas MC (2007) Advances in the immunobiology and treatment of inflammatory myopathies. Curr Rheumatol 9(4):291–297

Dekker M, Bonifati V, van Swieten J, Leenders N, Galjaard RJ, Snijders P, Horstink M, Heutink P, Oostra B, van Duijn C (2003) Clinical features and neuroimaging of PARK7-linked parkinsonism. Mov Disord 18:751–757

Dekker MCJ, Eshuis SA, Maguire RP, Veenma-van der Duijn L, Pruim J, Snijders PJ, Oostra BA, van Duijn CM, Leenders KL (2004) PET neuroimaging and mutations in the DJ-1 gene. J Neural Transm 111:1575–1581

Engel WK, Askanas V (2006) Inclusion-body myositis: clinical, diagnostic, and pathologic aspects. Neurology 66:20–29

Galter D, Westerlund M, Belin AC, Olson L (2007) DJ-1 and UCH-L1 gene activity patterns in the brains of controls, Parkinson and schizophrenia patients and in rodents. Physiol Behav 92:46–53

Garcia-Ruiz PJ, Chaudhuri KR, Martinez-Martin P (2014) Non-motor symptoms of Parkinson's disease a review from the past. J Neurol Sci 338:30–33

Girotto S, Cendron L, Bisaglia M, Tessari I, Mammi S, Zanotti G, Bubacco L (2014) DJ-1 is a copper chaperone acting on SOD1 activation. J Biol Chem 289:10887–10899

Heo HY, Park JM, Kim CH, Han BS, Kim KS, Seol W (2010) LRRk2 enhances oxidative stress-induced neurotoxicity via its kinase activity. Exp Cell Res 316(4):649–656

Hering R, Strauss KM, Tao X, Bauer A, Woitalla D, Mietz EM, Petrovic S, Bauer P, Schaible W, Müller T, Schöls L, Klein C, Berg D, Meyer PT, Schulz JB, Wollnik B, Tong L, Krüger R, Riess O (2004) Novel homozygous p.E64D mutation in DJ1 in early onset Parkinson disease (PARK7). Hum Mutat 24:321–329

Hernandez DG, Reed X, Singleton AB (2016) Genetics in Parkinson's disease: mendelian versus non-mendelian inheritance. J Neurochem 139:59. https://doi.org/10.1111/jnc.13593

Im JY, Lee KW, Woo JM, Junn E, Mouradian MM (2012) DJ-1 induces thioredoxin 1 expression through the Nrf2 pathway. Hum Mol Gen 21:3013–3024

Ishikawa S, Taira T, Niki T, Takahashi-Niki K, Maita C, Maita H, Ariga H, Iguchi-Ariga SM (2009) Oxidative status of DJ-1-dependent activation of dopamine synthesis through interaction of tyrosine hydroxylase and 4-dihydroxy-L-phenylalanine (L-DOPA) decarboxylase with DJ-1. J Biol Chem 284:28832–28844

Jellinger KA (2001) The pathology of Parkinson's disease. Adv Neurol 86:55–72

Jellinger KA (2012) Neuropathology of sporadic Parkinson's disease: evaluation and changes of concepts. Mov Disorders 27(1):8–30

Jeong HJ, Kim DW, Kim MJ, Woo SJ, Kim HR, Kim SM, Jo HS, Hwang HS, Kim DS, Cho SW, Won MH, Han KH, Park JS, Eum WS, Choi SY (2012a) Protective effects of transduced Tat-DJ-1 protein against oxidative stress and ischemic brain injury. Exp Mol Med 44:586–593

Jeong HJ, Kim DW, Woo SJ, Kim HR, Kim SM, Jo HS, Park M, Kim DS, Kwon OS, Hwang IK, Han KH, Park J, Eum WS, Choi SY (2012b) Transduced Tat-DJ-1 protein protects against oxidative stress-induced SH-SY5Y cell death and Parkinson disease in a mouse model. Mol Cells 33:471–478

Jin J, Meredith GE, Chen L, Zhou Y, Xu J, Shie FS, Lockhart P, Zhang J (2005) Quantitative proteomic analysis of mitochondrial proteins: relevance to Lewy body formation and Parkinson's disease. Brain Res Mol Brain Res 134:119–138

Kahle PJ, Waak J, Gasser T (2009) DJ-1 and prevention of oxidative stress in Parkinson's disease and other age-related disorders. Free Rad Biol Med 47:1354–1361

Kato I, Maita H, Takahashi-Niki K, Saito Y, Noguchi N, Iguchi-Ariga SM, Ariga H (2013) Oxidized DJ-1 inhibits p53 by sequestering p53 from promoters in a DNA-binding affinity-dependent manner. Mol Cell Biol 33:340–359

Kim RH, Peters M, Jang Y, RH K, Peters M, Jang Y, Shi W, Pintilie M, Fletcher GC, DeLuca C, Liepa J, Zhou L, Snow B, Binari RC, Manoukian AS, Bray MR, Liu FF, Tsao MS, Mak TW (2005) DJ-1, a novel regulator of the tumor suppressor PTEN. Cancer Cell 7:263–273

Kim YC, Kitaura H, Taira T, Iguchi-Ariga SM, Ariga H (2009) Oxidation of DJ-1dependent cell transformation through direct binding of DJ-1 to PTEN. Int J Oncol 35:1331–1341

Kinumi T, Kimata J, Taira T, Ariga H, Niki E (2004) Cysteine-106 of DJ-1 is the most sensitive cysteine residue to hydrogen peroxide-mediated oxidation in vivo in human umbilical vein endothelial cells. Biochem Biophys Res Commun 317:722–728

Knippenberg S, Sipos J, Thau-Habermann N, Körner S, Rath KJ, Dengler R, Petri S (2013) Altered expression of DJ-1 and PINK1 in sporadic ALS and in the SOD1(G93A) ALS mouse model. J Neuropathol Exp Neurol 72:1052–1061

Krebiehl G, Ruckerbauer S, Burbulla LF, Kieper N, Maurer B, Waak J, Wolburg H, Gizatullina Z, Gellerich FN, Woitalla D, Reiss O, Kahle PJ, Proikas-Cezanne T, Kruger R (2010) Reduced basal autophagy and impaired mitochondrial dynamics due to loss of Parkinson's disease-associated protein DJ-1. PLos ONE 5(2):e9367

Kumaran R, Kingsbury A, Coulter I, Lashley T, Williams D, de Silva R, mann D, Revesz T, Lees A, Bandopadhyay R (2007) DJ-1 (PARK7) is associated with 3R and 4R tau neuronal and glial inclusions in neurodegenerative disorders. Neurobiol Dis 28:122Y32

Kumaran R, Vandrovcova J, Luk C, Sharma S, Renton A, Wood NW, Hardy JA, Lees AJ, Bandopadhyay R (2009) Differential DJ-1 gene expression in Parkinson's disease. Neurobiol Dis 36:393–400

Larsen NJ, Ambrosi G, Mullett SJ, Berman SB, Hinkle DA (2011) DJ-1 knock-down impairs astrocyte mitochondrial function. Neuroscience 196:251Y64

Lee JY, Song J, Kwon K, Jang S, Kim C, Baek K, Kim J, Park C (2012) Human DJ-1 and its homologs are novel glyoxalases. Hum Mol Genet 21(14):3215–3225

Lev N, Barhum Y, Pilosof NS, Ickowicz D, Cohen HY, Melamed E, Offen D (2013) DJ-1 protects against dopamine toxicity: implications for Parkinson's disease and aging. J Gerontol A Biol Sci Med Sci 68(3):215–225. https://doi.org/10.1093/gerona/gls147. Epub 2012 Aug 10

Lev N, Barhum Y, Ben-Zur T, Aharony I, Trifonov L, Regev N, Melamed E, Gruzman A, Offen D (2015) A DJ-1 based peptide attenuates dopaminergic degeneration in mice models of Parkinson's disease via enhancing Nrf2. PLoS One 10(5):e0127549. https://doi.org/10.1371/journal.pone.0127549. eCollection 2015

Li HM, Niki T, Taira T, Iguchi-Ariga SMM, Ariga H (2005) Association of DJ-1 with chaperones and enhanced association and colocalisation with mitochondrial Hsp70 by oxidative stress. Free Radic Res 39:1091–1099

Lin X, Cook TJ, Zabetian CP, Leverenz JB, Peskind ER, Hu SC, Cain KC, Pan C, Edgar JS, Dr G, Racette BA, Checkoway H, Montine TJ, Shi M, Zhang J (2012) DJ-1 isoforms in whole blood as potential biomarkers of Parkinson disease. Sci Rep 2:954

Luth HJ, Ogunlade V, Kuhla B, Kientsch_Engel R, Sthal P, Webster J, Arendt T, Munch G (2005) Age- and stage-dependent accumulation of advanced glycation end products in intracellular deposits in normal and Alzheimer's disease brains. Cereb Cortex 15:211–220

Manning BD, Cantley LC (2007) AKT/PKB signaling: navigating downstream. Cell 129:1261–1274

Meulener MC, Graves CL, Sampathu DM (2005) DJ-1 is present in a large molecular complex in human brain tissue and interacts with alpha-synuclein. J Neurochem 93(6):1524–1532

Milani P, Ambrosi G, Gammoh O, Blandini F, Cereda C (2013) SOD1 and DJ-1 converge at Nrf2 pathway: a clue for antioxidant therapeutic potential in neurodegeneration. Oxidative Med Cell Longev 2013:836760

Miller DW, Ahmad R, Hague S, Baptista MJ, Canet-Aviles R, Mc Lendon C, Carter DM, Zhu P-P, Stadler J, Chandran J, Klinefelter GR, Blackstone C, Cookson MR (2003) L166P mutant DJ-1, causative for recessive Parkinson's disease, is degraded through the ubiquitin-proteasome system. J Biol Chem 278:36588–36595

Miyazaki S, Yanagida T, Nunome K, Ishikawa S, Inden M, Kitamura Y, Nakagawa S, Taira T, Hirota K, Niwa M, Iguchi-Ariga SM, Ariga H (2008) DJ-1-binding compounds prevent

oxidative stress-induced cell death and movement defect in Parkinson's disease model rats. J Neurochem 105:2418–2434
Moore DJ, Zhang L, Troncoso J, Lee MK, Hattori N, Mizuno Y, Dawson TM, Dawson VL (2005) Association of DJ-1 and parkin mediated by pathogenic DJ-1 mutations and oxidative stress. Hum Mol Genet 14:71–84
Mullett SJ, Hamilton RL, Hinkle DA (2009) DJ-1 immunoreactivity in human brain astrocytes is dependent on infarct presence and infarct age. Neuropathology 29:125Y31
Mullett SJ, Di Maio R, Greenamyre JT, Hinkle DA (2013) DJ-1 expression modulates astrocyte-mediated protection against neuronal oxidative stress. J Mol Neurosci 49:507Y11
Munch G, Luth HJ, Wong A, Arnedt T, Hirsch E, Ravid R, Riederer P (2000) Crosslinking of alpha-synuclein by advanced glycation endproducts—an early pathophysiological step in Lewy body formation? J Chem Neuroanat 20:253–257
Nagakubo D, Taira T, Kitaura H, Ikeda M, Tamai K, Iguchi-Ariga SM, Ariga H (1997) DJ-1, a novel oncogene which transforms mouse NIH3T3 cells in cooperation with ras. Biochem Biophys Res Commun 231:509–513
Natale M, Bonino D, Consoli P, Alberio T, Ravid RG, Fasano M, Bucci EM (2010) A meta-analysis of two-dimensional electrophoresis pattern of the Parkinson's disease-related protein DJ-1. Bioinformatics 26:946–952
Needham M, Mastaglia FL (2007) Inclusion body myositis: current pathogenetic concepts and diagnostic and therapeutic approaches. Lancet Neurol 6:620–631
Neumann M, Müller V, Görner K, Kretzschmar HA, Haass C, Kahle PJ (2004) Pathological properties of the Parkinson's disease-associated protein DJ-1 in alpha-synucleinopathies and tauopathies: relevance for multiple system atrophy and Pick's disease. Acta Neuropathol 107:489–496
Nogalska A, Wojcik S, Engel WK, McFerrin J, Askanas V (2007) Endoplasmic reticulum stress induces myostatin precursor protein and NF-kappaB in cultured human muscle fibers: relevance to inclusion body myositis. Exp Neurol 204:610–618
Oldfors A, Moslemi AR, Jonasson L, Jonasson L, Ohlsson M, Kollberg G, Lindberg C (2006) Mitochondrial abnormalities in inclusion-body myositis. Neurology 66:49–55
Olzmann JA, Bordelon JR, Muly EC, Rees HD, Levey AI, Li L, Chin LS (2007) Selective enrichment of DJ-1 protein in primate striatal neuronal processes: implications for Parkinson's disease. J Comp Neurol 500:585–599
Quinn N, Critchley P, Marsden CD (1987) Young onset Parkinson's disease. Mov Disord 2:73–91
Rizzu P, Hinkle DA, Zhukareva V, Bonifati V, Severijinen LA, Martinez D, Ravid R, Kamphorst W, Eberwine JH, Lee VM, Trojanowski JQ, Heutink P (2004) DJ-1 colocalizes with tau inclusions: a link between parkinsonism and dementia. Ann Neurol 55:113–118
Saito Y (2014) Oxidized DJ-1 as a possible biomarker of Parkinson's disease. J Clin Biochem Nutr 54:138–144
Saito Y, Hamakubo T, Yoshida Y, Ogawa Y, Hara Y, Fujimura H, Imai Y, Iwanari H, Mochizuki Y, Shichiri M, Nishio K, kinumi T, Noguchi N, Kodama T, Niki E (2009) Preparation and application of monoclonal antibodies against oxidized DJ-1. Significant elevation of oxidized DJ-1 in erythrocytes of early-stage Parkinson disease patients. Neurosci Lett 465:1–5
Saito Y, Miyasaka T, Hatsuta H, Takahashi-Niki K, Mita Y, Kusano-Arai O, Iwanari H, ariga H, Hamakubo T, Yoshida Y, Niki E, Murayama S, Ihara Y, Noguchi N (2014) Immunostaining of oxidized DJ-1 in human and mouse brains. J Neuropathol Exp Neurol 73:714–728
Sajjad MU, Green EW, Miller-Fleming L (2014) DJ-1 modulates aggregation and pathogenesis in models of Huntington's disease. Hum Mol Genet 23:755–766
Salmena L, Carracedo A, Pandolfi PP (2008) Tenets of PTEN tumor suppression. Cell 133:403–414
Schweitzer KJ, Brüssel T, Leitner P (2007) Transcranial ultrasound in different monogenetic subtypes of Parkinson's disease. J Neurol 254:613–616
Sheng C, Heng X, Zhang G, Xiong R, Li H, Zhang S, Chen S (2013) DJ-1 deficiency perturbs microtubule dynamics and impairs striatal neurite outgrowth. Neurobiol Aging 34:489Y98

Taipa R, Pereira C, Reis I, Alonso I, Baston-Lima A, MetoPieres M, Magalhaes M (2016) DJ-1 linked parkinsonism (PARK7) is associated with LB. Pathology 139(6):1680–1687

Taira T, Saito Y, Niki T, Iguchi-Ariga SM, Takahashi K, Ariga H (2004) DJ-1 has a role in antioxidative stress to prevent cell death. EMBO Rep 5:213–218

Tateyama M, Takeda A, Onodera Y, Matsuzaki M, Hasegawa T, Munomura A, Hirai K, Perry G, Smith MA, Itoyama Y (2003) Oxidative stress and predominant Abeta42(43) deposition in myopathies with rimmed vacuoles. Acta Neuropathol 105:581–585

Terracciano C, Nogalska A, Engel WK, Wojcik S, Askanas V (2008) In inclusion-body myositis muscle fibers Parkinson-associated DJ-1 is increased and oxidized. Free Radic Biol Med 45:773–779

Thornalley PJ, Langborg A, Minhas HS (1999) Formation of glyoxal, methylglyoxal and 3-deoxyglucosone in the glycation of proteins by glucose. Biochem J 344:109–116

Toyoda Y, Erkut C, Pan-Montojo F, Boland S, Stewart MP, Muller DJ, Wurst W, Hyman AA, Kurzchalia TV (2014) Products of the Parkinson's disease-related glyoxalase DJ-1, D-lactate and glycolate, support mitochondrial membrane potential and neuronal survival. Biol Open 3:777–784

Van der Merwe C, Dashti ZJS, Christoffels A, Loos B, Bardien S (2015) Evidence for a common biological pathway linking three Parkinson's disease-causing genes: parkin, PINK1 and DJ-1. Eur J Neurosci 41:1113–1125

Wang Z, Zhang Y, Zhang S, Guo Q, Tan Y, Wang X, Xiong R, Ding J, Chen S (2011a) DJ-1 can inhibit microtubule associated protein 1 B formed aggregates. Mol Neurodegener 6:38

Wang Z, Liu J, Chen S, Wang Y, Cao L, Zhang Y, Kang W, Li H, Gui Y, Chen S, Ding J (2011b) DJ-1 modulates the expression of Cu/Zn-superoxide dismutase-1 through the Erk1/2-Elk1 pathway in neuroprotection. Ann Neurol 70:591–599

Wilson MA (2011) The role of cysteine oxidation in DJ-1 function and dysfunction. Antioxid Redox Signal 15:111–122

Xiong H, Wang D, Chen L, Choo YS, M H TC, Xia K, Jiang W, Ronal Z, Zhuang X, Zhang Z (2009) Parkin, PINK1 and DJ-1 form a ubiquitin E3 ligase complex promoting unfolded protein degradation. J Clin Investig 119(3):650–660

Yamane K, Kitamura Y, Yanagida T, Takata K, Yanagisawa D, Taniguchi T, Taira T, Ariga H (2009) Oxidative neurodegeneration is prevented by UCP0045037, an allosteric modulator for the reduced form of DJ-1, a wild-type of familial Parkinson's disease-linked PARK7. Int J Mol Sci 10:4789–4804

Yamashita S, Mori A, Kimura E, Mita S, Maeda Y, Hirano T, Uchino M (2010) DJ-1 forms complexes with mutant SOD1 and ameliorates its toxicity. J Neurochem 113:860–870

Yanagisawa D, Kitamura Y, Inden M, Takata K, Taniguchi T, Morikawa S, Morita M, Inubushi T, Tooyama I, Taira T, Iguchi-Ariga SM, Akaike A, Ariga H (2008) DJ-1 protects against neurodegeneration caused by focal cerebral ischemia and reperfusion in rats. J Cereb Blood Flow Metab 28:563–578

Yang CC, Alvarez RB, Engel WK, Askanas V (1996) Increase of nitric oxide synthases and nitrotyrosine in inclusion-body myositis. Neuroreport 8:153–158

Yang CC, Askanas V, Engel WK, Alvarez RB (1998) Immunolocalization of transcription factor NF-kappaB in inclusion-body myositis muscle and at normal human neuromuscular junctions. Neurosci Lett 254:77–80

Yokota T, Sugawara K, Ito K, Takahashi R, Ariga H, Mizusawa H (2003) Down regulation of DJ-1 enhances cell death by oxidative stress, ER stress, and proteasome inhibition. Biochem Bioph Res Co 312:1342–1348

Zhang L, Shimoji M, Thomas B, Moore DJ, S-W Y, Marupudi NI, Torp R, Torgner IA, Otterson OP, Dawson TM, Dawson VL (2012) Mitochondrial localization of the Parkinson's disease related protein DJ-1: implications for pathogenesis. Hum Mol Genet 14:2063–2073

Zhong N, Kim CY, Rizzu P, Geula C, Porter DR, Pothos EN, Squitieri F, Heutink P, Xu J (2006) DJ-1 transcriptionally up-regulates the human tyrosine hydroxylase by inhibiting the sumoylation of pyrimidine tract-binding protein-associated splicing factor. J Biol Chem 281:20940–20948

Zhou W, Freed CR (2005) DJ-1 up-regulates glutathione synthesis during oxidative stress and inhibits A53T alpha-synuclein toxicity. J Biol Chem 280:43150–43158

Zhou W, Zhu M, Wilson MA, Petsko GA, Fink AL (2006) The oxidation state of DJ-1 regulates its chaperone activity toward alpha-synuclein. J Mol Biol 356:1036–1048

Zondler L, Miller-Fleming L, Repici M, Gonçalves S, Tenreiro S, Rosado-Ramos R, Betzer C, Straatman KR, Jensen PH, Giorgini F, Outeiro TF (2014) DJ-1 interactions with α-synuclein attenuate aggregation and cellular toxicity in models of Parkinson's disease. Cell Death Dis 5:e1350

Chapter 4
Expression of DJ-1 in Cancer Cells: Its Correlation with Clinical Significance

Takahiko Kawate, Benio Tsuchiya, and Keiichi Iwaya

Abstract Upregulation of *DJ-1* mRNA is commonly observed in various human cancers such as ductal carcinoma of the breast, non-small cell carcinoma of the lung, pancreatic duct adenocarcinoma, urinary transitional cell carcinoma, and gynecologic carcinoma. At the protein level, intensity and intracellular localization of DJ-1 expression is varied, and the DJ-1 protein regulates cancer progression, clinical aggressiveness, differentiation, cancer cell morphology, and drug sensitivity. Thus, DJ-1 plays a critical role in cancer. Although DJ-1 has an important role within cancer cells, cancer cells secrete DJ-1 outside the cells. DJ-1 may serve as a tumor marker that can be detected from an early stage in the blood, secretory fluids, ascites, or pleural effusion.

Keywords DJ-1 • Cancer • Oxidative stress • In situ hybridization • Prognosis • Drug resistance • Secretion • Tumor marker

4.1 Introduction

dj-1 was discovered as an oncogene that promotes neoplastic transformation in collaboration with *Ras*. It is located in locus 1p36.12-13 in the human genome, and in this region, loss of heterozygosity has been confirmed in various kinds of cancers such as gastrointestinal carcinoma, lung cancer, and breast cancer, suggesting that it acts as a tumor suppressor gene.

T. Kawate
Department of Breast Oncology, Tokyo Medical University, Shinjuku-ku, Tokyo 160-0023, Japan

B. Tsuchiya
Department of Pathology, School of Allied Health Sciences, Kitasato University, Minami-ku, Sagamihara, Kanagawa 252-0373, Japan

K. Iwaya (✉)
Department of Pathology, Kyoundo Hospital, Sasaki Institute, 101-0062 2-2 Kandasurugadai, Chiyoda-ku, Tokyo, Japan
e-mail: iwaya@po.kyoundo.jp

H. Ariga, S.M.M. Iguchi-Ariga (eds.), *DJ-1/PARK7 Protein*, Advances in Experimental Medicine and Biology 1037,
https://doi.org/10.1007/978-981-10-6583-5_4

Whether *dj-1* acts as oncogene or suppressor gene is not determined partly due to the fact that DJ-1 has pleiotropic functions. As for DJ-1 expression, DJ-1 is expressed in the majority of cancer cells, and it may be easy to think *dj-1* as an oncogene. We review and show variety of expression patterns of DJ-1 in the main carcinoma tissues (breast cancer, lung cancer, pancreatic cancer, bladder cancer, gynecologic cancer) in this chapter and comment on the clinicopathological significance.

4.2 Breast Cancer

4.2.1 DJ-1 Expression in Normal Breast Tissue

DJ-1 protein expression is seen in the breast tissue. DJ-1 protein is expressed in both types of epithelial cells: glandular epithelial cells and myoepithelial cells (Fig. 4.1). DJ-1 is also expressed in vascular smooth muscle cells, endothelial cells, and nerve cells in the breast tissue. DJ-1 protein is located both in nucleus and cytoplasm and in varying nucleocytoplasmic ratios. There is no detectable expression that is apparent to the fibroblasts around the mammary duct by immunohistochemistry.

4.2.2 DJ-1 Expression in Breast Cancer Tissue

Both invasive and noninvasive ductal carcinomas display stronger expression of *DJ-1* mRNA than non-cancerous breast duct epithelium. For the expression level of DJ-1 protein by immunohistochemistry, Naour et al. first reported low expression compared to normal breast ductal or lobular epithelium, particularly in the nuclei (Le Naour et al. 2001). On the other hand, Kim et al. described cytoplasmic

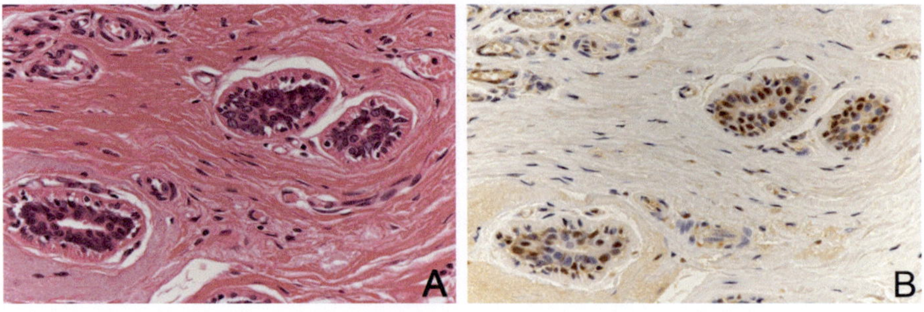

Fig. 4.1 DJ-1 expression in normal breast tissue. (**a**) Three mammary ducts are seen, which are composed of inner layer of glandular epithelium and outer layer of myoepithelium (H & E stain, original ×400). (**b**) DJ-1 protein is expressed in both kinds of epithelial cells (DJ-1 immunostaining, original ×400)

staining for DJ-1 in cancer cells (Kim et al. 2005). In our study, the intensity of DJ-1 immunohistochemical staining in cancer cells was compared to that of adjacent normal ductal cells without distinguishing nuclear and cytoplasmic distribution (Tsuchiya et al. 2012). As a result, about half of cancer cells displayed higher DJ-1 staining than normal ductal cells. (Fig. 4.2). In the breast cancer cases with low DJ-1 protein expression, upregulation of *DJ-1* mRNA was detected in almost all the cases, and thus, there was a discordance in DJ-1 expression between mRNA and protein levels in about half of breast cancer cases (Fig. 4.3).

4.2.3 Correlation of DJ-1 Expression and Clinicopathological Factors and Prognosis

The correlation between low DJ-1 protein expression and clinicopathological factors was examined. Lower levels of DJ-1 were found to be associated with larger tumor size and higher histological grade. Additionally, low DJ-1 protein expression was positively correlated with hormone receptor-negative status (estrogen receptor, progesterone receptor), HER2 score 3+, and high Ki-67 labeling index. All these

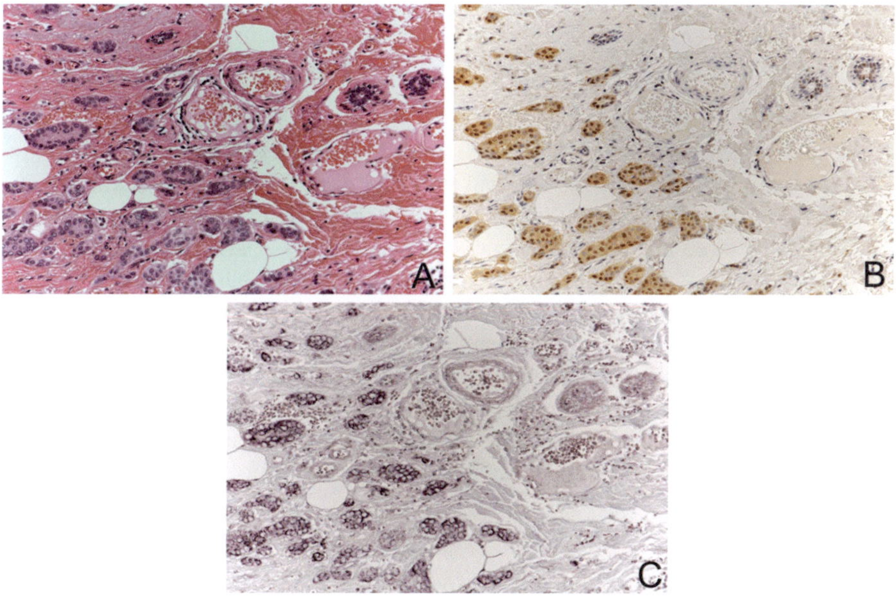

Fig. 4.2 DJ-1 expression in invasive ductal cancer tissue. (**a**) Invasive ductal carcinoma cells (*left side*) and normal mammary ducts and blood vessels (*right side*) (H & E stain, original ×200). (**b**) High DJ-1 protein expression is detected in cancer cells compared with that in normal mammary ductal cells and vascular endothelial cells (DJ-1 immunostaining, original ×200). (**c**) Stronger expression of *DJ-1* mRNA in cancer cells than in normal cells and vascular endothelial cells (in situ hybridization of DJ-1, original ×200)

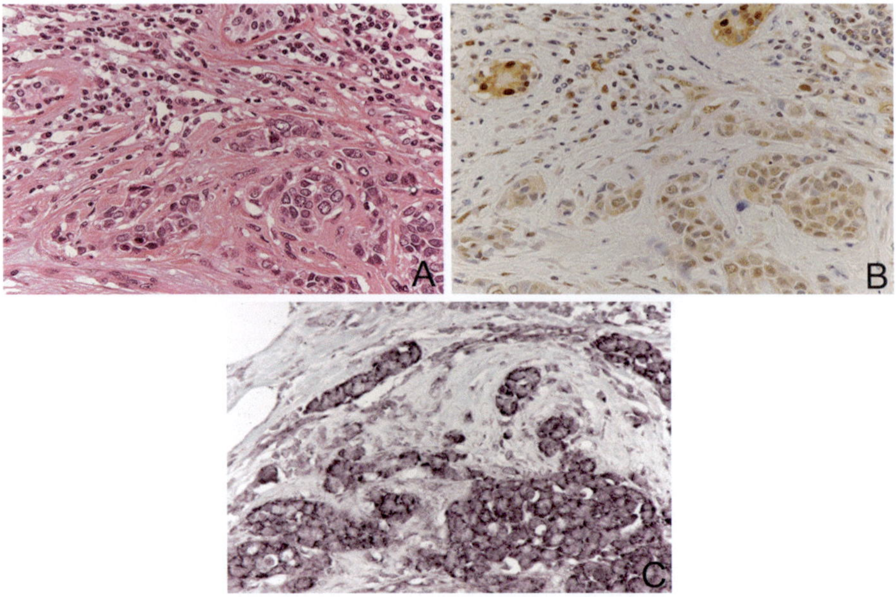

Fig. 4.3 DJ-1 expression in invasive ductal cancer tissue. (**a**) Invasive ductal carcinoma cells are seen. Normal mammary ducts are seen in the *upper side* (H & E stain, original ×400). (**b**) Lower DJ-1 protein expression is detected in cancer cells compared to that in normal mammary ductal cells (DJ-1 immunostaining, original ×400). (**c**) High expression of *DJ-1* mRNA in cancer cells (in situ hybridization of DJ-1, original ×400)

factors are indicators of poor prognoses. Indeed, in breast cancer patients with low DJ-1 protein level, disease-free survival time and overall survival time was significantly shorter than in those with high DJ-1 protein level. In addition, we found no correlation between nucleocytoplasmic localization of DJ-1 and prognosis.

To test the efficacy of DJ-1 as predictor for preoperative chemotherapy for breast cancer, DJ-1 protein expression and pathological complete remission (pCR) were studied in three therapy groups—anthracycline (EC) + docetaxel (DOC), in EC + paclitaxel (Pacli), and trastuzumab combination (Kawate et al. 2013). Breast cancer cases with low DJ-1 expression became pCR frequently in all the three therapy groups. Among the subtypes, low DJ-1 expression was associated with a high pCR rate in luminal A, luminal B, triple-negative breast cancer.

4.2.4 *DJ-1 as a Secretory Protein*

Breast cancer cells secrete DJ-1 protein. In a study examining sera of breast cancer patients using the ELISA method, Naour et al. detected higher levels of DJ-1 protein than those of non-cancer patients for the first time (Le Naour et al. 2001). In addition, in the nipple discharge, which is the aggregate of the secreted material,

DJ-1 protein levels were higher in patients with breast cancer than in non-cancer patients (Oda et al. 2012). Higher DJ-1 levels were detected in the sera of breast cancer cases showing low DJ-1 expression by cancer cells than those showing high DJ-1 expression (Kawate et al. 2015).

Furthermore, we examined DJ-1 isoforms in serum of breast cancer patients using 2-D gel electrophoresis (Kawate et al. 2015). Among several isoforms, the isoform of pI: 6.3 was found in all breast cancer cases. In contrast, a more acidic isoform was detected in the non-cancerous patients. It has been suggested that pI: 6.3 isoform may serve as cancer-specific marker.

4.3 Lung Cancer

4.3.1 Expression of DJ-1 in the Lung Carcinoma Tissue and the Clinicopathological Significance

Expression of *DJ-1* mRNA is enhanced in the non-small cell lung carcinoma as compared to normal lung tissue. Expression levels of *DJ-1* mRNA and protein were well correlated, and DJ-1 protein is overexpressed in cancer tissue as determined by western blot and immunostaining (Kim et al. 2005). DJ-1 protein was found to be localized mainly in the cytoplasm and sparsely in the nucleus of non-small cell lung carcinomas. In normal lung tissue, DJ-1 protein was hardly detected by immunohistochemistry (Bai et al. 2012).

High expression of *DJ-1* mRNA is a predictor of recurrence. As for DJ-1 protein expression level and cancer progression, the non-small cell lung carcinoma with high DJ-1 protein frequently showed lymph node metastases (Bai et al. 2012) and shorter overall survival time (Zeng et al. 2011). High expression of DJ-1 protein was correlated with cisplatin resistance. The combination treatment of paclitaxel and MEK inhibitor uniquely altered DJ-1 protein (MacKeigan et al. 2003).

4.3.2 DJ-1 Levels in the Blood and Pleural Effusion in Patients with Lung Carcinoma

Serum DJ-1 protein levels in non-small cell lung carcinoma patients were significantly higher than those in the healthy controls. High DJ-1 level in the serum was correlated with positive lymph node metastasis, advanced clinical stage, and distant metastasis (Fan et al. 2016).

Furthermore, significantly higher DJ-1 levels were detected in the malignant pleural effusion than those of the reactive pleural effusion (Vavougios et al. 2015).

4.4 Pancreatic Cancer

4.4.1 DJ-1 Expression of Normal Pancreatic Tissue

Negative or equivocal staining of DJ-1 protein was detected in pancreatic duct cells, acinar cells, or the islet cells by immunohistochemistry. Additionally, we found that the islet cells sometimes express DJ-1 protein, localized mainly in the nucleus in the atrophic pancreatic tissue (Fig. 4.4).

4.4.2 DJ-1 Expression of Pancreatic Cancer Tissue and Pancreatic Cancer Cell Line

DJ-1 protein is overexpressed frequently in pancreatic duct carcinoma cells (Tian M et al. 2008). Overexpression of DJ-1 protein is commonly detected in well-differentiated adenocarcinoma (Fig. 4.5). However, about half of poorly differentiated adenocarcinomas also express high levels of DJ-1 protein (Fig. 4.6). Intensity of DJ-1 expression seems to be associated with cancer cell differentiation in pancreatic ductal carcinoma cells.

In the pancreatic cancer cell line that expressed low levels of *DJ-1* mRNA, apoptosis was induced in response to gemcitabine. By contrast, gemcitabine resistance was induced when DJ-1 was overexpressed (Chen et al. 2012).

4.4.3 Clinical Correlation of DJ-1 Protein Levels in the Blood and in the Pancreatic Juice

DJ-1 protein levels in the pancreatic juice are significantly higher in pancreatic cancer patients than in healthy controls (Tian et al. 2008). Similarly in the blood, serum DJ-1 protein level is an independent prognostic factor in pancreatic cancer patients (He et al. 2011). Significantly higher DJ-1 levels were detected in pancreatic cancer patients than those in healthy controls (Chen et al. 2012) and patients suffering from chronic pancreatitis patients (He et al. 2011). In pancreatic cancer, patients with high serum DJ-1 protein levels had poor prognoses. Such patients were associated with advanced TNM stage and abnormally high CA19-9 level.

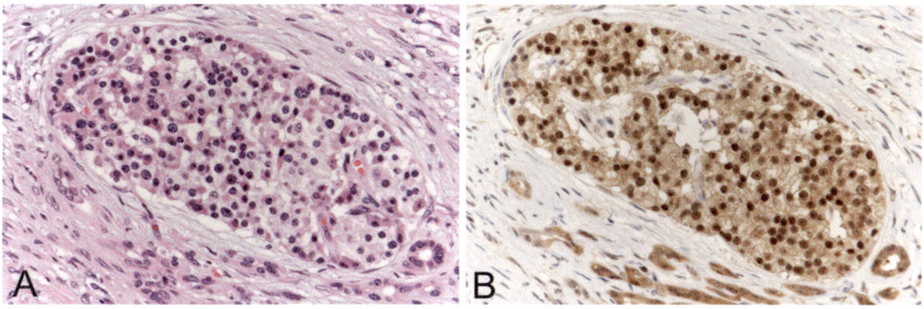

Fig. 4.4 DJ-1 expression in normal pancreatic tissue. (**a**) Staining of Langerhans island in the pancreatic tissue (H & E stain, original ×400). (**b**) The islet cells sometimes express DJ-1 protein mainly in the nuclei (DJ-1 immunostaining, original ×400)

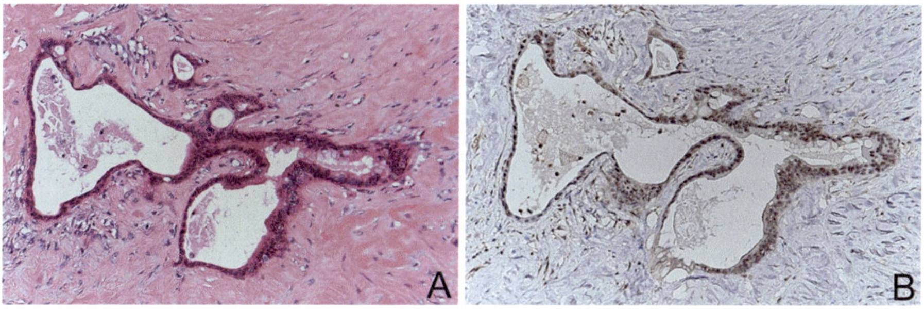

Fig. 4.5 DJ-1 protein expression in well-differentiated pancreatic adenocarcinoma. (**a**) Well-differentiated adenocarcinoma cells form irregular ducts (H & E stain, original ×400). (**b**) B DJ-1 protein is strongly expressed in the nuclei of ductal cancer cells (DJ-1 immunostaining, original ×400)

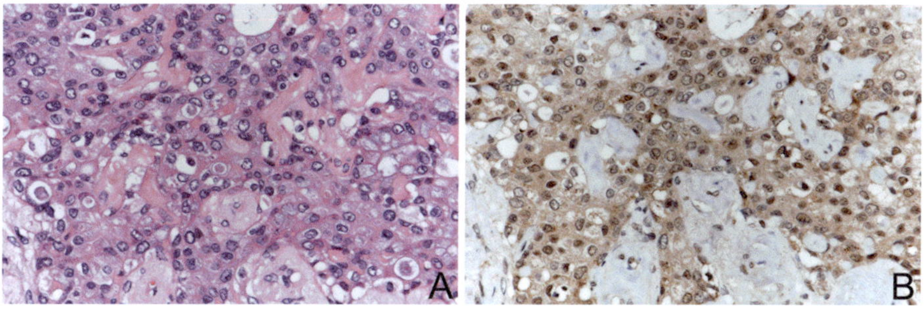

Fig. 4.6 DJ-1 protein expression in poorly differentiated pancreatic adenocarcinoma. (**a**) Poorly differentiated adenocarcinoma cells form irregular sheets (H & E stain, original ×400). (**b**) DJ-1 protein is strongly expressed in both the nucleus and cytoplasm of ductal cancer cells (DJ-1 immunostaining, original ×400)

4.5 Bladder Cancer

4.5.1 Expression of DJ-1 in Normal Bladder Tissue

DJ-1 protein was found to be localized mainly in the nucleus, but was also detected in the cytoplasm and along the membrane of urothelial cells (Lee et al. 2012).

4.5.2 Expression of DJ-1 Protein in the Bladder Cancer Tissue and the Clinicopathological Significance

Lee et al. performed immunohistochemistry on urothelial carcinoma (UC) tissue obtained from operative specimens by transurethral resection or radical cystectomy and compared the staining intensity of DJ-1 protein expression with nonneoplastic bladder tissue (Lee et al. 2012). DJ-1 expression was detected in 84.3% of the cases, and strong expression was observed in 66.7% cases. DJ-1 protein was mostly localized in the cytoplasm of cancer cells, and little expression in the nucleus and the cell membrane was also observed. Strong expression was detected in non-papillary, high-grade, invasive UC (Fig.4.7); however, DJ-1 staining was faint in papillary, low-grade, or noninvasive UC (Fig. 4.8). Increased of DJ-1 protein was detected more frequently in T1–T3 case than in Ta cases, suggesting the association with the invasiveness of cancer cells.

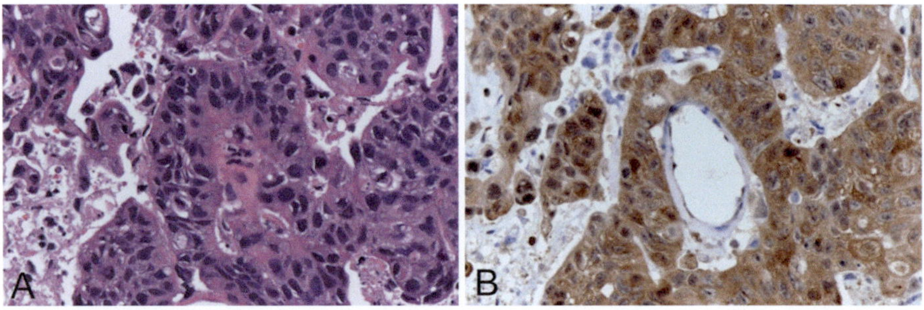

Fig. 4.7 DJ-1 protein expression in high-grade papillary urothelial carcinoma. (**a**) Invasive high-grade urothelial carcinoma cells (H & E stain, original ×400). (**b**) High expression of DJ-1 protein is detected in cancer cells (DJ-1 immunostaining, original ×400)

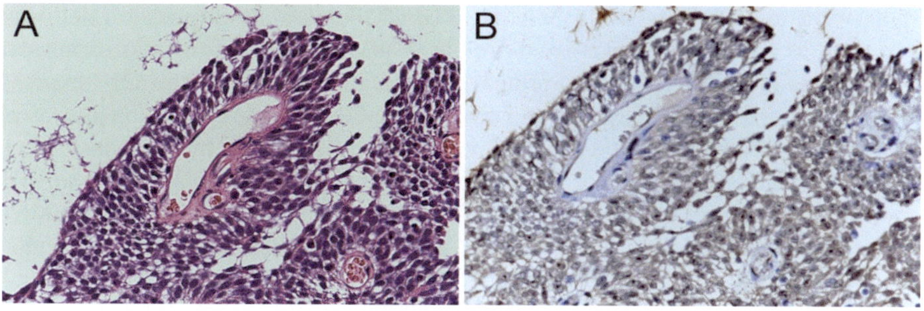

Fig. 4.8 DJ-1 protein expression in low-grade papillary urothelial carcinoma. (**a**) Low-grade urothelial carcinoma cells with papillary growth (H & E stain, original ×400). (**b**) Low expression of DJ-1 protein is detected in cancer cells (DJ-1 immunostaining, original ×400)

4.5.3 DJ-1 Protein Level as a Possible Tumor Marker in the Urine

Kumar et al. compared the protein profile of urine in bladder cancer patients with that of healthy subjects. The quantitative proteomic analysis of urine was performed by high-resolution mass spectrometry in order to construct multiplex biomarker panel of bladder cancer (Kumar et al. 2015).

Five proteins were identified as possible tumor markers by mass spectrometry whose secretion in urine was two-fold or higher in pooled urine of bladder cancer patients. These proteins were validated by RT-PCR and western blot analyses. DJ-1 protein was one of these five proteins. ELISA assay of DJ-1 protein with the other markers successfully identified Ta/T1 bladder cancer patients by AUC 0.92, overall accuracy 85.3% (sensitivity 79.2%, specificity 100%), and T2/T3 patients by AUC 0.94, overall accuracy 90.6% (sensitivity 86.4%, specificity 100%).

Furthermore, there was little-to-negligible contribution from the plasma cells, leukocytes, or erythrocytes in urinary DJ-1, which was confirmed by the immunostaining of the bladder cancer tissue that revealed strong expression of DJ-1 protein only in cancer cells. Therefore, the effect of hematuria on the quantity of urinary DJ-1 can be ignored.

4.6 Gynecologic Cancer

4.6.1 Cervical Cancer

Normal tissue, high-grade squamous intraepithelial lesion (HSIL) and squamous cell carcinomatous lesion were separately biopsied from same patients with cervical cancer. Each biopsied specimen was processed by microdissection and applied to two-dimensional fluorescence difference gel electrophoresis (2D DIGE). Differential spots were identified by using MALDI TOF/TOF mass spectrometry. DJ-1 was identified as a differential expressed protein. Squamous cell carcinomas expressed significantly higher levels of DJ-1 protein as compared to normal tissue. DJ-1 expression was variable in HSIL lesions (Wei et al. 2008). DJ-1 protein expression increased progressively from normal to CIN and invasive cancer as studied by immunostaining further supporting the results of the proteome analysis (Choi et al. 2015). In normal squamous epithelium, DJ-1 is localized in the nuclei of basal cells (Fig. 4.9). In low-grade intraepithelial lesion, faint staining of DJ-1 protein was detected in the cytoplasm of slightly atypical epithelial cells (Fig. 4.10). In high-grade intraepithelial lesion, strong staining of DJ-1 was seen in both the cytoplasm and nuclei of atypical epithelial cells (Fig. 4.11), while invasive squamous cell carcinoma cells revealed strong DJ-1 staining in both the cytoplasm and nuclei (Fig. 4.12).

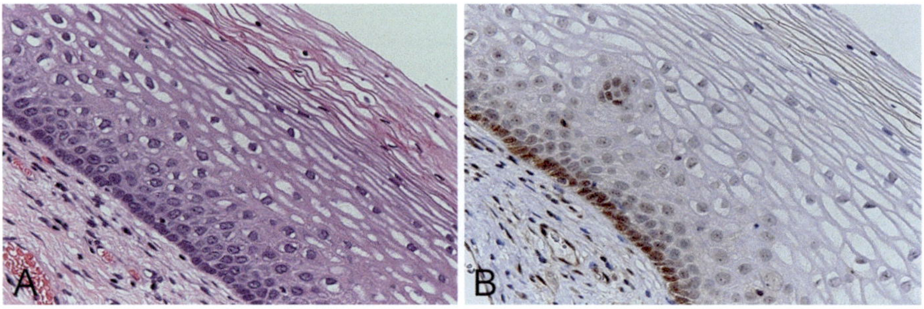

Fig. 4.9 DJ-1 expression in normal uterine cervix. (**a**) Normal stratified squamous epithelial cells (H & E stain, original ×400). (**b**) DJ-1 protein is localized in the nuclei of basal cells (DJ-1 immunostaining, original ×400)

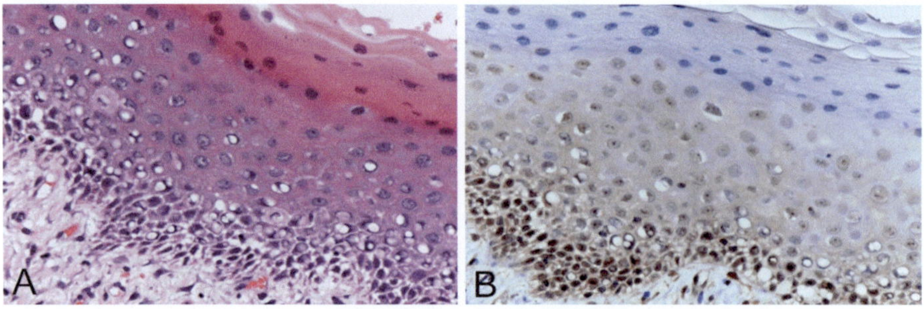

Fig. 4.10 DJ-1 protein expression in low-grade intraepithelial lesion. (**a**) Slightly atypical epithelial cells are seen in the lower layer of squamous epithelium (H & E stain, original ×400). (**b**) Faint staining of DJ-1 protein was detected in the cytoplasm of slightly atypical epithelial cells (DJ-1 immunostaining, original ×400)

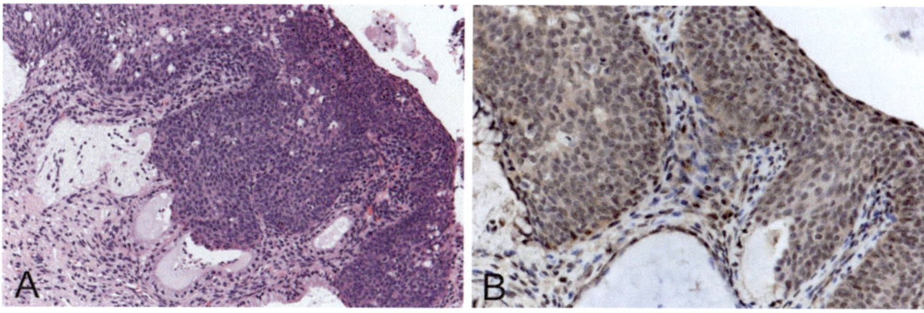

Fig. 4.11 DJ-1 protein expression in high-grade intraepithelial lesion. (**a**) Highly atypical epithelial cells are seen in the squamous epithelium (H & E stain, original ×400). (**b**) Strong staining of DJ-1 protein is detected mainly in the cytoplasm atypical epithelial cells (DJ-1 immunostaining, original ×400)

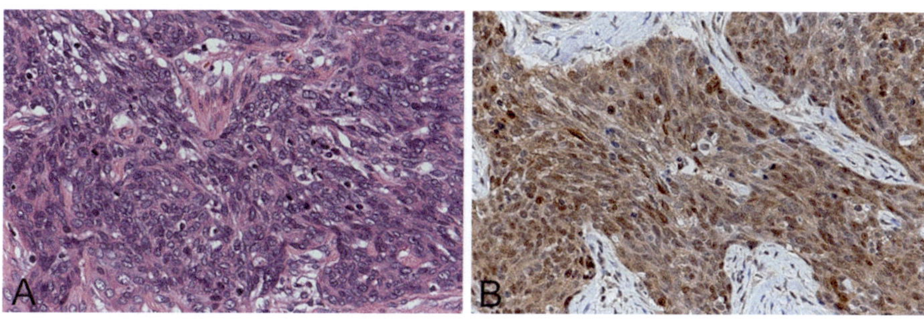

Fig. 4.12 DJ-1 protein expression in invasive squamous cell carcinoma cells. (**a**) Squamous cell carcinoma cells invade into submucosal tissue (H & E stain, original ×400). (**b**) Strong staining of DJ-1 protein is detected in both the cytoplasm and nuclei of squamous cell carcinomas (DJ-1 immunostaining, original ×400)

4.6.2 Uterine Cancer

Carcinoma tissue and adjacent non-cancerous tissue was separately obtained from uterine body cancer patients, and DJ-1 protein and mRNA levels were compared by western blot and RT-PCR, respectively. DJ-1 was shown to be highly expressed in carcinoma tissues. At the transcript level, higher expression was detected in poorly differentiated cancers rather than in well-differentiated cancers and in patients who had lymph node metastasis (Shu et al. 2013).

Another study by 2D DIGE + MALDI TOF/TOF mass spectrometry revealed higher expression of DJ-1 in poorly differentiated endometrioid carcinomas (Fig. 4.13) vs. well/moderately differentiated cancers (Fig. 4.14), in serous and clear cell adenocarcinomas vs. endometrioid adenocarcinoma. DJ-1 protein was localized in both the nucleus and the cytoplasm of the cancer cells and was particularly strongly stained in the muscle-invasive cancer cells (Morelli et al. 2014).

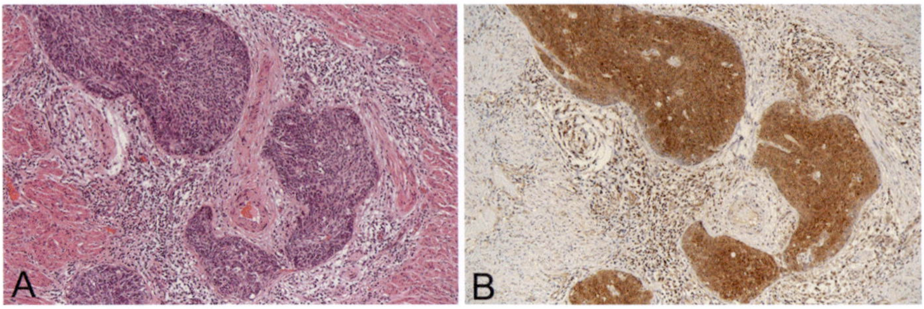

Fig. 4.13 DJ-1 protein expression in poorly differentiated endometrioid adenocarcinoma. (**a**) Poorly differentiated endometrioid adenocarcinoma cells invading into muscle (H & E stain, original ×400). (**b**) Strong staining of DJ-1 protein was detected in both the cytoplasm and nuclei of carcinoma cells (DJ-1 immunostaining, original ×400)

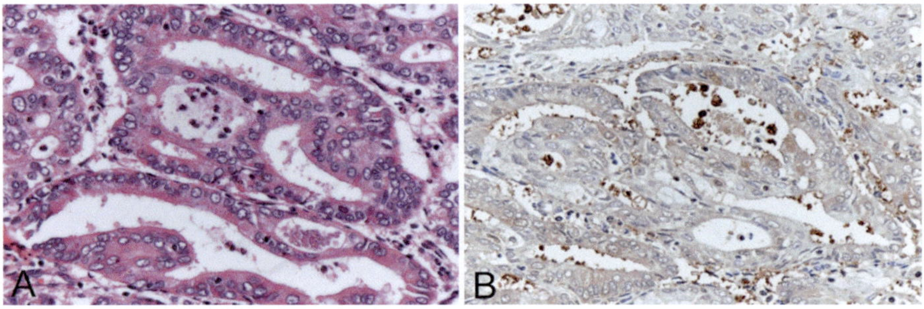

Fig. 4.14 DJ-1 protein expression in well-differentiated endometrioid adenocarcinoma. (**a**) Well-differentiated endometrioid adenocarcinoma cells form atypical glands (H & E stain, original ×400). (**b**) Weak staining of DJ-1 protein was detected in the cytoplasm of carcinoma cells (DJ-1 immunostaining, original ×400)

4.6.3 Ovarian Cancer

Various ovarian cancers (serous, mucinous, endometrioid, mixed epithelial, undifferentiated) overexpress *DJ-1* mRNA, frequently at advanced stages. Ovarian cancer patients with high levels of DJ-1 protein in post-chemotherapy ascites had poor prognoses (Davidson et al. 2008).

4.7 Summary

Majority of human cancers overexpress *DJ-1* mRNA. In contrast, protein levels of DJ-1 vary according to cancer subtype. DJ-1 is a prognostic factor, although differential expression of DJ-1 is critical for accurate prognosis. Visualization of DJ-1 expression suggests that distinct DJ-1 expression patterns are involved in cancer

Table 4.1 Expression of DJ-1 and drug resistance

Type of cancer	DJ-1 situation	Methods	Clinical significance	References
Non-small cell lung cancer	33/67, Overexpression	Western blot	Cisplatin resistance	Zeng et al. (2011)
Prostatic cancer cell line (PC-3)	High protein expression (8–13-fold)	Western blot	Cytotoxic agents resistance	Hod (2004)
Invasive ductal carcinoma, breast	Lower DJ-protein in 39/49 of pathological complete remission (pCR) cases	Immunohistochemistry, in situ hybridization	Pathological complete remission anthracycline + (docetaxel or paclitaxel) combination with Trastuzumab	Kawate et al. (2013)
Breast cancer cell line (MCF-7)	High (compared with DJ-1 silence)	Western blot	Adriamycin resistance	Zhang et al. (2015)
Pancreatic duct adenocarcinoma	Overexpression of both m RNA and protein DJ-1	RT-PCR, Western blot	Gemcitabine resistance	Chen et al. (2012)
Pancreatic carcinoma	Overexpression	Immunohistochemistry	Gemcitabine resistance	Tsiaousidou et al. (2013)
Clear cell renal cell carcinoma	High (compared with DJ-1 silence)	RT-PCR	Cisplatin resistance	Trivedi et al. (2016)
Ovarian cancer cell line	High (compared with DJ-1 silence)	Quantitative PCR and Western blot	Cisplatin resistance	Schumann et al. (2016)
Cervical cancer cell line (HeLa cell)	High (compared with DJ-1 silence)	RT-PCR	Dihydroartemisinin resistance	Zhu et al. (2014)

progression, differentiation, and morphology. Regarding drug sensitivity, high expression of DJ-1 mRNA is a predictor of drug resistance irrespective of drugs with different mechanisms of action in various cancers (Table 4.1). This may be because DJ-1 regulates multiple apoptotic pathways and, hence, is a promising target for cancer therapy.

Moreover, DJ-1 is a possible predictive marker during chemotherapy in various cancers. As non-cancerous cells also secrete DJ-1 under stress conditions, it will be of clinical significance to determine if DJ-1 protein secreted by cancer cells has any cancer-specific role or function.

The effects of oxidative stress experienced by cells in cancer or diseased such as neurodegenerative disorders are not yet fully understood. The expression of DJ-1 protein and its clinical significance outlined in this chapter may provide a hint in this direction.

References

Bai J, Guo C, Sun W, Li M, Meng X, Yu Y, Jin Y, Tong D, Geng J, Huang Q, Qi J, Fu S (2012) DJ-1 may contribute to metastasis of non-small cell lung cancer. Mol Biol Rep 39(3):2697–2703

Chen Y, Kang M, Lu W, Guo Q, Zhang B, Xie Q, Wu Y (2012) DJ-1, a novel biomarker and a selected target gene for overcoming chemoresistance in pancreatic cancer. J Cancer Res Clin Oncol 138(9):1463–1474

Choi SK, Hong YO, Lee WM, Kim EK, Joo JE, Kim DW, Lee H (2015) Overexpression of PI3K-p110α in the progression of uterine cervical neoplasia and its correlation with pAkt and DJ-1. Eur J Gynaecol Oncol 36(4):389–393

Davidson B, Hadar R, Schlossberg A, Sternlicht T, Slipicevic A, Skrede M, Risberg B, Flørenes VA, Kopolovic J, Reich R (2008) Expression and clinical role of DJ-1, a negative regulator of PTEN, in ovarian carcinoma. Hum Pathol 39(1):87–95

Fan J, Yu H, Lv Y, Yin L (2016) Diagnostic and prognostic value of serum thioredoxin and DJ-1 in non-small cell lung carcinoma patients. Tumor Biol 37(2):1949–1958

He XY, Liu BY, Yao WY, Zhao XJ, Zheng Z, Li JF, BQ Y, Yuan YZ (2011) Serum DJ-1 as a diagnostic marker and prognostic factor for pancreatic cancer. J Dig Dis 12(2):131–137

Hod Y (2004) Differential control of apoptosis by DJ-1 in prostate benign and cancer cells. J Cell Biochem 92(6):1221–1233

MacKeigan JP, Clements CM, Lich JD, Pope RM, Hod Y, Ting JP (2003) Proteomic profiling drug-induced apoptosis in non-small cell lung carcinoma: identification of RS/DJ-1 and RhoGDI. Cancer Res 63:6928–6934

Kawate T, Iwaya K, Kikuchi R, Kaise H, Oda M, Sato E, Hiroi S, Matsubara O, Kohno N (2013) DJ-1 protein expression as a predictor of pathological complete remission after neoadjuvant chemotherapy in breast cancer patients. Breast Cancer Res Treat 139(1):51–59

Kawate T, Iwaya K, Koshikawa K, Moriya T, Yamasaki T, Hasegawa S, Kaise H, Fujita T, Matsuo H, Nakamura T, Ishikawa T, Hiroi S, Iguchi-Ariga SMM, Ariga H, Murota K, Fujimori M, Yamamoto J, Matsubara O, Kohno N (2015) High levels of DJ-1 protein and isoelectric point 6.3 isoform in sera of breast cancer patients. Cancer Sci 106(7):938–943

Kim RH, Peters M, Jang Y, Shi W, Pintilie M, Fletcher GC, DeLuca C, Liepa J, Zhou L, Snow B, Binari RC, Manoukian AS, Bray MR, Liu FF, Tsao MS, Mak TW (2005) DJ-1, a novel regulator of the tumor suppressor PTEN. Cancer Cell 7(3):263–273

Kumar P, Nandi S, Tan TZ, Ler SG, Chia KS, Lim WY, Bütow Z, Vordos D, De la Taille A, Al-Haddawi M, Raida M, Beyer B, Ricci E, Colombel M, Chong TW, Chiong E, Soo R, Park

MK, Ha HK, Gunaratne J, Thiery JP (2015) Highly sensitive and specific novel biomarkers for the diagnosis of transitional bladder carcinoma. Oncotarget 6(15):13539–13549
Le Naour F, Misek DE, Krause MC, Deneux L, Giordano TJ, Scholl S, Hanash SM (2001) Proteomics-based identification of RS/DJ-1 as a novel circulating tumor antigen in breast cancer. Clin Cancer Res 7(11):3328–3335
Lee H, Choi SK, Ro JY (2012) Overexpression of DJ-1 and HSP90α, and loss of PTEN associated with invasive urothelial carcinoma of urinary bladder: possible prognostic markers. Oncol Lett 3(3):507–512
Morelli M, Scumaci D, Di Cello A, Venturella R, Donato G, Faniello MC, Quaresima B, Cuda G, Zullo F, Costanzo F (2014) DJ-1 in endometrial cancer: a possible biomarker to improve differential diagnosis between subtypes. Int J Gynecol Cancer 24(4):649–658
Oda M, Makita M, Iwaya K, Akiyama F, Kohno N, Tsuchiya B, Iwase T, Matsubara O (2012) High levels of DJ-1 protein in nipple fluid of patients with breast cancer. Cancer Sci 103(6):1172–1176
Trivedi R, Dihazi GH, Eltoweissy M, Mishra DP, Mueller GA, Dihazi H (2016) The antioxidant protein PARK7 plays an important role in cell resistance to Cisplatin-induced apoptosis in case of clear cell renal cell carcinoma. Eur J Pharmacol 784:99–110
Schumann C, Chan S, Khalimonchuk O, Khal S, Moskal V, Shah V, Alani AW, Taratula O, Taratula O (2016) Mechanistic nanotherapeutic approach based on siRNA-mediated DJ-1 protein suppression for platinum-resistant ovarian cancer. Mol Pharm 13:2070–2083
Shu K, Xiao Z, Long S, Yan J, Yu X, Zhu Q, Mei T (2013) Expression of DJ-1 in endometrial cancer: close correlation with clinicopathological features and apoptosis. Int J Gynecol Cancer 23(6):1029–1035
Tian M, Cui YZ, Song GH, Zong MJ, Zhou XY, Chen Y, Han JX (2008) Proteomic analysis identifies MMP-9, DJ-1 and A1BG as overexpressed proteins in pancreatic juice from pancreatic ductal adenocarcinoma patients. BMC Cancer 8. https://doi.org/10.1186/1471-2407-8-241
Trivedi R, Dihazi GH, Eltoweissy M, Mishra DP, Mueller GA, Dihazi H (2016) The antioxidant protein PARK7 plays an important role in cell resistance to Cisplatin-induced apoptosis in case of clear cell renal cell carcinoma. European Journal of Pharmacology 784:99-110
Tsiaousidou A, Lambropoulou M, Chatzitheoklitos E, Tripsianis G, Tsompanidou C, Simopoulos C, Tsaroucha AK (2013) B7H4, HSP27 and DJ-1 molecular markers as prognostic factors in pancreatic cancer. Pancreatology 13(6):564–569
Tsuchiya B, Iwaya K, Kohno N, Kawate T, Akahoshi T, Matsubara O, Mukai K (2012) Clinical significance of DJ-1 as a secretory molecule: retrospective study of DJ-1 expression at mRNA and protein levels in ductal carcinoma of the breast. Histopathology 61(1):69–77
Vavougios G, Kerenidi T, Tsilioni I, Zarogiannis SG, Gourgoulianis KI (2015) Pleural effusion levels of DJ-1 are increased in elderly lung cancer patients with malignant pleural effusions. Redox Rep 20(6):254–258
Wei L, Wang Z, Cui T, Yi F, Bu Y, Ding S, Ma Y, Song F (2008) Proteomic analysis of cervical cancer cells treated with adenovirus-mediated MDA-7. Cancer Biol Ther 7(4):510–516
Zeng HZ, YQ Q, Zhang WJ, Xiu B, Deng AM, Liang AB (2011) Proteomic analysis identified DJ-1 as a cisplatin resistant marker in non-small cell lung cancer. Int J Mol Sci 12(6):3489–3499
Zhang GQ, He C, Tao L, Liu F (2015) Role of DJ-1 siRNA in reverse sensitivity of breast cancer cells of chemotherapy and its possible mechanism. Int J Clin Exp Pathol 8(6):6944–6951
Zhu H, Liao SD, Shi JJ, Chang LL, Tong YG, Cao J, YY F, Chen XP, Ying MD, Yang B, He QJ, JJ L (2014) DJ-1 mediates the resistance of cancer cells to dihydroartemisinin through reactive oxygen species removal. Free Radical Biol Med 71:121–132

Chapter 5
Role of DJ-1 in Fertilization

An Chunna and Xiao-Ping Pu

Abstract Neither a sperm nor an egg can develop into an individual alone. Only when the sperm and egg bind and fuse, which is known as fertilization, can they acquire the ability of developing into new individuals. DJ-1 was reported to be involved in the process of fertilization.

Keywords DJ-1 • Fertilization • Asthenozoospermia

5.1 DJ-1 and Sperm Motility

Whyard et al. (2000) demonstrate that DJ-1/RS co-localized with β-tubulin, a major component of the axoneme, suggesting that DJ-1 is also associated with the flagellum axoneme and may play some role in flagellar motility. They also identified an infertile individual with immotile sperm of which the level of DJ-1/RS was undetectable. But they do did not know yet whether the reduced level of DJ-1/RS is the primary reason for the sperm abnormality or a consequence of another aberration.

5.2 DJ-1 and Sperm Capacitation

Freshly ejaculated sperm are unable to fertilize. Under physiological conditions, sperm must reside in the female reproductive tract for a period of time before they fertilize an egg. This process is called capacitation. Sperm that have undergone capacitation display hyperactivated motility, and they are said to become

A. Chunna
School of Basic Medical Sciences, North China University of Science and Technology, Tangshan, China

X.-P. Pu (✉)
Department of Molecular and Cellular Pharmacology, School of Pharmaceutical Science, Peking University Health Science Center, Beijing 100191, China
e-mail: puxiaopingbj@163.com

H. Ariga, S.M.M. Iguchi-Ariga (eds.), *DJ-1/PARK7 Protein*,
Advances in Experimental Medicine and Biology 1037,
https://doi.org/10.1007/978-981-10-6583-5_5

hyperactivated. The sperm of many mammals, including humans, can also be capacitated by incubation with capacitation medium in vitro.

Contraception-associated protein 1 (CAP1) is a homolog of DJ-1. Siva and the colleagues (Siva et al. 2004) observed the fate of DJ-1/CAP1 during capacitation by in vitro capacitation. Immunofluorescence showed that DJ-1/CAP1 protein locates in the principal piece and the mid piece of the tails in noncapacitated sperm, while some of the capacitated spermatozoa showed a loss of DJ-1/CAP1 from the mid piece. Loss of DJ-1/CAP1 was also confirmed by immunoblot analysis.

The exact role of DJ-1 in capacitation was not explained, but the loss of DJ-1 in capacitated sperm was confirmed by others (Choi et al. 2008). It must be noticed that most DJ-1 protein still remained after capacitation, indicating the DJ-1 has important roles in the following process of fertilization.

5.3 DJ-1 and Sperm-Oocyte Binding

Klinefelter et al. (2002) studied the role of DJ-1 protein in fertility using polyclonal and monoclonal DJ-1 antibodies. They found that when cauda epididymal sperm from the rat were incubated with polyclonal or monoclonal DJ-1 antibodies, the in vitro fertilization rates were inhibited. Likewise, decreased fertility of in utero insemination was observed when cauda epididymal sperm from the rat were incubated with polyclonal or monoclonal DJ-1 antibodies. That polyclonal and monoclonal antibodies were each equally capable of inhibiting fertilization both in vitro and in vivo, in the absence of sperm agglutination or deficits in sperm motility, which suggests that binding of these antibodies altered the function of DJ-1, causing fertility to be compromised. To further investigate the role of DJ-1 in fertilization, they did the in vitro fertilization experiment using hamster zona-free oocytes. It turned out that DJ-1 antibody inhibited fertilization rates only at high concentrations. This suggests that DJ-1 antibody inhibits fertilization at the level of the zona pellucida, indicating that DJ-1 protein assists sperm pass though zona pellucida. On the other hand, DJ-1 antibody inhibits fertilization even when zona pellucida is absent, suggesting that DJ-1 protein may function in the process of sperm-egg fusion.

Okada et al. (2002) found that the amount of DJ-1 in sperm and the efficiency of in vitro fertilization decreased in mice fed with ornidazole. The efficiency of fertilization in vitro decreased when an anti-mouse DJ-1 serum was added to sperm solution before the in vitro fertilization reaction with eggs. They thought that DJ-1 directly participates in the fertilization reaction.

5.4 DJ-1 and Male Infertility

Klinefelter et al. (1997) first reported the relationship between DJ-1/SP22 and male infertility. They fed rats with a series of reproductive toxicities for 4 days; then proximal cauda epididymal sperm were inseminated in utero into females. They found the fertility and the amount of cauda epididymal sperm DJ-1/SP22 were both decreased. Besides, DJ-1/SP22 was highly correlated with fertility. They indicated that the amount of DJ-1/SP22 in a cauda epididymal sperm sample may be a useful predictor of fertility in toxicant-treated animals. The relationship between DJ-1 and male infertility has been reported by many others.

While many sperm proteins now have been purported to be pivotal to male fertility, DJ-1/SP22 is the only sperm protein that has been demonstrated in multiple studies to be a consistent biomarker of effect, i.e., fertility (Klinefelter 2008).

Except animals, DJ-1 is also associated with human male infertility. We (An et al. 2011) collected sperm sample from113 asthenozoospermia patients and 58 age-matched controls (Table 5.1). Quantitative ELISA revealed that sperm DJ-1 concentration in asthenozoospermia patients is lower than that in control subjects (Fig. 5.1). DJ-1 mRNA level was also decreased (Fig. 5.2). Besides, DJ-1 with a more acidic pI was increased in asthenozoospermia patients (Fig. 5.3). Furthermore, DJ-1 concentration is positively correlated with sperm motility and sperm SOD activity indicated by partial correlation analysis (Table 5.2).

Table 5.1 Demographic and clinical data of subjects

		Asthenozoospermia		
Group	Controls	Total	Mild	Moderate
N	58	113	70	43
Age	32.1 ± 5.6	32.3 ± 5.2	32.3 ± 4.8	32.1 ± 5.9
pH	7.2 ± 0.3	7.2 ± 0.3	7.2 ± 0.3	7.2 ± 0.3
Ejaculate volume (ml)	3.3 ± 0.9	2.9 ± 1.1*	2.9 ± 1.2*	3.0 ± 0.9
Sperm density (10^6/ml)	86.2 ± 46.3	75.2 ± 52.1	74.9 ± 50.1	75.6 ± 56.0
Grade A sperm (%)				
Average ± SD	40.8 ± 10.2	12.8 ± 7.1***	17.4 ± 4.46***	5.3 ± 3.0***
Range	25.15–66.38	0.24–24.46	10.2–24.6	0.24–9.93
Grade (A + B) sperm (%)				
Average ± SD	52.5 ± 11.1	18.2 ± 9.7***	24.4 ± 6.3***	8.2 ± 4.3***
Range	37.92–75.50	0.71–39.27	14.13–39.27	0.71–18.22

$^*P < 0.05$, $^{***}P < 0.001$ versus controls by student's unpaired t-test

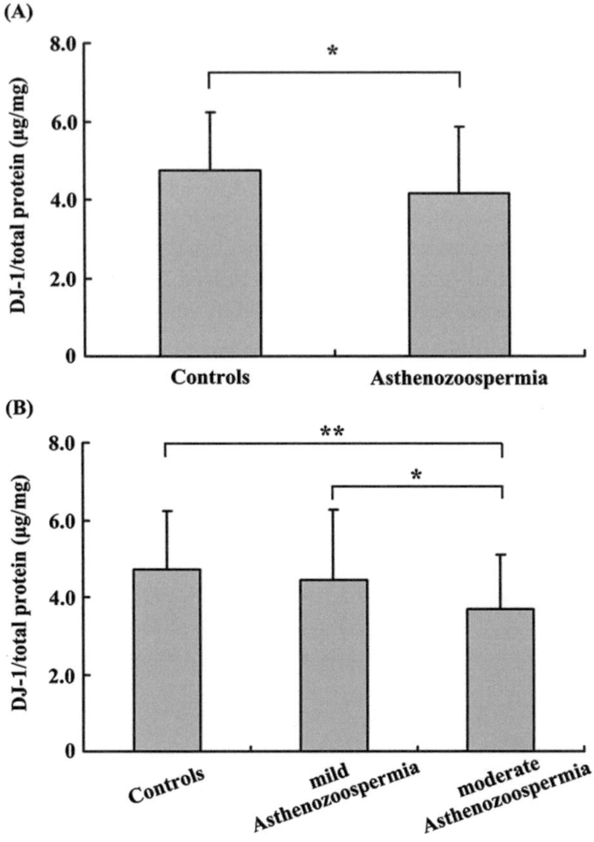

Fig. 5.1 Quantitative ELISA analysis of DJ-1 levels in ejaculated spermatozoa. (a) Sperm DJ-1 content in controls ($n = 58$) and asthenozoospermia patients ($n = 113$). $*P < 0.05$ by student's unpaired t-test. (b) Sperm DJ-1 content in controls ($n = 58$), mild asthenozoospermia patients ($n = 70$), and moderate asthenozoospermia patients ($n = 43$). $*P < 0.05$, $**P < 0.01$ by ANOVA with post hoc Bonferroni's test

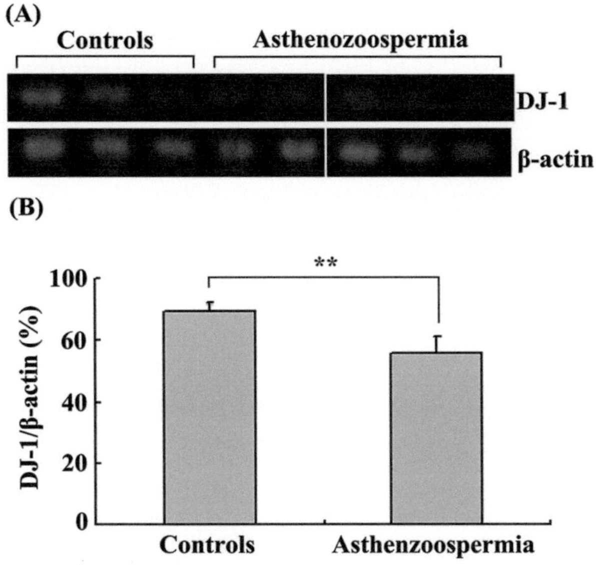

Fig. 5.2 Sperm DJ-1 mRNA expression in controls ($n = 3$) and asthenozoospermia patients ($n = 5$). (**a**) Sperm DJ-1 mRNA detection by RT-PCR. (**b**) Levels of DJ-1 mRNA were normalized against β-actin. $**P < 0.01$ by student's unpaired t-test

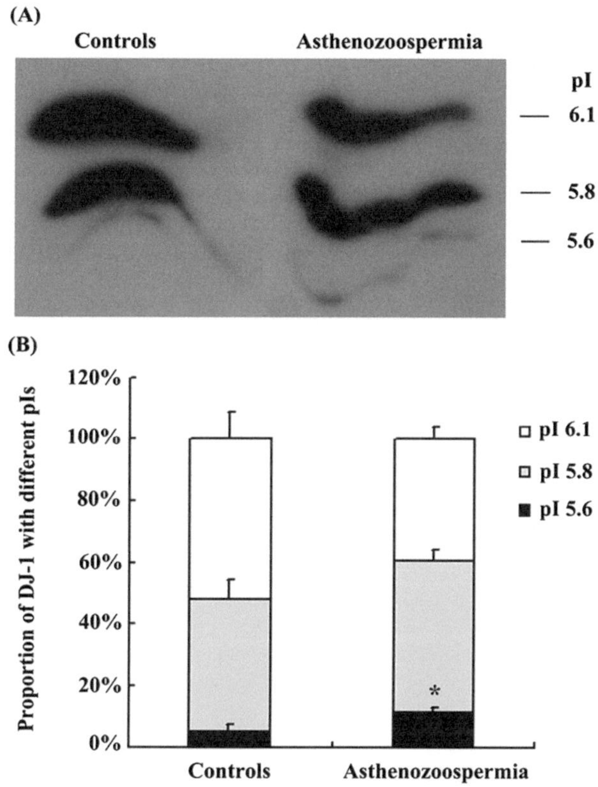

Fig. 5.3 Detection of pI values of DJ-1 protein in spermatozoa from controls and asthenozoospermia patients. (**a**) Representative result of immunoblot. (**b**) Proportion of DJ-1 with different pIs. Data were representative of three independent experiments. $*P < 0.05$ versus controls by student's t-test

Table 5.2 Correlation between two variables analyzed by partial correlation analysis controlling for age ($n = 171$)

Variables	Grade A sperm %	Grade (A + B) Sperm %	SOD activity
DJ-1 content	$P < 0.01$, R = 0.200	$P < 0.01$, $R = 0.222$	$P < 0.001$, $R = 0.352$
SOD activity	$P > 0.05$	$P > 0.05$	–

References

An CN, Jiang H, Wang Q, Yuan RP, Liu JM, Shi WL, Zhang ZY, XP P (2011) Downregulation of DJ-1 protein in the ejaculated spermatozoa from Chinese asthenozoospermia patients. Fertil Steril 96(1):19–23

Choi YJ, Uhm SJ, Song SJ et al (2008) Cytochrome c upregulation during capacitation and spontaneous acrosome reaction determines the fate of pig sperm cells: linking proteome analysis. J Reprod Dev 54(1):68–83

Klinefelter GR (2008) Saga of a sperm fertility biomarker. Anim Reprod Sci 105(1–2):90–103

Klinefelter GR, Laskey JW, Ferrell J et al (1997) Discriminant analysis indicates a single sperm protein (SP22) is predictive of fertility following exposure to epididymal toxicants. J Androl 18(2):139–150

Klinefelter GR, Welch JE, Perreault SD et al (2002) Localization of the sperm protein SP22 and inhibition of fertility in vivo and in vitro. J Androl 23(1):48–63

Okada M, Matsumoto K, Niki T et al (2002) DJ-1, a target protein for an endocrine disrupter, participates in the fertilization in mice. Biol Pharm Bull 25(7):853–856

Siva AB, Sundareswaran VR, Yeung CH et al (2004) Hamster contraception associated protein 1 (CAP1). Mol Reprod Dev 68(3):373–383

Whyard TC, Cheung W, Sheynkin Y et al (2000) Identification of RS as a flagellar and head sperm protein [J]. Mol Reprod Dev 55(2):189–196

Chapter 6
The Multifaceted Roles of DJ-1 as an Antioxidant

Prahlad V. Raninga, Giovanna Di Trapani, and Kathryn F. Tonissen

Abstract The DJ-1 protein was originally linked with Parkinson's disease and is now known to have antioxidant functions. The protein has three redox-sensitive cysteine residues, which are involved in its dimerisation and functional properties. A mildly oxidised form of DJ-1 is the most active form and protects cells from oxidative stress conditions. DJ-1 functions as an antioxidant through a variety of mechanisms, including a weak direct antioxidant activity by scavenging reactive oxygen species. DJ-1 also regulates a number of signalling pathways, including the inhibition of apoptosis signal-regulating kinase 1 (ASK1)-induced apoptosis under oxidative stress conditions. Other proteins regulated by DJ-1 include enzymes, chaperones, the 20S proteasome and transcription factors, including Nrf2. Once activated by oxidative stress, Nrf2 upregulates antioxidant gene expression including members of the thioredoxin and glutathione pathways, which in turn mediate an antioxidant protective function. Crosstalk between DJ-1 and both the thioredoxin and glutathione systems has also been identified. Thioredoxin reduces a cysteine residue on DJ-1 to modulate its activity, while glutaredoxin1 de-glutathionylates DJ-1, preventing degradation of DJ-1 and resulting in its accumulation. DJ-1 also regulates the activity of glutamate cysteine ligase, which is the rate-limiting step for glutathione synthesis. These antioxidant functions of DJ-1 are key to its role in protecting neurons from oxidative stress and are hypothesised to protect the brain from the development of neurodegenerative diseases such as Parkinson's disease (PD) and to protect cardiac tissues from ischaemic-reperfusion injury. However, DJ-1, as an antioxidant, also protects cancer cells from undergoing oxidative stress-induced apoptosis.

Keywords DJ-1 • Antioxidant • Thioredoxin • Glutathione • Oxidative stress • Reactive oxygen species

---

P.V. Raninga • K.F. Tonissen (✉)
School of Natural Sciences, Griffith University, Nathan, QLD, 4111, Australia

Griffith Institute for Drug Discovery, Griffith University, Nathan, QLD, 4111, Australia
e-mail: Prahlad.Raninga@griffithuni.edu.au; K.Tonissen@griffith.edu.au

G. Di Trapani
School of Natural Sciences, Griffith University, Nathan, QLD, 4111, Australia
e-mail: G.DiTrapani@griffith.edu.au

H. Ariga, S.M.M. Iguchi-Ariga (eds.), *DJ-1/PARK7 Protein*,
Advances in Experimental Medicine and Biology 1037,
https://doi.org/10.1007/978-981-10-6583-5_6

6.1 Cellular Antioxidants

Redox homeostasis is an important regulatory process that ensures cell survival and function in the presence of an oxygen environment by balancing reactive oxygen species (ROS) with the help of the antioxidant systems. ROS are made as by-products of oxygen metabolism, primarily by the mitochondria, and play some vital roles in cell signalling. However when levels of ROS are increased in cells, oxidative stress occurs, which can lead to detrimental cellular consequences such as cancer or other human disease conditions. Examples of ROS are shown in Fig. 6.1, including superoxide anions (O_2^-), which are formed from the reduction of molecular oxygen (Nordberg and Arner 2001). Superoxide can then undergo dismutation to form hydrogen peroxide, which can be partially reduced to the hydroxyl radical (•OH) or fully reduced to form water. Reactive nitrogen species (RNS) are also regarded as free radicals and can cause cellular damage. Nitric oxide (NO), for example, can react with the superoxide anion to form $OONO^-$ (peroxynitrite), which is highly reactive and damaging to proteins (Nordberg and Arner 2001).

The cell responds to high levels of ROS by expressing and activating antioxidant systems. The two most important antioxidant systems are the thioredoxin system and the glutathione system. The cytoplasmic thioredoxin system consists of a 12kDa redox-active protein, thioredoxin 1 (Trx1), thioredoxin reductase 1 (TrxR1) and NADPH (Holmgren 1985; Lu and Holmgren 2014). There is also a corresponding system in the mitochondria comprised of Trx2, TrxR2 and NADPH. Trx contains two cysteine residues in its active site that form a disulphide bond upon reduction of its substrate. It is then recycled to a reduced active form by the action of TrxR and NADPH (Fig. 6.2). Its many substrates include the peroxiredoxins, transcription factors and many other regulatory proteins (Arner and Holmgren 2000). The Trx system can also directly quench ROS (Das and Das 2000).

Glutathione (GSH) is the most abundant thiol-based antioxidant within a cell and can act as an electron donor to detoxify compounds either directly or indirectly through the action of glutathione S-transferases (Nordberg and Arner 2001; Lushchak 2012) (Fig. 6.3). Glutathione peroxidases reduce hydrogen peroxide through the oxidation of two GSH molecules, forming GSSG. Oxidised GSSG can

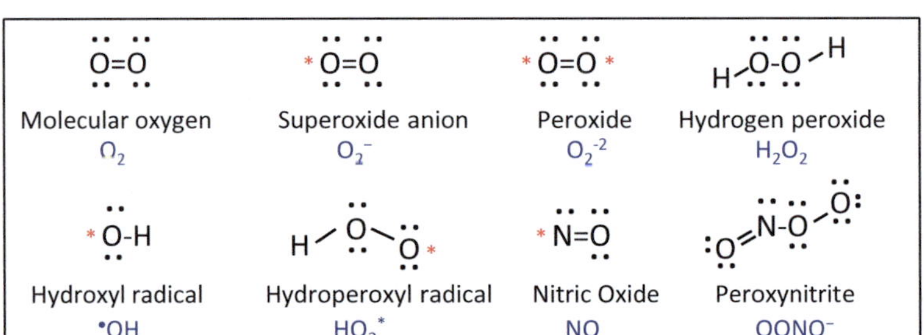

Fig. 6.1 Examples of reactive oxygen species and reactive nitrogen species (The red stars represent unpaired electrons)

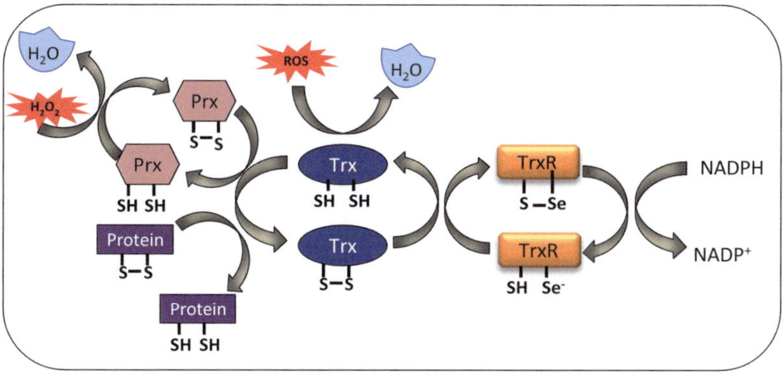

Fig. 6.2 The thioredoxin (*Trx*) system. Trx reduces the disulphide bonds in other proteins, including peroxiredoxins (*Prdx*). Trx can also reduce ROS. The oxidised form of Trx is recycled to a reduced state by the action of TrxR

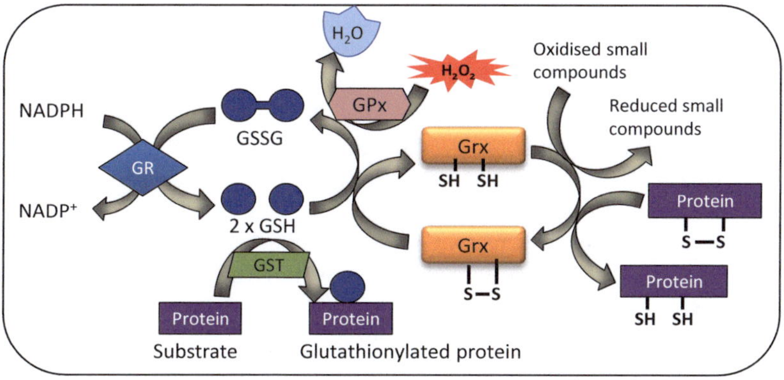

Fig. 6.3 The glutathione (*GSH*) system. The GSH system acts to reduce small compounds or other proteins through the action of glutaredoxins (*Grx*), glutathione peroxidases (*GPx*) and glutathione S-transferases (*GST*)

be recycled to GSH through the action of glutathione reductase. GSH can also reduce glutaredoxin enzymes (Grx), which in turn reduce other oxidised substrates. The glutathione system has distinct substrates from that of the thioredoxin system, although there is some overlap in their functions (Le Moan et al. 2006; Go et al. 2013).

These systems maintain the intracellular redox homeostasis and their disruption results in diseased conditions. In addition to the Trx and GSH systems, there are also a number of other proteins in the cell with antioxidant function, and these also play a role in maintaining redox homeostasis. In recent years it has become increasingly recognised that DJ-1 has antioxidant functions and provides cytoprotection against increased oxidative stress (Taira et al. 2004; Lev et al. 2009). DJ-1 is upregulated by various oxidative stress stimuli and exerts its antioxidant function in different cell types and tissues including neuronal (Lev et al. 2008; Goldberg et al. 2012), renal (Eltoweissy et al. 2011) and cardiac cells (Yu et al. 2013).

While its exact functions are not fully understood, DJ-1 has been shown to play a significant regulatory role in various biological processes including chaperone activity (Shendelman et al. 2004; Zhou et al. 2006), transcriptional regulation (Clements et al. 2006; Fan et al. 2008; McNally et al. 2011), antioxidative stress activity (Waak et al. 2009; Im et al. 2010) and proteasome function (Moscovitz et al. 2015). While these roles are not yet fully defined, consistent findings demonstrate the antioxidative stress function of the DJ-1 protein. The next sections of the chapter examine the specific antioxidant functions of DJ-1 in more detail.

6.2 Redox-Sensitive Residues of DJ-1 and Their Role in Antioxidant Function

The oxidation state of DJ-1 is key to its functionality as an antioxidant and to its role in the disease state. For example, the levels of oxidised DJ-1 are significantly increased in the brain of Parkinson's disease (PD) and Alzheimer's patients compared to that of healthy individuals (Choi et al. 2006). Many studies have focussed on assessing how oxidation alters the function or structure of DJ-1 and how these different oxidation states could be potentially targeted in therapies. The cysteine residues have therefore been a specific focus for many research groups. The DJ-1 protein contains three redox-sensitive cysteine residues, Cys46, Cys53 and Cys106 (Fig. 6.4). Of these the Cys106 residue has a low thiol pKa (approximately 5.4), and at physiological pH, it exists predominantly in the reactive thiolate anion form (Witt et al. 2008; Wilson 2011). It is therefore not a coincidence that many studies have shown Cys106 to be essential for DJ-1 to exert its cytoprotective action against oxidative stress (Canet-Aviles et al. 2004; Kim et al. 2009; Waak et al. 2009; Im et al. 2010; Fernandez-Caggiano et al. 2016). The Cys106 residue is the most sensitive to oxidative stress and is susceptible to undergo oxidation to SOH (sulfenic acid form), to SO_2H (sulfinic acid form) and then to the SO_3H (sulphonic acid form), as shown in Fig. 6.4 (Canet-Aviles et al. 2004; Kinumi et al. 2004; Zhou et al. 2006; Wilson 2011; Ariga et al. 2013). The DJ-1 protein exhibits different properties depending on the Cys106 oxidation state. Mild oxidation of Cys106 to SO_2H is required for DJ-1 localisation and mitochondrial function, to protect cells from oxidative stress (Blackinton et al. 2009) and also to inhibit the fibrillation of α-synuclein (Zhou et al. 2006). However, extensive oxidation of Cys106 to SO_3H causes the DJ-1 protein to aggregate and to become inactive (Zhou et al. 2006). Subsequently Cys46 and Cys53 may also become oxidised (Zhou et al. 2006) and, unlike Cys106, can also be subjected to S-nitrosylation (Ito et al. 2006).

While most focus has been on the Cys106 residue as being a functionally essential residue, the other cysteine residues have also been shown to be important for DJ-1 function. DJ-1 exists as a homodimer, and the two cysteine residues at

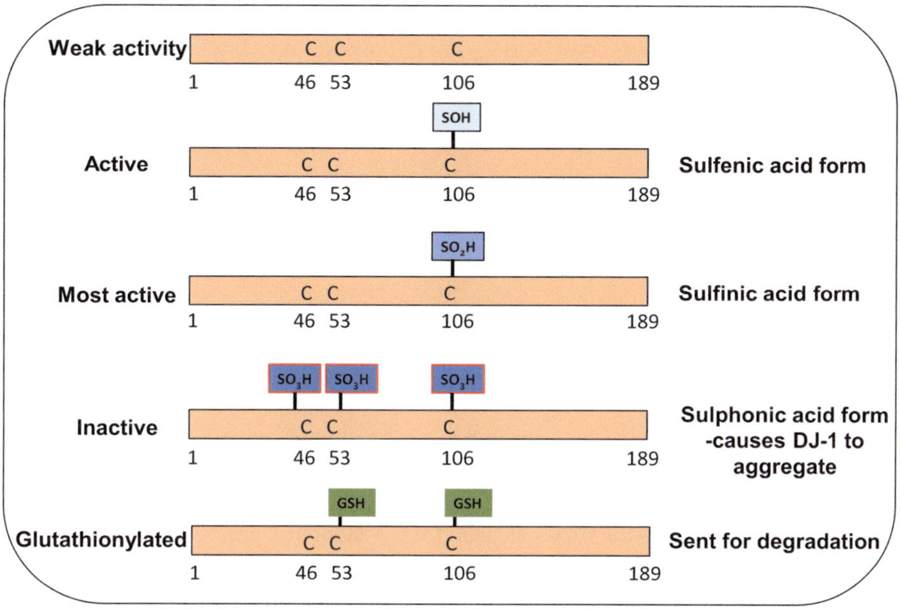

Fig. 6.4 Cysteine residue oxidation and activation of DJ-1. DJ-1 contains three cysteine residues at amino acid numbers 46, 53 and 106. Cys106 is sequentially oxidised to SOH, SO_2H and SO_3H, and then Cys46 and Cys53 are oxidised and may also be nitrosylated (not shown). Cys53 and Cys106 can also be glutathionylated (*GSH*) (Figure is modified from that previously published by Ariga and colleagues 2013)

positions 46 and 53 are close to the dimerisation interface, which involves mainly hydrophobic interactions (Honbou et al. 2003). There are also reports that implicate these Cys residues as important for dimerisation, which has downstream effects on DJ-1 function (Ito et al. 2006; Girotto et al. 2012; Fernandez-Caggiano et al. 2016). While Cys46 was initially found to be important for dimerisation (Ito et al. 2006), other researchers have since reported that the Cys53 residue can form a disulphide bond with a Cys53 residue in another DJ-1 monomer (Girotto et al. 2012; Fernandez-Caggiano et al. 2016).

DJ-1 has also been reported to have some functional capacity related to the Cys53 residue. DJ-1 was shown to have chaperone activity by preventing the aggregation of α-synuclein, a protein implicated in PD (Shendelman et al. 2004). This reported chaperone activity did not require the presence of Cys106. However mutation of Cys53 to an alanine prevented aggregation of α-synuclein and resulted in cell sensitivity to oxidative stress. However this mutation did not prevent formation of DJ-1 dimers (Shendelman et al. 2004). Solely relying on mutants may not necessarily reveal the functions of different cysteine residues since Zhou and colleagues subsequently showed using mass spectrometry that the oxidation of Cys106 to the sulfinic acid form is required for DJ-1 to prevent the aggregation of α-synuclein (Zhou et al. 2006). There is evidence for both Trx and peroxiredoxin-2 (Prdx-2) to bind to the Cys53 residue of DJ-1 (Fu et al. 2009; Fernandez-Caggiano et al. 2016). Other reports have shown Cys46 and Cys53 to play an important regulatory role for DJ-1 function. The Cys53 and Cys46 residues were shown to regulate the ability of DJ-1 to bind to the apoptosis signal-regulating kinase 1 (ASK1) protein through the Cys106 residue (Waak et al. 2009). There is also evidence that the redox state of the Cys53 residue affects the ability of Cys106 to be oxidised or to form heterodisulfides with other proteins (Fernandez-Caggiano et al. 2016).

6.3 Molecular Mechanisms Underlying the Antioxidative Stress Activity of DJ-1

DJ-1 functions as an antioxidant through a variety of mechanisms, including a direct antioxidant activity by scavenging ROS. It achieves this by acting as an atypical peroxiredoxin-like peroxidase through the oxidation of Cys106 but has only weak activity that is a thousand times slower than a standard peroxiredoxin (Andres-Mateos et al. 2007). In addition since DJ-1 is not catalytically converted back to the reduced state, as occurs for Trx for example, it has been suggested that this ROS-scavenging activity might not be its most effective mode of acting as an antioxidant (Andres-Mateos et al. 2007; Wilson 2011).

It is likely that the role of DJ-1 as an antioxidant primarily lies with its ability to bind to and to regulate other proteins that are involved with protecting cells from oxidative stress-induced cell death. A summary of these binding partners is shown in Table 6.1 including ASK1 (Waak et al. 2009; Im et al. 2010), p38-regulated/activated kinase (PRAK) (Tang et al. 2014), paraoxonase-2 (Parsanejad et al. 2014a), α-synuclein (Shendelman et al. 2004; Zhou et al. 2006), the 20S proteasome (Moscovitz et al. 2015), peroxiredoxin-2 (Fernandez-Caggiano et al. 2016) and Trx (Fu et al. 2009). These pathways and functions show the breadth of DJ-1's role to protect cells from oxidative stress.

One of DJ-1's key cytoprotective activities is as a direct binding partner for ASK1. ASK1 is a mitogen-activated protein kinase kinase kinase (MAPKKK) family member that is activated by oxidative stress stimuli, stimulating the activation of downstream pathways that result in apoptosis (Ichijo et al. 1997). DJ-1 functions to inhibit ASK1-induced apoptosis under oxidative stress conditions (Waak et al. 2009; Im et al. 2010; Mo et al. 2010) (Fig. 6.5).

DJ-1 binds to ASK1 through the Cys106 residue, which prevents activation of ASK1 in two different ways. First, DJ-1 prevents the homo-dimerisation of ASK1, which prevents ASK1 from activating the apoptotic pathway (Mo et al. 2010). Second, DJ-1 also prevents activation of ASK1 by regulating the binding of another inhibitor, Trx1. Under normal conditions ASK1 is bound by the reduced form of Trx1, which keeps ASK1 in an inactive state (Saitoh et al. 1998). Under oxidative stress conditions, Trx1 is oxidised and releases ASK1, allowing ASK1 to promote

Table 6.1 Key DJ-1 protein-binding partners for its antioxidant function

Binding partner	Function of binding partner	DJ-1 residue involved	Other DJ-1-relevant information	References
ASK1	Apoptosis (oxidative stress)	Cys106	Requires mild oxidation of Cys106 for binding	Waak et al. (2009) and Im et al. (2010)
PRAK	Apoptosis (oxidative stress)	Unknown	Binding to PRAK keeps DJ-1 in nucleus, which also keeps Daxx in nucleus	Tang et al. (2014)
Paraoxonase-2 (PON2)	PON2 is an enzyme that protects neurons from oxidative stress	Unknown	Direct interaction and modulation of PON2 activity in neuronal cells	Parsanejad et al. (2014a)
α-Synuclein	Chaperone	Cys53	Mutation of cys53 to ala abrogates chaperone activity and prevents DJ-1 from inhibiting the aggregation of α-synuclein	Shendelman et al. (2004)
	The interaction is weak and transitory	Cys106	Sulfinic acid form of Cys106 is required	Zhou et al. (2006)
20S proteasome	Role in protein degradation	Cys106	DJ-1 inhibits 20S proteasome degradation activity regardless of the Cys106 oxidation state	Moscovitz et al. (2015)
Peroxiredoxin- 2	Antioxidant	Cys53	DJ-1 forms a trimeric complex with peroxiredoxin-2 before the formation of DJ-1 disulphide-linked dimers in cardiac cells	Fernandez-Caggiano et al. (2016)
Thioredoxin	Antioxidant	Cys53	Cys53 is redox sensitive and can be reduced by Trx	Fu et al. (2009)

apoptosis (Saitoh et al. 1998). DJ-1 binds to ASK1 and prevents Trx1 from releasing ASK1, thus providing the cell with protection from apoptosis (Im et al. 2010).

DJ-1 also appears to regulate ASK1 via regulation of the death-associated protein 6 (Daxx), which usually acts as an activator of ASK1 (Chang et al. 1998). In unstimulated cells Daxx is usually located in the nucleus. But under oxidative stress conditions, it translocates into the cytoplasm where it can interact with and activate ASK1, leading to apoptosis (Song and Lee 2003). During oxidative stress DJ-1 was

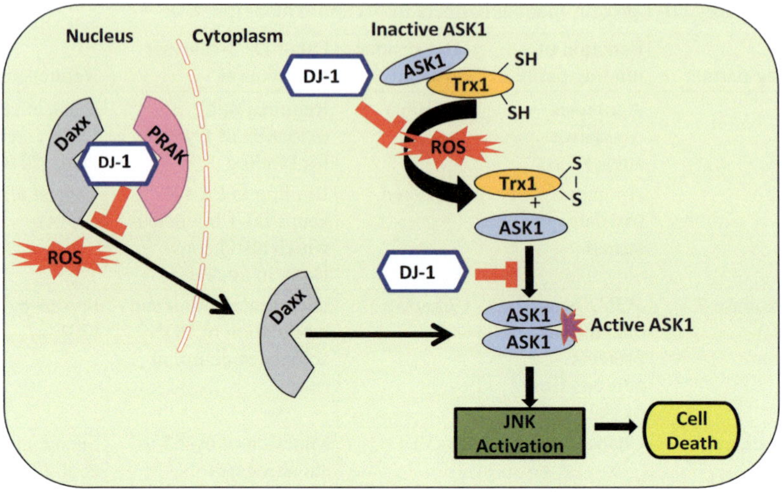

Fig. 6.5 Regulation of ASK1 activity by DJ-1. DJ-1 regulates ASK1 by inhibiting the dissociation of Trx from ASK1 during an oxidative stress stimulus. DJ-1 also prevents the dimerisation of ASK1. Under oxidative stress Daxx usually moves into the cytoplasm to activate ASK1. The nuclear protein PRAK binds to DJ-1, which retains Daxx in the nucleus preventing it from moving into the cytoplasm during oxidative stress

shown to sequester Daxx and keep it in the nucleus preventing it from activating ASK1 in the cytoplasm (Junn et al. 2005; Waak et al. 2009). The retention of DJ-1 in the nucleus is thus required, and since DJ-1 does not contain a nuclear localisation sequence (Taira et al. 2001), it most likely interacts with nuclear proteins. Using a yeast two-hybrid screen, DJ-1 was identified as a potential interacting partner with PRAK (Tang et al. 2014). This interaction was verified to occur in vitro and in vivo and to prevent oxidative stress-induced cell death. Following H_2O_2 stimulation, endogenous DJ-1 is retained in the nucleus of $PRAK^{+/+}$ cells but is absent in the nucleus in $PRAK^{-/-}$ cells, indicating that the interaction with PRAK allows DJ-1 to remain in the nucleus (Tang et al. 2014). By remaining in the nucleus, DJ-1 sequesters Daxx and prevents it from entering the cytoplasm to trigger oxidative stress-induced cell death through the ASK1 pathway (Fig. 6.5) (Waak et al. 2009).

6.4 DJ-1 as a Regulator of Antioxidant Gene Expression

DJ-1 also functions as an antioxidant by upregulating the expression of genes that encode proteins with an antioxidant function. DJ-1 regulates the activity of a number of transcription factors including p53 (Fan et al. 2008) and NF-κB (McNally et al. 2011), which in turn regulate the expression of multiple genes involved in the response to oxidative stress. One of the most relevant transcription factors in the cell for upregulating antioxidant gene expression is Nrf2. The Nrf2 transcription factor binds to an antioxidant-responsive element (ARE) present in the promoters of many genes that encode antioxidant proteins, including many members of the thioredoxin and glutathione systems (Hayes and Dinkova-Kostova 2014). Thus, activating Nrf2 contributes to a coordinated antioxidant response, and currently there is evidence to suggest that DJ-1 may play a role in regulating Nrf2 (Clements et al. 2006; Im et al. 2012; Cuevas et al. 2015).

Regulation of Nrf2 occurs at both the transcriptional and post-translational level. It has most recently been appreciated how complicated these levels of control are, involving multiple layers of regulation (Hayes and Dinkova-Kostova 2014). One mode of control is through inhibitors, such as Kelch-like ECH-associated protein 1 (Keap1), which bind to Nrf2 in the cytoplasm and prevent Nrf2 from moving into the nucleus by sending it for degradation. Some electrophiles and oxidants cause conformational changes in Keap1 and prevent Nrf2 ubiquitination and subsequent degradation. This allows Nrf2 to accumulate quickly and translocate into the nucleus (Hayes and Dinkova-Kostova 2014).

DJ-1 was first implicated as a regulator of Nrf2 activity through a study that showed DJ-1 had a role in stabilising Nrf2 in a number of different cell types. Knockdown of DJ-1 expression resulted in a significant decrease in expression of the Nrf2 target gene NQQ1 (Clements et al. 2006). DJ-1 did not alter Nrf2 expression levels nor was it detected to bind with Nrf2 on ARE sequences. DJ-1 was however shown to reduce the ubiquitination levels of Nrf2, thus leading to higher protein levels of Nrf2 accumulating in the cells (Clements et al. 2006).

Other reports have shown that Nrf2 can be activated independently of DJ-1 (Gan et al. 2010), and therefore it has been suggested that Nrf2 activation occurs downstream of DJ-1 function (Im et al. 2012). Overexpression of DJ-1 was shown to increase Nrf2 protein levels, promoting its transport into the nucleus, resulting in increased Trx gene expression via Nrf2 binding to the ARE sequence in the Trx gene promoter (Im et al. 2012). This study showed that DJ-1 did not function through a direct interaction with either Nrf2 or Keap1 and in their studies they did not find a change in Nrf2 ubiquitination. Their data suggests that DJ-1 is functioning through a non-Keap1-dependent mechanism (Im et al. 2012). This conclusion was supported by an in vivo mouse kidney study, which showed that DJ-1 was required to prevent degradation of Nrf2 (Cuevas et al. 2015). As with earlier studies, they found a connection with DJ-1 and Nrf2 ubiquitination, since DJ-1 $^{-/-}$ mice exhibited increased ubiquitination. However other studies show that DJ-1 does not always exert its antioxidant function through Nrf2 activation, suggesting there are multiple pathways regulated by DJ-1 (Gan et al. 2010; Ismail et al. 2015).

6.5 Crosstalk with Other Antioxidant Systems

The cellular response to oxidative stress involves a coordinated activation of antioxidant systems that may be cell type and stimulus specific. The interaction between DJ-1 and the other major antioxidant systems, such as the Trx system (Fu et al. 2009;

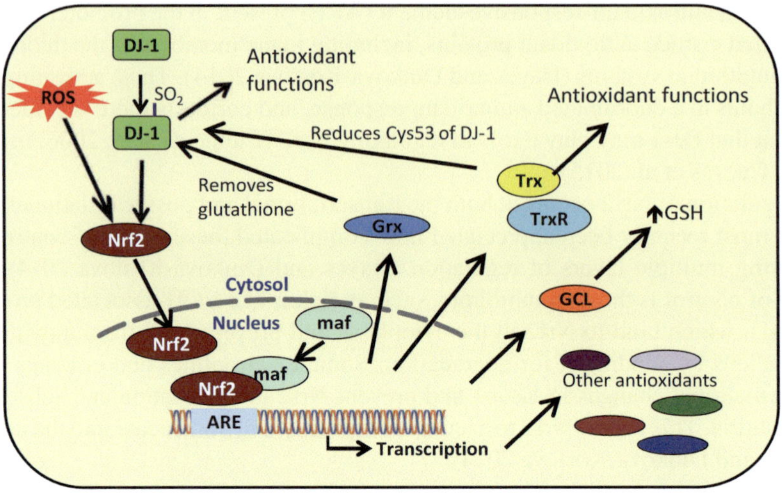

Fig. 6.6 Crosstalk between DJ-1 and the thioredoxin and glutathione systems. Under oxidative stress DJ-1 activates Nrf2, which upregulates the expression of ARE-containing genes, including Trx, TrxR, GCL and Grx1, and other antioxidants. Upregulation of Trx results in antioxidant function in the cell as well as being able to reduce the Cys53 residue of DJ-1. Upregulation of GCL results in increased GSH synthesis. Grx1 is responsible for removing the glutathione moiety from DJ-1 to maintain high levels of DJ-1 in the cell

Im et al. 2012; Mahmood et al. 2013; Raninga et al. 2014; Fernandez-Caggiano et al. 2016) and glutathione systems (Zhou and Freed 2005; Johnson et al. 2016), is likely to be complex but important for optimum responsiveness to an oxidative stress stimulus.

The connection between Trx and DJ-1 can be either through a direct interaction or via upregulation of Trx gene expression through the Nrf2 pathway (Fig. 6.6). A mass spectrometry study showed that the Cys53 residue of DJ-1 can be reduced by Trx1 both in vitro and in vivo (Fu et al. 2009). In addition, when TrxR is inhibited by auranofin, the disulphide dimer of DJ-1 is formed, explained to be because Trx is not available to reduce the Cys53 residue of DJ-1 (Fernandez-Caggiano et al. 2016). As well as a target for Trx, the Cys53 residue is also the site where DJ-1 binds another oxidative stress-responsive protein, peroxiredoxin-2 (Prdx-2) (Fernandez-Caggiano et al. 2016), which itself is a target for Trx. It is suggested that the binding of Prdx-2 influences the activity of DJ-1 through the formation of homodimers, which induces a conformation change. This in turn regulates the binding of other proteins to the Cys106 residue and protects it from hyperoxidation (Fernandez-Caggiano et al. 2016).

Overexpression of DJ-1 was shown to upregulate Trx levels (Im et al. 2012) suggesting that DJ-1 may act indirectly on the Trx pathway by enhancing its expression levels. It has been suggested that the regulation of Trx expression occurs by DJ-1-activating Nrf2, but as described above, there is still some debate over this mechanism. Once upregulated, Trx functions to reduce oxidised proteins, and Trx could therefore in some situations be mediating the antioxidant function of DJ-1. It is therefore interesting that several of DJ-1's target pathways such as NF-κB (McNally et al. 2011), p53 (Fan et al. 2008), ASK1 (Waak et al. 2009) and PTEN (Kim et al. 2005a) are also subject to redox regulation by Trx (Matthews et al. 1992; Saitoh et al. 1998; Ueno et al. 1999; Meuillet et al. 2004).

There is also crosstalk between the glutathione system and DJ-1, which as with Trx can be either through a direct interaction or through a connection with Nrf2 (Fig. 6.6). Under oxidative stress conditions, DJ-1 was shown to upregulate intracellular glutathione synthesis by increasing both the expression and activity of the glutamate cysteine ligase (GCL) enzyme, the rate-limiting step for glutathione synthesis. This leads to the protection of primary dopaminergic neurons against H_2O_2 and 6-hydroxydopamine (6-OHDA)-induced oxidative stress (Zhou and Freed 2005). Subsequently oxidative stress-induced expression from the GCL gene was identified to be mediated by Nrf2 through an ARE promoter element (Mani et al. 2013), suggesting that DJ-1 may upregulate GCL expression via Nrf2 activation.

A further connection with the glutathione pathway is through glutaredoxin1 (Grx1). Grx1 is the main deglutathionylating enzyme in the cell and has been shown to regulate DJ-1 protein levels in vivo. DJ-1 can be glutathionylated on the Cys53 and Cys106 residues (Fig. 6.4), which leads to its degradation. The action of Grx1 is to reverse the glutathionylation of DJ-1 and hence stabilises DJ-1 leading to increased overall levels, which contribute to the survival of the neuronal cells in vitro and in vivo (Johnson et al. 2016).

6.6 DJ-1: Antioxidant Functions in Neuronal Cells

The degeneration of the dopaminergic neurons in the substantia nigra pars compacta is the leading cause of Parkinson's disease (PD) (Sulzer and Surmeier 2013), and increased oxidative stress and mitochondrial dysfunction are major factors contributing to this degeneration (Jenner and Olanow 2006; Zhu and Chu 2010; Schapira and Jenner 2011). The link between oxidative stress and dopaminergic neuron degeneration is supported by various studies. The post-mortem brain analysis of human PD patients showed increased accumulation of a lipid peroxidation product 4-hydroxyl-2-nonenal (Yoritaka et al. 1996), carbonyl modification of soluble proteins (Floor and Wetzel 1998) and DNA and RNA oxidation products (Zhang et al. 1999). In addition, several studies have shown that inducing an oxidative stress insult by using various oxidative stress-inducing toxins including rotenone, paraquat and 6-OHDA leads to PD features in experimental mouse models (Perier et al. 2003; Callio et al. 2005; Richardson et al. 2005). High levels of ROS can be generated in the neurons and glia from various sources including the electron transport chain in the mitochondria (Dumont and Beal 2011), monoamine oxidase (MAO), NADPH oxidase (NOX) and flavoenzymes (Johnson et al. 2012). Such increased generation of ROS leads to dopaminergic neuron degeneration leading to PD and other neurodegenerative disorders. DJ-1 has been implicated as an antioxidant capable of protecting neurons from oxidative stress-induced damage.

DJ-1 protein expression was increased in SH-SY5Y neuroblastoma cells treated with hydrogen peroxide (H_2O_2). The pI of DJ-1 was shifted to a more acidic point, indicating a modification had occurred to produce the more oxidised DJ-1 forms (Taira et al. 2004). ROS exposure also resulted in a decrease in the pI of DJ-1 (Canet-Aviles et al. 2004; Kinumi et al. 2004; Choi et al. 2006). As noted previously a change in the oxidative state leads to active binding of other proteins, which forms part of the DJ-1 antioxidant function. Further in vitro studies also supported the hypothesis that DJ-1 acted to reduce intracellular ROS levels and protect neuroblastoma cells and neurons from undergoing cell death in response to increased oxidative stress caused by H_2O_2, rotenone and 6-OHDA (Yokota et al. 2003; Taira et al. 2004; Lev et al. 2009; Kim et al. 2012b).

The role of DJ-1 in protecting neurons from oxidative stress may be through a myriad of specific functions and interactions with other antioxidants. As described above DJ-1 can function to increase glutathione levels, which protect dopaminergic neurons against H_2O_2- and 6-OHDA-induced oxidative stress (Zhou and Freed 2005). DJ-1 also interacts with and enhances the activity of the enzyme paraoxonase-2, which has a role in protecting neurons from oxidative stress (Parsanejad et al. 2014a). As described earlier, under oxidative stress conditions, the sulfinic acid form of DJ-1 prevents the aggregation of α-synuclein and thus acts as a chaperone (Shendelman et al. 2004; Zhou et al. 2006). Further oxidation of DJ-1 reduces its chaperone activity and is therefore proposed to lead to oxidation-induced damage and possibly to neurological disease, including PD (Zhou et al. 2006).

DJ-1 has also been shown to exert antioxidative and cytoprotective functions in vivo. It was reported that DJ-1 knockout mice have significantly increased mitochondrial oxidative stress levels (Goldberg et al. 2012). Moreover, DJ-1 knockout mice were more susceptible to 1-methyl-4-phenyl-1,2,3,6-tetrahydropyridine (MPTP) and 6-OHDA compared to wild-type mice. Overexpression of DJ-1 in DJ-1 knockout mice reduced MPTP-induced neurotoxicity and neuron degeneration (Kim et al. 2005b; Paterna et al. 2007; Kim et al. 2012a). Moreover, microinjection of DJ-1 protein into the rat 6-OHDA model of PD improved motor behaviour and dopamine content, reduced neuron degeneration and improved the neuron count (Sun et al. 2012). It is clear that DJ-1 has an in vivo role to protect neuronal cells from oxidative stress-induced cell death, and the downstream targets identified in the in vitro studies provide possible mechanisms for this function.

6.7 DJ-1: Antioxidant Function in Cancer Cells

Increased intrinsic oxidative stress is one of the hallmarks of cancer cells, and it plays a crucial role in tumour development and progression (Pelicano et al. 2004). In response to increased ROS levels, cancer cells upregulate antioxidant molecules and stress response proteins to counteract the toxic effects. Moreover, antioxidant upregulation is also correlated with tumour proliferation, resistance to undergo apoptosis and drug resistance (Lincoln et al. 2003; Fernandes and Holmgren 2004; Kim et al. 2005c; Park et al. 2006; Nerini-Molteni et al. 2008; Raninga et al. 2015). Inhibition of antioxidant molecules has been shown to reduce tumour growth, to induce apoptosis and to overcome chemoresistance by making cancer cells more susceptible to ROS-induced cell death (Li et al. 2012; He et al. 2013; Raninga et al. 2016b).

DJ-1 expression is also increased in many human cancer types (MacKeigan et al. 2003; Liu et al. 2008; Zhang et al. 2008; Shu et al. 2013; Wang et al. 2014), and DJ-1 plays an important antioxidant function in cancer cells. DJ-1 has been shown to play a significant role in chemoresistance by eliminating the intracellular ROS produced in response to drug treatment and therefore providing cytoprotection. Increased DJ-1 expression was observed in dihydroartemisinin (DHA)-resistant HeLa cells compared to the DHA-sensitive HeLa cells. Increased mitochondrial translocation of oxidised DJ-1 was observed in DHA-resistant HeLa cells, and its inhibition increased DHA-induced intracellular ROS levels and sensitised resistant HeLa cells to DHA (Zhu et al. 2014). Another study reported the role of DJ-1 in protecting breast cancer cells, MCF-7, from the oxidative stress induced in response to 2′-benzoyloxycinnamaldehyde (BCA) treatment. BCA treatment increased DJ-1 expression in MCF-7 cells, and its inhibition increased mitochondrial damage in response to BCA-induced oxidative stress in these cells (Ismail et al. 2012). Thus, these results support the role of DJ-1 in protecting cancer cells against oxidative stress-induced damage and cell death, with DJ-1 likely to play either a direct role as an antioxidant by binding to the key targets discussed above (Table 6.1) or an indirect role by activating Nrf2 to upregulate antioxidant gene expression (Fig. 6.6).

6.8 Antioxidant Function of DJ-1 Against Hypoxia-Induced Stress

Hypoxia is a condition that arises due to an imbalance between the availability of oxygen and its consumption by cells. Reduced blood flow can also limit oxygen availability and is known as ischaemia. Both hypoxia and ischaemia play a significant role in the development and progression of various pathological conditions including neurological disorders, cardiac disease, stroke and cancer (Michiels 2004). When cells are exposed to a hypoxic microenvironment, it results in the stabilisation of hypoxia-inducible transcription factor-1 (HIF-1). Paradoxically HIF-1 requires ROS for activity (Chandel et al. 2000) but is also regulated by the combined action of Trx and Ref-1, which reduce specific cysteine residues (Ema et al. 1999). Correspondingly, antioxidant levels are also upregulated by hypoxia. Exposure of oral squamous cell carcinoma to a hypoxic microenvironment resulted in increased expression of peroxiredoxin-1 (Prdx-1) (Zhang et al. 2014), and we showed that exposure of multiple myeloma cells to an hypoxic microenvironment increased TrxR1 protein expression (Raninga et al. 2016b) and, most recently, also DJ-1 (Raninga et al. 2016a).

There is also evidence that DJ-1 is an upstream activator of HIF-1 under hypoxia and it also protects cancer cells against hypoxia-induced cell death (Vasseur et al. 2009). DJ-1 likely achieves this through the regulation of the Von Hippel-Lindau (VHL) protein, which causes ubiquitination of HIF-1α, leading to its degradation. DJ-1 was shown to bind to the VHL protein in vitro while in mouse neurons DJ-1 acted by negatively regulating the ubiquitination activity of VHL, leading to HIF-1 stabilisation and protection of the neuronal cells against hypoxic stress (Parsanejad et al. 2014b).

Other pathological conditions arise from the damage caused by either hypoxia or ischaemia followed by a rapid reoxygenation called reperfusion. Under such conditions, cells encounter a sudden increase in intracellular ROS levels, and a highly relevant example is myocardial ischaemic-reperfusion (sI/R) injury. DJ-1 was shown to protect cardiac myocytes against sI/R-induced oxidative stress. Stable overexpression of DJ-1 in H9c2 myocytes enhanced the antioxidant capacity of these cells, inhibited the sI/R-induced increase of intracellular ROS levels and prevented sI/R-induced cell viability loss (Yu et al. 2013). Myocardial preconditioning is a process where cardiac myocytes are subjected to brief episodes of sublethal ischaemia, which enhances the resistance of these cells to subsequent ischaemic injuries (Murry et al. 1986). Hypoxic preconditioning upregulated DJ-1 protein expression in cardiac myocytes, decreased intracellular ROS levels and attenuated sI/R-induced cell viability loss in these cells (Lu et al. 2012). The mechanism by which DJ-1 achieves this cardioprotection is still not fully understood but is likely to involve the regulation of key signalling and transcription factor pathways, including Nrf2 and the upregulation of other antioxidants (Yan et al. 2015). Of note, a similar function in protecting cardiac cells from sI/R has been attributed to Trx (Turoczi et al. 2003), so it will be interesting to ascertain if DJ-1 and Trx work cooperatively to protect cells from sI/R injury and if other specific antioxidants are involved.

6.9 Conclusions

The antioxidant functions of DJ-1 offer an explanation for its role in protecting cells from oxidative stress-induced damage that could otherwise lead to neurodegenerative diseases such as PD or to other pathologies. While DJ-1 has a weak peroxidase-like activity, it is clear that it interacts with other antioxidant systems and signalling pathways to mediate most of its antioxidant functions (Fig. 6.7). DJ-1 acts either by directly binding to its target proteins or by regulating specific transcription factors that increase expression of the Trx and GSH redox systems, which in turn can modulate DJ-1 activity. There are still many conflicting reports in the literature regarding the precise function of DJ-1 in these interlinked pathways, and there are without doubt more targets yet to be identified. However it is evident that DJ-1 has an

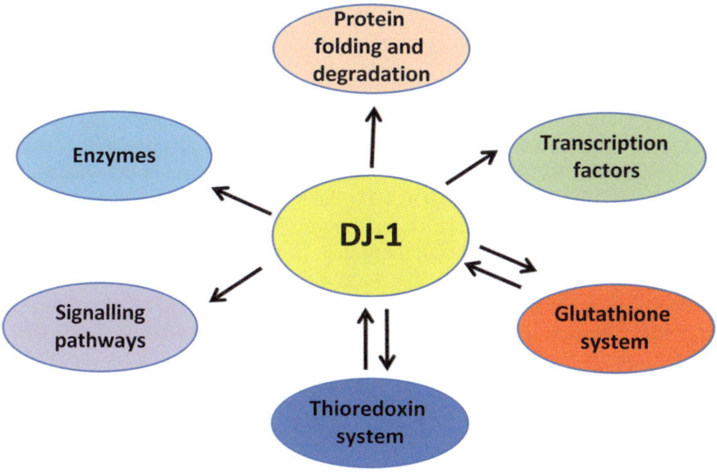

Fig. 6.7 The connection of DJ-1 with its many antioxidant functions. DJ-1 regulates the function of many other proteins, antioxidant systems and transcription factors. Both the thioredoxin and glutathione systems also regulate the activity of DJ-1

integral role to play in protecting cells from a broad range of oxidative stress insults. A thorough understanding of how DJ-1 exerts antioxidant functions in the cells and its crosstalk with other cellular antioxidant systems may help in better understanding the role of DJ-1 in many human diseases including PD and cancers. This knowledge may also help in designing therapies targeting DJ-1 to treat PD and cancers.

References

Andres-Mateos E, Perier C, Zhang L, Blanchard-Fillion B, Greco TM, Thomas B, Ko HS, Sasaki M, Ischiropoulos H, Przedborski S, Dawson TM, Dawson VL (2007) DJ-1 gene deletion reveals that DJ-1 is an atypical peroxiredoxin-like peroxidase. Proc Natl Acad Sci U S A 104(37):14807–14812. https://doi.org/10.1073/pnas.0703219104

Ariga H, Takahashi-Niki K, Kato I, Maita H, Niki T, Iguchi-Ariga SM (2013) Neuroprotective function of DJ-1 in Parkinson's disease. Oxidative Med Cell Longev 2013:683920. https://doi.org/10.1155/2013/683920

Arner ES, Holmgren A (2000) Physiological functions of thioredoxin and thioredoxin reductase. Eur J Biochem 267(20):6102–6109

Blackinton J, Lakshminarasimhan M, Thomas KJ, Ahmad R, Greggio E, Raza AS, Cookson MR, Wilson MA (2009) Formation of a stabilized cysteine sulfinic acid is critical for the mitochondrial function of the parkinsonism protein DJ-1. J Biol Chem 284(10):6476–6485. https://doi.org/10.1074/jbc.M806599200

Callio J, Oury TD, Chu CT (2005) Manganese superoxide dismutase protects against 6-hydroxydopamine injury in mouse brains. J Biol Chem 280(18):18536–18542. https://doi.org/10.1074/jbc.M413224200

Canet-Aviles RM, Wilson MA, Miller DW, Ahmad R, McLendon C, Bandyopadhyay S, Baptista MJ, Ringe D, Petsko GA, Cookson MR (2004) The Parkinson's disease protein DJ-1 is neuroprotective due to cysteine-sulfinic acid-driven mitochondrial localization. Proc Natl Acad Sci U S A 101(24):9103–9108. https://doi.org/10.1073/pnas.0402959101

Chandel NS, McClintock DS, Feliciano CE, Wood TM, Melendez JA, Rodriguez AM, Schumacker PT (2000) Reactive oxygen species generated at mitochondrial complex III stabilize hypoxia-inducible factor-1 alpha during hypoxia: a mechanism of O2 sensing. J Biol Chem 275(33):25130–25138. https://doi.org/10.1074/jbc.M001914200

Chang HY, Nishitoh H, Yang X, Ichijo H, Baltimore D (1998) Activation of apoptosis signal-regulating kinase 1 (ASK1) by the adapter protein Daxx. Science 281(5384):1860–1863

Choi J, Sullards MC, Olzmann JA, Rees HD, Weintraub ST, Bostwick DE, Gearing M, Levey AI, Chin LS, Li L (2006) Oxidative damage of DJ-1 is linked to sporadic Parkinson and Alzheimer diseases. J Biol Chem 281(16):10816–10824. https://doi.org/10.1074/jbc.M509079200

Clements CM, McNally RS, Conti BJ, Mak TW, Ting JP (2006) DJ-1, a cancer- and Parkinson's disease-associated protein, stabilizes the antioxidant transcriptional master regulator Nrf2. Proc Natl Acad Sci U S A 103(41):15091–15096. https://doi.org/10.1073/pnas.0607260103

Cuevas S, Yang Y, Konkalmatt P, Asico LD, Feranil J, Jones J, Villar VA, Armando I, Jose PA (2015) Role of nuclear factor erythroid 2-related factor 2 in the oxidative stress-dependent hypertension associated with the depletion of DJ-1. Hypertension 65(6):1251–1257. https://doi.org/10.1161/HYPERTENSIONAHA.114.04525

Das KC, Das CK (2000) Thioredoxin, a singlet oxygen quencher and hydroxyl radical scavenger: redox independent functions. Biochem Biophys Res Commun 277(2):443–447. https://doi.org/10.1006/bbrc.2000.3689

Dumont M, Beal MF (2011) Neuroprotective strategies involving ROS in Alzheimer disease. Free Radic Biol Med 51(5):1014–1026. https://doi.org/10.1016/j.freeradbiomed.2010.11.026

Eltoweissy M, Muller GA, Bibi A, Nguye PV, Dihazi GH, Muller CA, Dihazi H (2011) Proteomics analysis identifies PARK7 as an important player for renal cell resistance and survival under oxidative stress. Mol BioSyst 7(4):1277–1288. https://doi.org/10.1039/c0mb00116c

Ema M, Hirota K, Mimura J, Abe H, Yodoi J, Sogawa K, Poellinger L, Fujii-Kuriyama Y (1999) Molecular mechanisms of transcription activation by HLF and HIF1alpha in response to hypoxia: their stabilization and redox signal-induced interaction with CBP/p300. EMBO J 18(7):1905–1914. https://doi.org/10.1093/emboj/18.7.1905

Fan J, Ren H, Jia N, Fei E, Zhou T, Jiang P, Wu M, Wang G (2008) DJ-1 decreases Bax expression through repressing p53 transcriptional activity. J Biol Chem 283(7):4022–4030. https://doi.org/10.1074/jbc.M707176200

Fernandes AP, Holmgren A (2004) Glutaredoxins: glutathione-dependent redox enzymes with functions far beyond a simple thioredoxin backup system. Antioxid Redox Signal 6(1):63–74. https://doi.org/10.1089/152308604771978354

Fernandez-Caggiano M, Schroder E, Cho HJ, Burgoyne J, Barallobre-Barreiro J, Mayr M, Eaton P (2016) Oxidant-induced Interprotein disulfide formation in cardiac protein DJ-1 occurs via an interaction with Peroxiredoxin 2. J Biol Chem 291(19):10399–10410. https://doi.org/10.1074/jbc.M115.699850

Floor E, Wetzel MG (1998) Increased protein oxidation in human substantia nigra pars compacta in comparison with basal ganglia and prefrontal cortex measured with an improved dinitrophenylhydrazine assay. J Neurochem 70(1):268–275

Fu C, Wu C, Liu T, Ago T, Zhai P, Sadoshima J, Li H (2009) Elucidation of thioredoxin target protein networks in mouse. Mol Cell Proteomics 8(7):1674–1687. https://doi.org/10.1074/mcp.M800580-MCP200

Gan L, Johnson DA, Johnson JA (2010) Keap1-Nrf2 Activation in the presence and absence of DJ-1. Eur J Neurosci 31(6):967–977. https://doi.org/10.1111/j.1460-9568.2010.07138.x

Girotto S, Sturlese M, Bellanda M, Tessari I, Cappellini R, Bisaglia M, Bubacco L, Mammi S (2012) Dopamine-derived quinones affect the structure of the redox sensor DJ-1 through modifications at Cys-106 and Cys-53. J Biol Chem 287(22):18738–18749. https://doi.org/10.1074/jbc.M111.311589

Go YM, Roede JR, Walker DI, Duong DM, Seyfried NT, Orr M, Liang Y, Pennell KD, Jones DP (2013) Selective targeting of the cysteine proteome by thioredoxin and glutathione redox systems. Mol Cell Proteomics 12(11):3285–3296. https://doi.org/10.1074/mcp.M113.030437
Goldberg JA, Guzman JN, Estep CM, Ilijic E, Kondapalli J, Sanchez-Padilla J, Surmeier DJ (2012) Calcium entry induces mitochondrial oxidant stress in vagal neurons at risk in Parkinson's disease. Nat Neurosci 15(10):1414–1421. https://doi.org/10.1038/nn.3209
Hayes JD, Dinkova-Kostova AT (2014) The Nrf2 regulatory network provides an interface between redox and intermediary metabolism. Trends Biochem Sci 39(4):199–218. https://doi.org/10.1016/j.tibs.2014.02.002
He T, Banach-Latapy A, Vernis L, Dardalhon M, Chanet R, Huang ME (2013) Peroxiredoxin 1 knockdown potentiates beta-lapachone cytotoxicity through modulation of reactive oxygen species and mitogen-activated protein kinase signals. Carcinogenesis 34(4):760–769. https://doi.org/10.1093/carcin/bgs389
Holmgren A (1985) Thioredoxin. Annu Rev Biochem 54:237–271. https://doi.org/10.1146/annurev.bi.54.070185.001321
Honbou K, Suzuki NN, Horiuchi M, Niki T, Taira T, Ariga H, Inagaki F (2003) The crystal structure of DJ-1, a protein related to male fertility and Parkinson's disease. J Biol Chem 278(33):31380–31384. https://doi.org/10.1074/jbc.M305878200
Ichijo H, Nishida E, Irie K, ten Dijke P, Saitoh M, Moriguchi T, Takagi M, Matsumoto K, Miyazono K, Gotoh Y (1997) Induction of apoptosis by ASK1, a mammalian MAPKKK that activates SAPK/JNK and p38 signaling pathways. Science 275(5296):90–94
Im JY, Lee KW, Junn E, Mouradian MM (2010) DJ-1 protects against oxidative damage by regulating the thioredoxin/ASK1 complex. Neurosci Res 67(3):203–208. https://doi.org/10.1016/j.neures.2010.04.002
Im JY, Lee KW, Woo JM, Junn E, Mouradian MM (2012) DJ-1 induces thioredoxin 1 expression through the Nrf2 pathway. Hum Mol Genet 21(13):3013–3024. https://doi.org/10.1093/hmg/dds131
Ismail IA, Kang HS, Lee HJ, Kwon BM, Hong SH (2012) 2′-Benzoyloxycinnamaldehyde-mediated DJ-1 upregulation protects MCF-7 cells from mitochondrial damage. Biol Pharm Bull 35(6):895–902
Ismail IA, Abdel Shakor AB, Hong SH (2015) DJ-1 protects breast cancer cells against 2′-Benzoyloxycinnamaldehyde-induced oxidative stress independent of Nrf2. J Cell Physiol 230(9):2262–2269. https://doi.org/10.1002/jcp.24957
Ito G, Ariga H, Nakagawa Y, Iwatsubo T (2006) Roles of distinct cysteine residues in S-nitrosylation and dimerization of DJ-1. Biochem Biophys Res Commun 339(2):667–672. https://doi.org/10.1016/j.bbrc.2005.11.058
Jenner P, Olanow CW (2006) The pathogenesis of cell death in Parkinson's disease. Neurology 66(10 Suppl 4):S24–S36
Johnson WM, Wilson-Delfosse AL, Mieyal JJ (2012) Dysregulation of glutathione homeostasis in neurodegenerative diseases. Forum Nutr 4(10):1399–1440. https://doi.org/10.3390/nu4101399
Johnson WM, Golczak M, Choe K, Currran PL, Gorelenkova Miller O, Yao C, Wang W, Lin J, Milkovic NM, Ray A, Ravindranath V, Zhu X, Wilson MA, Wilson-Delfosse AL, Chen SG, Mieyal JJ (2016) Regulation of DJ-1 by glutaredoxin 1 in vivo – implications for Parkinson's disease. Biochemistry. https://doi.org/10.1021/acs.biochem.5b01132
Junn E, Taniguchi H, Jeong BS, Zhao X, Ichijo H, Mouradian MM (2005) Interaction of DJ-1 with Daxx inhibits apoptosis signal-regulating kinase 1 activity and cell death. Proc Natl Acad Sci U S A 102(27):9691–9696. https://doi.org/10.1073/pnas.0409635102
Kim SJ, Miyoshi Y, Taguchi T, Tamaki Y, Nakamura H, Yodoi J, Kato K, Noguchi S (2005a) High thioredoxin expression is associated with resistance to docetaxel in primary breast cancer. Clin Cancer Res 11(23):8425–8430. https://doi.org/10.1158/1078-0432.CCR-05-0449
Kim RH, Peters M, Jang Y, Shi W, Pintilie M, Fletcher GC, DeLuca C, Liepa J, Zhou L, Snow B, Binari RC, Manoukian AS, Bray MR, Liu FF, Tsao MS, Mak TW (2005b) DJ-1, a novel regulator of the tumor suppressor PTEN. Cancer Cell 7(3):263–273. https://doi.org/10.1016/j.ccr.2005.02.010

Kim RH, Smith PD, Aleyasin H, Hayley S, Mount MP, Pownall S, Wakeham A, You-Ten AJ, Kalia SK, Horne P, Westaway D, Lozano AM, Anisman H, Park DS, Mak TW (2005c) Hypersensitivity of DJ-1-deficient mice to 1-methyl-4-phenyl-1,2,3,6-tetrahydropyrindine (MPTP) and oxidative stress. Proc Natl Acad Sci U S A 102(14):5215–5220. https://doi.org/10.1073/pnas.0501282102

Kim YC, Kitaura H, Taira T, Iguchi-Ariga SM, Ariga H (2009) Oxidation of DJ-1-dependent cell transformation through direct binding of DJ-1 to PTEN. Int J Oncol 35(6):1331–1341

Kim SJ, Park YJ, Hwang IY, Youdim MB, Park KS, YJ O (2012a) Nuclear translocation of DJ-1 during oxidative stress-induced neuronal cell death. Free Radic Biol Med 53(4):936–950. https://doi.org/10.1016/j.freeradbiomed.2012.05.035

Kim SJ, Park YJ, YJ O (2012b) Proteomic analysis reveals a protective role for DJ-1 during 6-hydroxydopamine-induced cell death. Biochem Biophys Res Commun 422(1):8–14. https://doi.org/10.1016/j.bbrc.2012.04.063

Kinumi T, Kimata J, Taira T, Ariga H, Niki E (2004) Cysteine-106 of DJ-1 is the most sensitive cysteine residue to hydrogen peroxide-mediated oxidation in vivo in human umbilical vein endothelial cells. Biochem Biophys Res Commun 317(3):722–728. https://doi.org/10.1016/j.bbrc.2004.03.110

Le Moan N, Clement G, Le Maout S, Tacnet F, Toledano MB (2006) The Saccharomyces Cerevisiae proteome of oxidized protein thiols: contrasted functions for the thioredoxin and glutathione pathways. J Biol Chem 281(15):10420–10430. https://doi.org/10.1074/jbc.M513346200

Lev N, Ickowicz D, Melamed E, Offen D (2008) Oxidative insults induce DJ-1 upregulation and redistribution: implications for neuroprotection. Neurotoxicology 29(3):397–405. https://doi.org/10.1016/j.neuro.2008.01.007

Lev N, Ickowicz D, Barhum Y, Lev S, Melamed E, Offen D (2009) DJ-1 protects against dopamine toxicity. J Neural Transm 116(2):151–160. https://doi.org/10.1007/s00702-008-0134-4

Li C, Thompson MA, Tamayo AT, Zuo Z, Lee J, Vega F, Ford RJ, Pham LV (2012) Over-expression of Thioredoxin-1 mediates growth, survival, and chemoresistance and is a druggable target in diffuse large B-cell lymphoma. Oncotarget 3(3):314–326. 10.18632/oncotarget.463

Lincoln DT, Ali Emadi EM, Tonissen KF, Clarke FM (2003) The thioredoxin-thioredoxin reductase system: over-expression in human cancer. Anticancer Res 23(3B):2425–2433

Liu H, Wang M, Li M, Wang D, Rao Q, Wang Y, Xu Z, Wang J (2008) Expression and role of DJ-1 in leukemia. Biochem Biophys Res Commun 375(3):477–483. https://doi.org/10.1016/j.bbrc.2008.08.046

Lu J, Holmgren A (2014) The thioredoxin antioxidant system. Free Radic Biol Med 66:75–87. https://doi.org/10.1016/j.freeradbiomed.2013.07.036

Lu HS, Chen HP, Wang S, Yu HH, Huang XS, Huang QR, He M (2012) Hypoxic preconditioning up-regulates DJ-1 protein expression in rat heart-derived H9c2 cells through the activation of extracellular-regulated kinase 1/2 pathway. Mol Cell Biochem 370(1–2):231–240. https://doi.org/10.1007/s11010-012-1414-8

Lushchak VI (2012) Glutathione homeostasis and functions: potential targets for medical interventions. J Amino Acids 2012:736837. https://doi.org/10.1155/2012/736837

MacKeigan JP, Clements CM, Lich JD, Pope RM, Hod Y, Ting JP (2003) Proteomic profiling drug-induced apoptosis in non-small cell lung carcinoma: identification of RS/DJ-1 and RhoGDIalpha. Cancer Res 63(20):6928–6934

Mahmood DF, Abderrazak A, El Hadri K, Simmet T, Rouis M (2013) The thioredoxin system as a therapeutic target in human health and disease. Antioxid Redox Signal 19(11):1266–1303. https://doi.org/10.1089/ars.2012.4757

Mani M, Khaghani S, Gol Mohammadi T, Zamani Z, Azadmanesh K, Meshkani R, Pasalar P, Mostafavi E (2013) Activation of Nrf2-antioxidant response element mediated glutamate cysteine ligase expression in hepatoma cell line by homocysteine. Hepat Mon 13(5):e8394. https://doi.org/10.5812/hepatmon.8394

Matthews JR, Wakasugi N, Virelizier JL, Yodoi J, Hay RT (1992) Thioredoxin regulates the DNA binding activity of NF-kappa B by reduction of a disulphide bond involving cysteine 62. Nucleic Acids Res 20(15):3821–3830

McNally RS, Davis BK, Clements CM, Accavitti-Loper MA, Mak TW, Ting JP (2011) DJ-1 enhances cell survival through the binding of Cezanne, a negative regulator of NF-kappaB. J Biol Chem 286(6):4098–4106. https://doi.org/10.1074/jbc.M110.147371

Meuillet EJ, Mahadevan D, Berggren M, Coon A, Powis G (2004) Thioredoxin-1 binds to the C2 domain of PTEN inhibiting PTEN's lipid phosphatase activity and membrane binding: a mechanism for the functional loss of PTEN's tumor suppressor activity. Arch Biochem Biophys 429(2):123–133. https://doi.org/10.1016/j.abb.2004.04.020

Michiels C (2004) Physiological and pathological responses to hypoxia. Am J Pathol 164(6):1875–1882. https://doi.org/10.1016/S0002-9440(10)63747-9

Mo JS, Jung J, Yoon JH, Hong JA, Kim MY, Ann EJ, Seo MS, Choi YH, Park HS (2010) DJ-1 modulates the p38 mitogen-activated protein kinase pathway through physical interaction with apoptosis signal-regulating kinase 1. J Cell Biochem 110(1):229–237. https://doi.org/10.1002/jcb.22530

Moscovitz O, Ben-Nissan G, Fainer I, Pollack D, Mizrachi L, Sharon M (2015) The Parkinson's-associated protein DJ-1 regulates the 20S proteasome. Nat Commun 6:6609. https://doi.org/10.1038/ncomms7609

Murry CE, Jennings RB, Reimer KA (1986) Preconditioning with ischemia: a delay of lethal cell injury in ischemic myocardium. Circulation 74(5):1124–1136

Nerini-Molteni S, Ferrarini M, Cozza S, Caligaris-Cappio F, Sitia R (2008) Redox homeostasis modulates the sensitivity of myeloma cells to bortezomib. Br J Haematol 141(4):494–503. https://doi.org/10.1111/j.1365-2141.2008.07066.x

Nordberg J, Arner ES (2001) Reactive oxygen species, antioxidants, and the mammalian thioredoxin system. Free Radic Biol Med 31(11):1287–1312

Park JH, Kim YS, Lee HL, Shim JY, Lee KS, YJ O, Shin SS, Choi YH, Park KJ, Park RW, Hwang SC (2006) Expression of peroxiredoxin and thioredoxin in human lung cancer and paired normal lung. Respirology 11(3):269–275. https://doi.org/10.1111/j.1440-1843.2006.00849.x

Parsanejad M, Bourquard N, Qu D, Zhang Y, Huang E, Rousseaux MW, Aleyasin H, Irrcher I, Callaghan S, Vaillant DC, Kim RH, Slack RS, Mak TW, Reddy ST, Figeys D, Park DS (2014a) DJ-1 interacts with and regulates paraoxonase-2, an enzyme critical for neuronal survival in response to oxidative stress. PLoS One 9(9):e106601. https://doi.org/10.1371/journal.pone.0106601

Parsanejad M, Zhang Y, Qu D, Irrcher I, Rousseaux MW, Aleyasin H, Kamkar F, Callaghan S, Slack RS, Mak TW, Lee S, Figeys D, Park DS (2014b) Regulation of the VHL/HIF-1 pathway by DJ-1. J Neurosci 34(23):8043–8050. https://doi.org/10.1523/JNEUROSCI.1244-13.2014

Paterna JC, Leng A, Weber E, Feldon J, Bueler H (2007) DJ-1 and Parkin modulate dopamine-dependent behavior and inhibit MPTP-induced nigral dopamine neuron loss in mice. Mol Ther 15(4):698–704. https://doi.org/10.1038/sj.mt.6300067

Pelicano H, Carney D, Huang P (2004) ROS stress in cancer cells and therapeutic implications. Drug Resist Updat 7(2):97–110. https://doi.org/10.1016/j.drup.2004.01.004

Perier C, Bove J, Vila M, Przedborski S (2003) The rotenone model of Parkinson's disease. Trends Neurosci 26(7):345–346. https://doi.org/10.1016/S0166-2236(03)00144-9

Raninga PV, Di Trapani G, Tonissen KF (2014) Cross talk between two antioxidant systems, Thioredoxin and DJ-1: consequences for cancer. Oncoscience 1(1):95–110. 10.18632/oncoscience.12

Raninga PV, Di Trapani G, Vuckovic S, Bhatia M, Tonissen KF (2015) Inhibition of thioredoxin 1 leads to apoptosis in drug-resistant multiple myeloma. Oncotarget 6(17):15410–15424. 10.18632/oncotarget.3795

Raninga PV, Di Trapani G, Vuckovic S, Tonissen KF (2016a) Targeted knockdown of DJ-1 induces multiple myeloma cell death via KLF6 upregulation. Apoptosis 21(12):1422–1437. https://doi.org/10.1007/s10495-016-1303-9

Raninga PV, Di Trapani G, Vuckovic S, Tonissen KF (2016b) TrxR1 Inhibition overcomes both hypoxia-induced and acquired bortezomib resistance in multiple myeloma through NF-kB inhibition. Cell Cycle 15:559–572. https://doi.org/10.1080/15384101.2015.1136038

Richardson JR, Quan Y, Sherer TB, Greenamyre JT, Miller GW (2005) Paraquat neurotoxicity is distinct from that of MPTP and rotenone. Toxicol Sci 88(1):193–201. https://doi.org/10.1093/toxsci/kfi304

Saitoh M, Nishitoh H, Fujii M, Takeda K, Tobiume K, Sawada Y, Kawabata M, Miyazono K, Ichijo H (1998) Mammalian thioredoxin is a direct inhibitor of apoptosis signal-regulating kinase (ASK) 1. EMBO J 17(9):2596–2606. https://doi.org/10.1093/emboj/17.9.2596

Schapira AH, Jenner P (2011) Etiology and pathogenesis of Parkinson's disease. Mov Disord 26(6):1049–1055. https://doi.org/10.1002/mds.23732

Shendelman S, Jonason A, Martinat C, Leete T, Abeliovich A (2004) DJ-1 is a redox-dependent molecular chaperone that inhibits alpha-synuclein aggregate formation. PLoS Biol 2(11):e362. https://doi.org/10.1371/journal.pbio.0020362

Shu K, Xiao Z, Long S, Yan J, Yu X, Zhu Q, Mei T (2013) Expression of DJ-1 in endometrial cancer: close correlation with Clinicopathological features and apoptosis. Int J Gynecol Cancer 23(6):1029–1035. https://doi.org/10.1097/IGC.0b013e3182959182

Song JJ, Lee YJ (2003) Role of the ASK1-SEK1-JNK1-HIPK1 signal in Daxx trafficking and ASK1 oligomerization. J Biol Chem 278(47):47245–47252. https://doi.org/10.1074/jbc.M213201200

Sulzer D, Surmeier DJ (2013) Neuronal vulnerability, pathogenesis, and Parkinson's disease. Mov Disord 28(6):715–724. https://doi.org/10.1002/mds.25187

Sun SY, An CN, XP P (2012) DJ-1 protein protects dopaminergic neurons against 6-OHDA/MG-132-induced neurotoxicity in rats. Brain Res Bull 88(6):609–616. https://doi.org/10.1016/j.brainresbull.2012.05.013

Taira T, Takahashi K, Kitagawa R, Iguchi-Ariga SM, Ariga H (2001) Molecular cloning of human and mouse DJ-1 genes and identification of Sp1-dependent activation of the human DJ-1 promoter. Gene 263(1–2):285–292

Taira T, Saito Y, Niki T, Iguchi-Ariga SM, Takahashi K, Ariga H (2004) DJ-1 has a role in antioxidative stress to prevent cell death. EMBO Rep 5(2):213–218. https://doi.org/10.1038/sj.embor.7400074

Tang J, Liu J, Li X, Zhong Y, Zhong T, Liu Y, Wang JH, Jiang Y (2014) PRAK interacts with DJ-1 and prevents oxidative stress-induced cell death. Oxidative Med Cell Longev 2014:735618. https://doi.org/10.1155/2014/735618

Turoczi T, Chang VW, Engelman RM, Maulik N, Ho YS, Das DK (2003) Thioredoxin redox signaling in the ischemic heart: an insight with transgenic mice overexpressing Trx1. J Mol Cell Cardiol 35(6):695–704

Ueno M, Masutani H, Arai RJ, Yamauchi A, Hirota K, Sakai T, Inamoto T, Yamaoka Y, Yodoi J, Nikaido T (1999) Thioredoxin-dependent redox regulation of p53-mediated p21 activation. J Biol Chem 274(50):35809–35815

Vasseur S, Afzal S, Tardivel-Lacombe J, Park DS, Iovanna JL, Mak TW (2009) DJ-1/PARK7 is an important mediator of hypoxia-induced cellular responses. Proc Natl Acad Sci U S A 106(4):1111–1116. https://doi.org/10.1073/pnas.0812745106

Waak J, Weber SS, Gorner K, Schall C, Ichijo H, Stehle T, Kahle PJ (2009) Oxidizable residues mediating protein stability and cytoprotective interaction of DJ-1 with apoptosis signal-regulating kinase 1. J Biol Chem 284(21):14245–14257. https://doi.org/10.1074/jbc.M806902200

Wang B, Qin H, Wang Y, Chen W, Luo J, Zhu X, Wen W, Lei W (2014) Effect of DJ-1 overexpression on the proliferation, apoptosis, invasion and migration of laryngeal squamous cell carcinoma SNU-46 cells through PI3K/AKT/mTOR. Oncol Rep 32(3):1108–1116. https://doi.org/10.3892/or.2014.3286

Wilson MA (2011) The role of cysteine oxidation in DJ-1 function and dysfunction. Antioxid Redox Signal 15(1):111–122. https://doi.org/10.1089/ars.2010.3481

Witt AC, Lakshminarasimhan M, Remington BC, Hasim S, Pozharski E, Wilson MA (2008) Cysteine pKa depression by a protonated glutamic acid in human DJ-1. Biochemistry 47(28):7430–7440. https://doi.org/10.1021/bi800282d

Yan YF, Chen HP, Huang XS, Qiu LY, Liao ZP, Huang QR (2015) DJ-1 mediates the delayed Cardioprotection of hypoxic preconditioning through activation of Nrf2 and subsequent upregulation of Antioxidative enzymes. J Cardiovasc Pharmacol 66(2):148–158. https://doi.org/10.1097/FJC.0000000000000257

Yokota T, Sugawara K, Ito K, Takahashi R, Ariga H, Mizusawa H (2003) Down regulation of DJ-1 enhances cell death by oxidative stress, ER stress, and proteasome inhibition. Biochem Biophys Res Commun 312(4):1342–1348

Yoritaka A, Hattori N, Uchida K, Tanaka M, Stadtman ER, Mizuno Y (1996) Immunohistochemical detection of 4-hydroxynonenal protein adducts in Parkinson disease. Proc Natl Acad Sci U S A 93(7):2696–2701

Yu HH, Xu Q, Chen HP, Wang S, Huang XS, Huang QR, He M (2013) Stable overexpression of DJ-1 protects H9c2 cells against oxidative stress under a hypoxia condition. Cell Biochem Funct. https://doi.org/10.1002/cbf.2949

Zhang J, Perry G, Smith MA, Robertson D, Olson SJ, Graham DG, Montine TJ (1999) Parkinson's Disease is associated with oxidative damage to cytoplasmic DNA and RNA in substantia nigra neurons. Am J Pathol 154(5):1423–1429. https://doi.org/10.1016/S0002-9440(10)65396-5

Zhang HY, Wang HQ, Liu HM, Guan Y, ZX D (2008) Regulation of tumor necrosis factor-related apoptosis-inducing ligand-induced apoptosis by DJ-1 in thyroid cancer cells. Endocr Relat Cancer 15(2):535–544. https://doi.org/10.1677/ERC-07-0195

Zhang M, Hou M, Ge L, Miao C, Zhang J, Jing X, Shi N, Chen T, Tang X (2014) Induction of peroxiredoxin 1 by hypoxia regulates heme oxygenase-1 via NF-kappaB in oral cancer. PLoS One 9(8):e105994. https://doi.org/10.1371/journal.pone.0105994

Zhou W, Freed CR (2005) DJ-1 up-regulates glutathione synthesis during oxidative stress and inhibits A53T alpha-synuclein toxicity. J Biol Chem 280(52):43150–43158. https://doi.org/10.1074/jbc.M507124200

Zhou W, Zhu M, Wilson MA, Petsko GA, Fink AL (2006) The oxidation state of DJ-1 regulates its chaperone activity toward alpha-synuclein. J Mol Biol 356(4):1036–1048. https://doi.org/10.1016/j.jmb.2005.12.030

Zhu J, Chu CT (2010) Mitochondrial dysfunction in Parkinson's disease. J Alzheimers Dis 20(Suppl 2):S325–S334. https://doi.org/10.3233/JAD-2010-100363

Zhu H, Liao SD, Shi JJ, Chang LL, Tong YG, Cao J, YY F, Chen XP, Ying MD, Yang B, He QJ, JJ L (2014) DJ-1 mediates the resistance of cancer cells to dihydroartemisinin through reactive oxygen species removal. Free Radic Biol Med 71:121–132. https://doi.org/10.1016/j.freeradbiomed.2014.03.026

Chapter 7
Transcriptional Regulation of DJ-1

Kazuko Takahashi-Niki, Takeshi Niki, Sanae M. M. Iguchi-Ariga, and Hiroyoshi Ariga

Abstract *DJ-1* is an oncogene and also a causative gene for familial Parkinson's disease. DJ-1 has various functions, and the oxidative status of a cysteine residue at position 106 (C106) is crucial for determination of the activation level of DJ-1.

DJ-1 binds to many proteins, including various transcription factors, and acts as a coactivator or corepressor for regulating their target genes without direct binding to DNA, thereby affecting various cell functions. DJ-1-regulating transcription factors and their modified proteins are the androgen receptor and its regulatory proteins, p53; polypyrimidine tract-binding protein-associated splicing factor (PSF); Keap1, an inhibitor for nuclear factor erythroid2-related factor 2 (Nrf2); sterol regulatory element-binding protein (SREBP); Ras-responsive element-binding protein (RREB1); signal transducer and activator of transcription 1 (STAT1); and Nurr1. Considering oxidative stress response and dopamine synthesis, the regulation of Nrf2, p53, and PSF by DJ-1 is especially important. In addition, SREBP1 and RREB1 functions that are positively regulated by DJ-1 may participate in the onset and pathogenesis of metabolic syndrome.

DJ-1 is expressed ubiquitously with high levels in the testis and brain and moderate levels in other tissues. Furthermore, DJ-1 is translocated from the cytoplasm to nucleus during the cell cycle after mitogen stimulation, suggesting that DJ-1 has a growth-related function. In this review, we describe how DJ-1 regulates cell growth/death and dopamine synthesis by targeting various transcription factors.

Keywords DJ-1 • Transcription • Oxidative stress • Parkinson's disease • Cancer

---

K. Takahashi-Niki (✉) • H. Ariga
Faculty of Pharmaceutical Sciences, Hokkaido University,
Kita 12, Nishi 6, Kita-ku, Sapporo 060-0812, Japan
e-mail: nkazu@pharm.hokudai.ac.jp; hiro@pharm.hokudai.ac.jp

T. Niki • S.M.M. Iguchi-Ariga
Faculty of Agriculture, Hokkaido University,
Kita 9, Nishi 9, Kita-ku, Sapporo 060-8589, Japan
e-mail: takeshi.niki@cira.kyoto-u.ac.jp; myan@res.agr.hokudai.ac.jp

H. Ariga, S.M.M. Iguchi-Ariga (eds.), *DJ-1/PARK7 Protein*,
Advances in Experimental Medicine and Biology 1037,
https://doi.org/10.1007/978-981-10-6583-5_7

7.1 Androgen Receptor

The androgen receptor (AR) is a member of the nuclear hormone receptor superfamily, which is comprised of a large group of ligand-dependent transcription factors. After a ligand binds to the AR, the AR is translocated into the nucleus and binds to the androgen-responsive element (ARE) of androgen-activating genes that are essential for the development and maintenance of male reproductive functions. DJ-1 regulates AR function through indirect binding mechanisms in which DJ-1 impairs the binding of a protein inhibitor of activated STAT (PIASxα) to the AR. PIASxα inhibits AR transcription activity in the nucleus, and DJ-1 sequesters PIASxα from the AR by its binding, thereby resulting in recovery of AR transcription activity (Takahashi et al. 2001). Since cell background and promoter context alter the effect of PIAS family proteins on AR-dependent transcription, the effect of DJ-1 on AR transcription activity depends on the cell type. In addition, it has been reported that DJ-1 positively regulates AR activity via DJ-1-binding protein (DJBP). DJBP represses AR transcription activity by inhibiting DNA-binding of the AR through its testosterone-dependent binding to the AR DNA-binding region, and DJ-1 restores AR transcription activity by sequestering DJBP (Niki et al. 2003). It has been reported, on the other hand, that DJ-1 binds to the AR to regulate its activity in hormonally treated prostate cancer cells (Tillman et al. 2007).

7.2 Proteins in the Dopamine Synthesis Pathway

It has been reported that DJ-1 interacts with several dopamine-associated genes or proteins. Dopamine is synthesized from tyrosine by tyrosine hydroxylase (TH), which converts tyrosine to L-DOPA, and by L-DOPA carboxylase (DDC), which converts L-DOPA to dopamine. Subsequently, dopamine is packed into synaptic vesicles by vesicular monoamine transporter 2 (VMAT2). VMAT2 reuptakes excess dopamine into synaptic vesicles to protect neurons from oxidized dopamine-induced toxicity.

It was first reported that DJ-1 forms a ternary complex with PSF and p54nrb in dopaminergic neuronal cells (Xu et al. 2005), and it was later reported that DJ-1 regulates the human *TH* gene by inhibiting the sumoylation of PSF, which acts as repressor against *TH* gene transcription (Zhong et al. 2008). Furthermore, we reported that DJ-1 positively regulates human *TH* gene expression in a human-specific manner by sequestering PSF from the human *TH* gene promoter (Ishikawa et al. 2010). Since the mouse *TH* gene does not contain the PSF-recognition sequence, this transcriptional regulation is only observed in the human *TH* gene. DJ-1 also positively regulates enzymatic activities of TH and DDC by direct association with them (Ishikawa et al. 2009). Positive regulation of TH and DDC activities by DJ-1 depends on the level of oxidative status of C106: When the sum of SH (reduced) and SOH forms of C106 is more than 50% of total forms of C106, DJ-1 stimulates TH and DDC activities. Since DJ-1 is oxidized with aging, these findings suggest that the activity of DJ-1 toward TH and DDC is decreased with aging, which is one of the crucial factors for the onset of Parkinson's disease (PD).

DJ-1 also regulates both *VMAT2* gene expression and its activity by direct binding (Ishikawa et al. 2012). DJ-1 contributes to reuptake of dopamine into synaptic vesicles via VMAT2 and protects neurons from oxidized dopamine-induced damage. Pathogenic mutations of DJ-1, including both homozygous and heterozygous mutations, reduce stimulating activity against TH, DDC, and VMAT2 (Ishikawa et al. 2009, 2010, 2012).

7.3 Nrf2

Nrf2 is a master transcription factor for oxidative stress and anti-oxidative stress responses. Without such stresses, Nrf2 is localized in the cytoplasm in a complex with Keap1, a ubiquitin ligase, resulting in degradation of Nrf2 by the ubiquitin-proteasome system. Under an oxidative condition, DJ-1 binds to and sequesters Keap1 from Nrf2, leading to translocation of Nrf2 into the nucleus to activate various anti-oxidative stress-related genes, thereby decreasing the levels of reactive oxygen species (ROS) (Clements et al. 2006). Moreover, DJ-1 regulates the expression of superoxide dismutase (SOD3) and glutathione ligase genes, resulting in reduction of the levels of ROS (Nishinaga et al. 2005, Zhou and Freed 2005, Yan et al. 2015).

7.4 p53

p53 is a tumor suppressor and plays roles in induction of senescence and apoptosis in cells and in regulation of mitochondrial homeostasis against oxidative stress. DJ-1 directly binds to p53 and regulates p53 activity in various ways: p53 is activated by Topors-mediated sumoylation and inactivated by DJ-1 through inhibition of Topors activity (Shinbo et al. 2005), and DJ-1 binds to the DNA-binding region of p53 to inhibit p53 transcriptional activity when affinity of p53 to DNA is low, leading to cell cycle progression (Kato et al. 2013).

It has also been reported that DJ-1 inhibits the induction of apoptosis by p53-induced Bax expression (Fan et al. 2008). Furthermore, we reported that DJ-1 binds to SIRT1 and regulates p53 transcriptional activity via SIRT1 (Takahashi-Niki et al. 2016).

DJ-1 downregulates PTEN-induced kinase 1 (PINK1) expression by binding to forkhead box O3a (Foxo3a), a transcription factor directly interacting with the *pink1* promoter to stimulate its transcriptional activity. Moreover, DJ-1 regulates tumor suppressor PTEN (Kim et al. 2005) by direct interaction with PTEN and inhibits its phosphatase activity (Kim et al. 2009).

In addition, it has been reported that DJ-1 regulates Legumain expression. Legumain is an asparaginyl endopeptidase observed in the mouse kidney and in tumors, and important roles of Legumain in tumor growth/metastasis and development of carotid artery atherosclerosis have been suggested. Legumain degrades annexin A2, and its expression and protease activity are regulated by p53 through binding of p53 to the *legumain* gene (Yamane et al. 2013b). The expression and enzymatic activity of Legumain are increased in DJ-1-knockout cells, and transcription of the *legumain* gene is regulated by p53 through DJ-1 by binding to the p53-binding site of the mouse *legumain* gene (Yamane et al. 2015). These findings indicate that DJ-1 regulates cell metabolism and proliferation.

7.5 SREBP

DJ-1 positively regulates the expression of the low-density lipoprotein receptor (LDLR) gene through association with sterol regulatory element-binding protein (SREBP), which is an enhancer-binding protein binding to the sterol regulatory element (SRE) of the *LDLR* gene (Yamaguchi et al. 2012). Activation of the LDLR promoter by DJ-1 is enhanced by oxidative stress as in the case of DJ-1-activating tyrosine hydroxylase promoter activity (see above).

7.6 RREB1

DJ-1 positively regulates cholecystokinin (CCK) gene expression (Yamane et al. 2013a). CCK is a peptide hormone and has roles in contraction of the gallbladder and in promotion of secretion of pancreatic fluid. CCK expression is regulated by two active sites on its promoter, the Ras-responsive element (RRE) and Sp1 site, and DJ-1 stimulates *CCK* gene expression in association with RRE-binding protein 1 (RREBP1) but not with Sp1. CCK is co-localized with dopamine in the substantia nigra and regulates the release of dopamine. Since DJ-1 regulates the expression of dopamine-synthesizing enzymes as described above, these findings suggest that DJ-1 directly or indirectly regulates the synthesis and secretion of dopamine.

7.7 STAT1

DJ-1 negatively regulates STAT1 transcriptional activity by stimulating interaction of STAT1 with its phosphatase, SH2 domain-containing protein-tyrosine phosphatase-1 (SHP-1). DJ-1, therefore, downregulates IFN-γ-induced inflammation by decreasing p-STAT1 level. Since brain inflammation is associated with the pathogenesis of PD, these findings suggest that the loss of DJ-1 function increases the risk of PD (Kim et al. 2015).

7.8 Nurr1

DJ-1 positively regulates transcriptional activity of the orphan nuclear receptor Nurr1, a transcription factor essential for dopaminergic neuron development and maturation. DJ-1 enhances transcriptional activity of Nurr1 by regulating the Raf/MEK/ERK pathway (Lu et al. 2016). DJ-1 directly binds to Raf, thereby activating the ERK pathway (Takahashi-Niki et al. 2015).

7.9 Conclusions

In summary, DJ-1 binds to various transcription factors and regulates their transcriptional activity, resulting in various effects on dopamine synthesis, oxidative stress reaction, and signaling pathways. Increased expression of DJ-1 has been observed in several cancer cell lines and tissues. Moreover, increased expression levels of DJ-1 in cancers are parallel to the severity of cancer. From the viewpoint of DJ-1 transcription effect, increased expression of DJ-1 upregulates AR transcription activity and suppresses p53 activity, subsequently inducing the development of prostate cancer and several other cancers, respectively.

On the other hand, it has been reported that loss of DJ-1 is related to male fertility and to type 2 diabetes. From the viewpoint of DJ-1 transcription effect, loss of DJ-1 suppresses LDLR expression with LDL uptake and AR transcriptional activity, thereby leading to lipidosis and type 2 diabetes and to male fertility, respectively (Fig. 7.1). These results suggest that loss of and excess activation of DJ-1 lead to the onset of Parkinson's disease or type 2 diabetes and cancer, respectively.

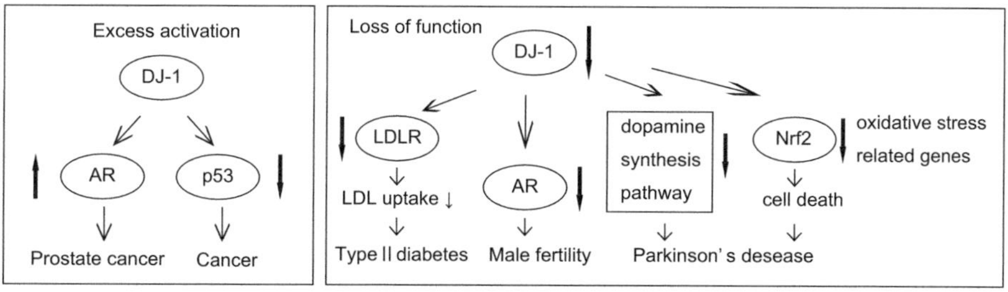

Fig. 7.1 DJ-1-mediated various diseases through cell growth and cell death pathways

References

Clements CM, McNally RS, Conti BJ, Mak TW, Ting JP (2006) DJ-1, a cancer- and Parkinson's disease-associated protein, stabilizes the antioxidant transcriptional master regulator Nrf2. Proc Natl Acad Sci U S A 103(41):15091–15096

Fan J, Ren H, Jia N, Fei E, Zhou T, Jiang P, Wu M, Wang G (2008) DJ-1 decreases Bax expression through repressing p53 transcriptional activity. J Biol Chem 283(7):4022–4030

Ishikawa S, Taira T, Niki T, Takahashi-Niki K, Maita C, Maita H, Ariga H, Iguchi-Ariga SMM (2009) Oxidative status of DJ-1-dependent activation of dopamine synthesis through interaction of tyrosine hydroxylase and 4-dihydroxy-L-phenylalanine (L-DOPA) decarboxylase with DJ-1. J Biol Chem 284:28832–28844

Ishikawa S, Taira T, Takahashi-Niki K, Niki T, Ariga H, Iguchi-Ariga SMM (2010) Human DJ-1-specific transcriptional activation of tyrosine hydroxylase gene. J Biol Chem 285:39718–39731

Ishikawa S, Tanaka Y, Takahashi-Niki K, Niki T, Ariga H, Iguchi-Ariga SMM (2012) Stimulation of vesicular monoamine transporter 2 activity by DJ-1 in SH-SY5Y cells. Biochem Biophys Res Commun 421(4):813–818

Kato I, Maita H, Takahashi-Niki K, Saito Y, Noguchi N, Iguchi-Ariga SMM, Ariga H (2013) Oxidized DJ-1 inhibits p53 by sequestering p53 from promoters in a DNA-binding affinity-dependent manner. Mol Cell Biol 33(2):340–359

Kim RH, Peters M, Jang Y, Shi W, Pintilie M, Fletcher GC, DeLuca C, Liepa J, Zhou L, Snow B, Binari RC, Manoukian AS, Bray MR, Liu FF, Tsao MS, Mak TW (2005) DJ-1, a novel regulator of the tumor suppressor PTEN. Cancer Cell 7(3):263–273

Kim YC, Kitaura H, Taira T, Iguchi-Ariga SMM, Ariga H (2009) Oxidation of DJ-1-dependent cell transformation through direct binding of DJ-1 to PTEN. Int J Oncol 35(6):1331–1341

Kim JH, Choi DJ, Jeong HK, Kim J, Kim DW, Choi SY, Park SM, Suh YH, Jou I, Joe EH (2015) DJ-1 facilitates the interaction between STAT1 and its phosphatase, SHP-1, in brain microglia and astrocytes: a novel anti-inflammatory function of DJ-1. Neurobiol Dis 60:1–10

Lu L, Zhao S, Gao G, Sun X, Zhao H, Yang H (2016) DJ-1/PARK7, but not its L166P mutant linked to autosomal recessive Parkinsonism, modulates the transcriptional activity of the orphan nuclear receptor Nurr1 in vitro and in vivo. Mol Neurobiol 53(10):7363–7374

Niki T, Takahashi-Niki K, Taira T, Iguchi-Ariga SMM, Ariga H (2003) DJBP: a novel DJ-1-binding protein, negatively regulates the androgen receptor by recruiting histone deacetylase complex, and DJ-1 antagonizes this inhibition by abrogation of this complex. Mol Cancer Res 1(4):247–261

Nishinaga H, Takahashi-Niki K, Taira T, Andreadis A, Iguchi-Ariga SMM, Ariga H (2005) Expression profiles of genes in DJ-1-knockdown and L 166 P DJ-1 mutant cells. Neurosci Lett 390(1):54–59

Shinbo Y, Taira T, Niki T, Iguchi-Ariga SMM, Ariga H (2005) DJ-1 restores p53 transcription activity inhibited by Topors/p53BP3. Int J Oncol 26(3):641–648

Takahashi K, Taira T, Niki T, Seino C, Iguchi-Ariga SMM, Ariga H (2001) DJ-1 positively regulates the androgen receptor by impairing the binding of PIASx alpha to the receptor. J Biol Chem 276(40):37556–37563

Takahashi-Niki K, Kato-Ose I, Murata H, Maita I-ASMM, Ariga H (2015) Epidermal growth factor-dependent activation of the ERK pathway by DJ-1 protein through its direct binding to c-Raf protein. J Biol Chem 290(29):17838–17847

Takahashi-Niki K, Ganaha Y, Niki T, Nakagawa S, Kato-Ose I, Iguchi-Ariga SMM, Ariga H (2016) DJ-1 activates SIRT1 through its direct binding to SIRT1. Biochem Biophys Res Commun 474(1):131–136

Tillman JE, Yuan J, Gu G, Fazli L, Ghosh R, Flynt AS, Gleave M, Rennie PS, Kasper S (2007) DJ-1 binds androgen receptor directly and mediates its activity in hormonally treated prostate cancer cells. Cancer Res 67(10):4630–4637

Xu J, Zhong N, Wang H, Elias JE, Kim CY, Woldman I, Pifl C, Gygi SP, Geula C, Yankner BA (2005) The Parkinson's disease-associated DJ-1 protein is a transcriptional co-activator that protects against neuronal apoptosis. Hum Mol Genet 14(9):1231–1241

Yamaguchi S, Yamane T, Takahashi-Niki K, Kato I, Niki T, Goldberg MS, Shen J, Ishimoto K, Doi T, Iguchi-Ariga SMM, Ariga H (2012) Transcriptional activation of low-density lipoprotein receptor gene by DJ-1 and effect of DJ-1 on cholesterol homeostasis. PLoS One 7(5):e38144

Yamane T, Suzui S, Kitaura H, Takahashi-Niki K, Iguchi-Ariga SMM, Ariga H (2013a) Transcriptional activation of the cholecystokinin gene by DJ-1 through interaction of DJ-1 with RREB1 and the effect of DJ-1 on the cholecystokinin level in mice. PLoS One 8(11):e78374

Yamane T, Murao S, Kato-Ose I, Kashima L, Yuguchi M, Kozuka M, Takeuchi K, Ogita H, Ohkubo I, Ariga H (2013b) Transcriptional regulation of the legumain gene by p53 in HCT116 cells. Biochem Biophys Res Commun 438(4):613–618

Yamane T, Yamamoto Y, Nakano Y, Nakagaki T, Ohkubo I, Ariga H (2015) Expression and protease activity of mouse legumain are regulated by the oncogene/transcription co-activator, DJ-1 through p53 and cleavage of annexin A2 is increased in DJ-1-knockout cells. Biochem Biophys Res Commun 467(3):472–477

Yan YF, Yang WJ, Xu Q, Chen HP, Huang XS, Qiu LY, Liao ZP, Huang QR (2015) DJ-1 upregulates anti-oxidant enzymes and attenuates hypoxia/re-oxygenation-induced oxidative stress by activation of the nuclear factor erythroid 2-like 2 signaling pathway. Mol Med Rep 12(3):4734–4742

Zhong N, Kim CY, Rizzu P, Geula C, Porter DR, Pothos EN, Squitieri F, Heutink P, Xu J (2008) DJ-1 transcriptionally up-regulates the human tyrosine hydroxylase by inhibiting the sumoylation of pyrimidine tract-binding protein-associated splicing factor. J Biol Chem 281(30):20940–20948

Zhou W, Freed CR (2005) DJ-1 up-regulates glutathione synthesis during oxidative stress and inhibits A53T alpha-synuclein toxicity. J Biol Chem 280(52):43150–43158

Chapter 8
Regulation of Signal Transduction by DJ-1

Stephanie E. Oh and M. Maral Mouradian

Abstract The ability of DJ-1 to modulate signal transduction has significant effects on how the cell regulates normal processes such as growth, senescence, apoptosis, and autophagy to adapt to changing environmental stimuli and stresses. Perturbations of DJ-1 levels or function can disrupt the equilibrium of homeostatic signaling networks and set off cascades that play a role in the pathogenesis of conditions such as cancer and Parkinson's disease.

DJ-1 plays a major role in various pathways. It mediates cell survival and proliferation by activating the extracellular signal-regulated kinase (ERK1/2) pathway and the phosphatidylinositol-3-kinase (PI3K)/Akt pathway. It attenuates cell death signaling by inhibiting apoptosis signal-regulating kinase 1 (ASK1) activation as well as by inhibiting mitogen-activated protein kinase kinase kinase 1 (MEKK1/MAP3K1) activation of downstream apoptotic cascades. It also modulates autophagy through the ERK, Akt, or the JNK/Beclin1 pathways. In addition, DJ-1 regulates the transcription of genes essential for male reproductive function, such as spermatogenesis, by relaying nuclear receptor androgen receptor (AR) signaling. In this chapter, we summarize the ways that DJ-1 regulates these pathways, focusing on how its role in signal transduction contributes to cellular homeostasis and the pathologic states that result from dysregulation.

Keywords DJ-1 • Signal transduction • Cell signaling • MAPK • ERK • MEK • Ras • Raf • PI3K • Akt • mTOR • MAPK • ASK1 • Daxx • Trx1 • JNK • p38 • MEKK1 • AR

S.E. Oh • M.M. Mouradian (✉)
Center for Neurodegenerative and Neuroimmunologic Diseases, Department of Neurology, Rutgers – Robert Wood Johnson Medical School, Piscataway, NJ 08854, USA
e-mail: m.mouradian@rutgers.edu

H. Ariga, S.M.M. Iguchi-Ariga (eds.), *DJ-1/PARK7 Protein*, Advances in Experimental Medicine and Biology 1037, https://doi.org/10.1007/978-981-10-6583-5_8

8.1 Introduction

In the past decade, research into DJ-1, which has varied activities involved in cellular homeostasis, has revealed another major function: its ability to modulate signal transduction. Cell signaling pathways convey, amplify, and translate the information transmitted from the plasma membrane to the nucleus, regulating normal cellular processes to adapt to changing environmental conditions. Investigating the function of DJ-1 in cell signaling has been crucial in understanding its role in the pathogenesis of cancer and Parkinson's disease (West et al. 2005; Devine et al. 2011). DJ-1 can tip the delicate balance between oncogenesis and cellular protection in either direction depending on cell type and extracellular stimuli. This equilibrium may be explained in the context of how DJ-1 influences the control processes that maintain important cellular signaling networks.

DJ-1, for example, can activate the extracellular signal-regulated kinase (ERK1/2) pathway and the phosphatidylinositol-3-kinase (PI3K)/Akt pathway to mediate cell survival and proliferation. It can attenuate cell death signaling by inhibiting apoptosis signal-regulating kinase 1 (ASK1) activation as well as mitogen-activated protein kinase kinase kinase 1 (MEKK1/MAP3K1) activation of downstream apoptotic cascades. It also appears to modulate autophagy through many signaling pathways, a process that can mediate either cell survival or cell death depending on the circumstances (Green and Llambi 2015). These pathways are regulated by DJ-1 in a multitude of ways. For instance, DJ-1 can bind directly to pathway effectors to modify their activity, and it can regulate a pathway indirectly by binding or modulating its co-activators or inhibitors.

In this chapter, we focus on the different ways that DJ-1 can affect major cell signaling pathways in diverse cell types to play a crucial role in cellular transformation, death, and protection against a variety of stressors (Fig. 8.1).

8.2 DJ-1 Activates the ERK1/2 Pathway

The extracellular signal-regulated kinase (ERK1/2) pathway is a classic mitogen-activated protein kinase (MAPK) signaling cascade that regulates cell proliferation, growth, autophagy, and differentiation. The core pathway members include Ras (small GTP-binding protein; activator), Raf (serine/threonine kinase; MAPKKK), MEK1/2 (mitogen-activated protein kinase/ERK kinase; MAPKK), and ERK1/2 (MAPK) (McCubrey et al. 2007; Cargnello and Roux 2011). This pathway is activated by various stimuli, including growth factors, polypeptide hormones, neurotransmitters, chemokines, and phorbol esters, which bind or activate a variety of receptors and proteins such as receptor tyrosine kinases (RTKs), G-protein-coupled receptors (GPCRs), and protein kinase C (PKC) (Yoshioka 2004).

ERK1/2, also known as p44/42 MAPK, are serine-threonine kinases that are positively regulated by MEK1/2-mediated phosphorylation. MEK1/2 are MAPKK proteins with ERK1/2 as their only known physiological substrates and can be spe-

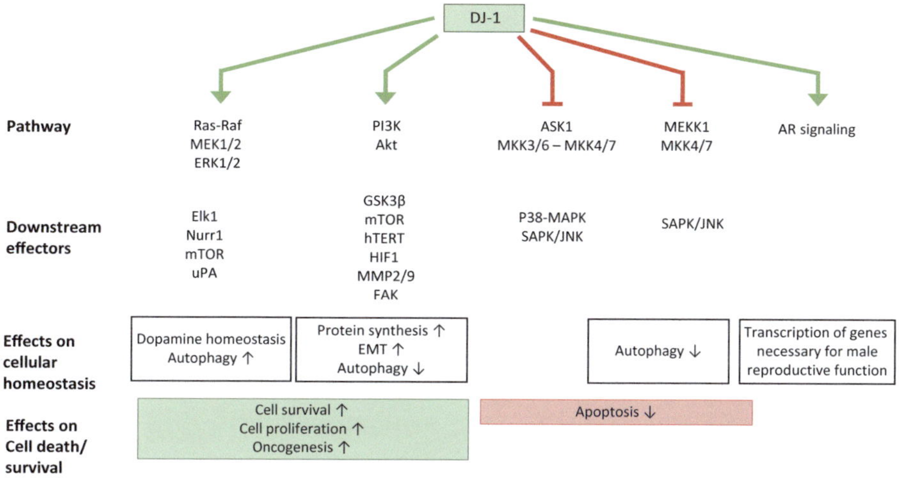

Fig. 8.1 Major signaling pathways regulated by DJ-1
DJ-1 activates the extracellular signal-regulated kinase (ERK1/2) pathway and the phosphatidylinositol-3-kinase (PI3K)/Akt pathway to mediate cell survival, proliferation, and autophagy. It can also activate androgen receptor (AR) signaling to induce transcription of genes necessary for male reproductive function. [*Activation is indicated by pointed green arrows*] DJ-1 can inhibit apoptosis signal-regulating kinase 1 (ASK1) activation as well as MEKK1 activation of downstream apoptotic cascades. [*Inhibition is indicated by blunted red arrows*]
EMT, epithelial-mesenchymal transition

cifically inhibited by small molecule inhibitors such as U0126 or PD98059 (Chang et al. 2003). On the other hand, ERK is negatively regulated by a family of dual-specificity (Thr/Tyr) MAPK phosphatases (DUSPs/MKPs) (Jeffrey et al. 2007).

ERK phosphorylates several downstream transcription factors such as AP-1, c-Jun, and c-Myc (Plotnikov et al. 2011). ERK can also activate ribosome S6 kinase (RSK) and inhibitor kappa-B kinase (IKK), which can lead to the respective activation of transcription factors cAMP response element-binding (CREB) and nuclear factor immunoglobulin kappa-chain enhancer-B-cell (NF-kappa-B) (Chang et al. 2003; Burotto et al. 2014). MEK/ERKs have also been reported to participate in a non-canonical signaling pathway that interacts with the mammalian target of rapamycin complex 1 (mTORC1) to regulate autophagy via Beclin-1 modulation (Wang et al. 2009).

DJ-1 was initially discovered and identified in 1997 by Nagakubo et al. as a putative oncogene capable of transforming cells in cooperation with H-Ras, a GTPase that can activate c-Raf in the ERK pathway (Nagakubo et al. 1997). Since then, many studies have shown DJ-1 to activate ERK1/2. The activation of this MAPK pathway could contribute to, or explain, many of the roles that DJ-1 plays in protecting cells from oxidative injury, in regulating gene transcription, and in activating autophagy (Fig. 8.2).

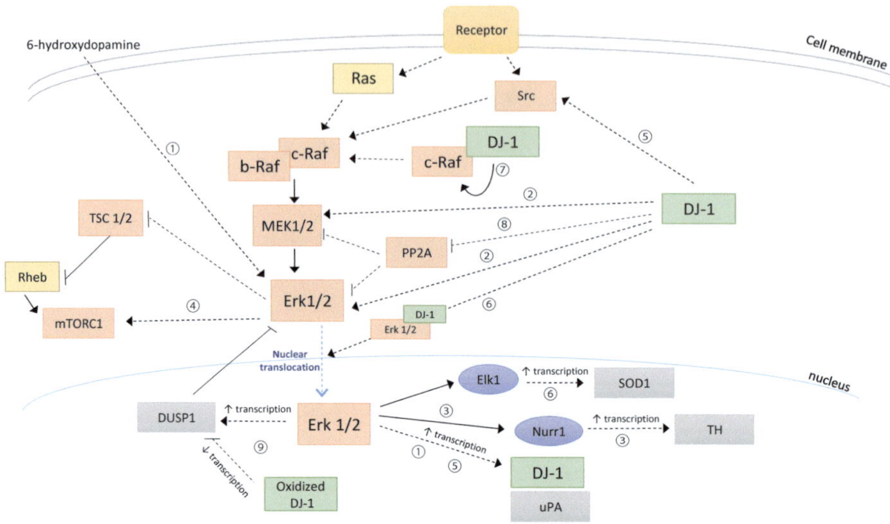

Fig. 8.2 DJ-1 activates ERK1/2 signaling pathway
1. 6-Hydroxydopamine-induced oxidative stress may upregulate DJ-1 through activation of ERK1/2 (Lev et al. 2009)
2. DJ-1, but not its L166P mutant, protects cells from oxidative injury by activating ERK1/2 and MEK1/2 (Gu et al. 2009)
3. DJ-1 mediates dopamine (DA) homeostasis through activation of ERK1/2 and the resulting nuclear translocation of the transcription factor Nurr1 (Lu et al. 2012; Lu et al. 2016)
4. DJ-1 activates the non-canonical MEK/ERK-mTOR pathway to activate autophagy (Gao et al. 2012; Krebiehl et al. 2010)
5. DJ-1 promotes oncogenic potential by activating the SRC/ERK/uPA axis (He et al. 2012)
6. DJ-1 interacts directly with ERK1/2 and may enhance the nuclear translocation of ERK1/2, where it can phosphorylate Elk1, leading to increased SOD expression (Wang et al. 2011)
7. DJ-1, but not its C106S mutant, can bind directly to and phosphorylate c-Raf, which can then activate MEK and ERK1/2 (Takahashi-Niki et al. 2015)
8. DJ-1, but not its L166P mutant, suppresses the expression of PP2A, an inhibitor of MEK1/2 and ERK1/2 family kinases (Gu et al. 2009)
9. DJ-1, but not its C106S mutant, may sequester p53 away from promoters in a DNA-binding affinity-dependent manner, resulting in downregulation of ERK1/2 inhibitor, DUSP1 (Kato et al. 2013)
[*positive regulation is indicated by pointed arrows, and negative regulation is indicated by blunted arrows. Direct or known regulation is indicated by solid lines; indirect or unknown regulation is indicated by dotted lines*]

8.2.1 Dopamine-Mediated Oxidative Stress May Upregulate DJ-1 Through Activation of ERK1/2

The involvement of DJ-1 with ERK was first reported in 2008, when it was demonstrated that ERK1/2 may upregulate DJ-1. As reactive oxygen species (ROS) generated from 6-hydroxydopamine (6-OHDA) could lead to upregulation of DJ-1 in SH-SY5Y human neuroblastoma cells (Lev et al. 2008), it was hypothesized that this augmentation may be mediated through the ERK pathway, which has been shown to play a role in forming a protective response against 6-OHDA stress (Lin et al. 2008). Phosphorylation/activation of ERK1/2 was seen to occur in both neuroblastoma cells treated with 6-OHDA and in the mouse striatum denervated with 6-OHDA. This ERK1/2 activation was shown in cellular models to precede the upregulation of DJ-1 mRNA, while inhibiting MEK1/2 by PD-98059 could attenuate ROS-induced upregulation of DJ-1 (Lev et al. 2009).

8.2.2 DJ-1 Protects Cells from Oxidative Injury by Activating ERK1/2 and MEK1/2

In contrast to studies suggesting that ERK pathway activation leads to upregulation of DJ-1, another experimental evidence has been reported suggesting that DJ-1 functions upstream of ERK1/2 phosphorylation (Gu et al. 2009; Letourneux et al. 2006; Kato et al. 2013; Wang et al. 2011). One such study found that overexpression of wild-type (WT) DJ-1 in COS-7 cells (African green monkey kidney fibroblast-like cell line) and MN9D cells (a fusion of mice embryonic ventral mesencephalic and neuroblastoma cells) upregulates ERK1/2 and MEK1/2 phosphorylation, whereas overexpression of a Parkinson's disease-linked mutant form of DJ-1, L166P, does not enhance ERK1/2 or MEK1/2 phosphorylation (Gu et al. 2009). Additionally, DJ-1 overexpression in various models improves the viability of cells stressed with hydrogen peroxide (H_2O_2) compared to control (Sekito et al. 2006; Kahle et al. 2009). Under these conditions, inhibiting ERK1/2 activation by pre-treating cells with the MEK1/2 inhibitor U0126 abrogates the protective effect of DJ-1 overexpression against oxidative injury, suggesting that WT DJ-1 may provide neuroprotection through activation of the ERK pathway (Gu et al. 2009).

8.2.3 DJ-1 Mediates Dopamine (DA) Homeostasis Through Activation of ERK1/2 and the Resulting Nuclear Translocation of the Transcription Factor Nurr1

In addition to protection from oxidative injury, another function of DJ-1-mediated ERK1/2 signaling may be the regulation of tyrosine hydroxylase (TH) expression. In both in vivo and in vitro models, DJ-1 has been found to modulate the transcription factor Nurr1. Nurr1 plays a major role in DA homeostasis and can regulate the expression of TH and L-dopa decarboxylase (DDC), both of which are involved in DA synthesis, as well as the expression of vesicular monoamine transporter 2 (VMAT-2), which is necessary for the transport of DA from the cytosol into synaptic vesicles (Iwawaki et al. 2000; Hermanson et al. 2003; Ishikawa et al. 2009). MN9D cells that overexpress WT DJ-1 exhibit an increase in the nuclear translocation of Nurr1 as well as an increase in the mRNA levels of Nurr1 targets. However, cells that overexpress the Parkinson-associated pathogenic L166P mutant DJ-1 do not show such an increase. Knocking down DJ-1 expression using RNAi attenuates the activity of Nurr1 and downregulates the expression of its target proteins, which can be rescued by subsequent overexpression of WT DJ-1 (Lu et al. 2012).

As ERK1/2 has been reported to increase Nurr1 transcriptional activity (Nordzell et al. 2004), it was hypothesized that DJ-1 may mediate Nurr1 activation through the ERK1/2 pathway. WT DJ-1, as compared to its L166P mutant, phosphorylates ERK1/2, while blocking ERK1/2 activation using U0126 prevents DJ-1-mediated nuclear translocation of Nurr1 and the induction of Nurr1 target protein levels (Lu et al. 2016). Similarly, overexpression of WT DJ-1 but not its L166P mutant in the rat substantia nigra using a lentiviral vector increases ERK activation, Nurr1 nuclear translocation, and Nurr1 target protein levels (Lu et al. 2016).

8.2.4 DJ-1 Activates the Non-canonical MEK/ERK-mTOR Pathway to Activate Autophagy

DJ-1 appears to impact autophagy and neuronal cell survival through the ERK pathway. In a rat model made to overexpress DJ-1 by injection of adeno-associated viral vector, ERK activation in the substantia nigra is significantly greater compared with control animals injected with a vector expressing only green fluorescent protein. Additionally, overexpression of DJ-1 protects against rotenone-induced injury and enhances autophagy markers in this rat model as well as in MN9D cells (Gao et al. 2012). Rotenone is an inhibitor of mitochondrial complex I that can induce oxidative stress and apoptosis, inhibit proteasome activity, and cause dopaminergic neuronal death in rodents (Shamoto-Nagai et al. 2003). Further, blocking autophagy in MN9D cells by inhibiting either MEK1/2 with U0126 or phosphoinositol 3-kinase (PI3K) with 3-methyladenine (3MA) reverses autophagic activation by DJ-1 and abrogates DJ-1-mediated protection against rotenone (Gao et al. 2012). This implies that the neuroprotective effect of DJ-1 may be at least partly due to its effects on two signaling pathways that have been shown to regulate autophagy – the canonical Akt/PI3K-mTOR pathway (Vasseur et al. 2009) and the non-canonical AMPK-MEK/ERK-TSC-mTOR pathway (Wang et al. 2009).

This notion is partially supported by data from another study which shows that DJ-1 knockout (KO) mouse embryonal fibroblasts (MEFs) exhibit reduced basal autophagic degradation, impaired lysosomal activity, and accumulation of defective mitochondria (Krebiehl et al. 2010). These DJ-1 KO MEFs also show a reduction in phosphorylated ERK2 in the mitochondrial fraction, but no effect on cytosolic fractions, suggesting that DJ-1-mediated ERK2 phosphorylation may be controlling autophagic and lysosomal function (Krebiehl et al. 2010).

8.2.5 DJ-1 Promotes Oncogenic Potential by Activating the SRC/ERK/uPA Axis in Pancreatic Cancer

DJ-1 is associated with a vast number of tumor types, with its overexpression and secretion found frequently in conjunction with abnormal cell transformation and tumor progression (Cao et al. 2015). Activation of the ERK pathway also plays a prominent role in oncogenesis by driving inappropriate cell proliferation and survival (Caunt et al. 2015). An examination of the role of DJ-1 in tumor invasion and metastasis in human pancreatic cancer cell lines (BxPC-3 and SW1990) showed that knockdown of DJ-1 can lead to cytoskeleton disruption as well as diminished urokinase plasminogen activator (uPA) activity and expression. Active uPA converts plasminogen to active plasmin, which can break down the extracellular matrix around the cell or activate growth factors to promote cancer cell migration (Blasi and Carmeliet 2002). Knockdown of DJ-1 in pancreatic cancer cell lines also decreases ERK1/2 and SRC kinase phosphorylation, which regulates the ERK pathway (Chang et al. 2003; Eichhorn et al. 2007). This decrease in ERK1/2 and SRC phosphorylation can be reversed by restoring DJ-1 expression. Inhibiting ERK has the same effect on pancreatic cancer invasion potential and cell migration as knocking down DJ-1; both lead to decreased uPA expression and activity. This suggests that the effect of DJ-1 on pancreatic cancer invasion and migration may be dependent on the SRC/ERK pathway (He et al. 2012).

8.2.6 DJ-1 Interacts Genetically with Ras/ERK Signaling Components

DJ-1 appears to play a definitive role in the activation of ERK1/2 and downstream effectors, but there is a differing range of data as to exactly how DJ-1 may be participating in this signaling cascade. While one group has shown DJ-1 can directly interact with ERK1/2 (Wang et al. 2011), other groups have shown that DJ-1 modulates upstream factors in the MAPK cascade, either through direct interaction or by affecting protein expression (Gu et al. 2009; Kato et al. 2013; Takahashi-Niki et al. 2015).

Initial studies found that aged mice lacking both DJ-1 and the receptor tyrosine kinase (RTK) receptor Ret lose more dopaminergic neurons in the substantia nigra compared to mice lacking only Ret, suggesting a possible cooperation between DJ-1 and Ret (Aron et al. 2010). Ret is upstream of the ERK pathway and is necessary for the neuronal survival activity of glial neurotrophic factor (GDNF) (Kramer et al. 2007).

To study the interaction between DJ-1 and Ret further, a developing *Drosophila* eye model system was employed, which is very sensitive to dosage changes in RTK signaling and downstream MAPK pathways. Overexpressing constitutively active Ret in this model led to the development of adult eyes with reduced size and rough morphology. However, overexpression of constitutively active versions of

Ret, Raf, ERK/rolled, or wild-type Akt1 did not affect endogenous DJ-1 levels. When flies expressing constitutively active Ret were crossed with flies expressing reduced DJ-1 levels (carrying DJ-1 microdeletions or DJ-1 loss-of-function alleles), the offspring exhibited normal eye phenotype, showing complete rescue of the eye defects. Conversely, when the flies expressing constitutively active Ret were crossed with flies overexpressing DJ-1, the offspring exhibited more severe eye defects. These findings indicate a genetic interaction between Ret and DJ-1 in controlling cell size and differentiation in the developing retinal photoreceptor neurons (Aron et al. 2010).

Similarly, DJ-1 interacts genetically with downstream RTK signaling component Ras and with ERK/*rolled* (rl), but not with PI3K/Akt. This suggests that DJ-1 does not modulate Akt activation in this model under normal circumstances, but it may synergize with Ret, Ras, and ERK during development, functioning either between Ras and ERK or in parallel to the Ras/ERK pathway to control cell differentiation and proliferation (Aron et al. 2010).

8.2.7 DJ-1 Interacts Directly with ERK1/2 and May Affect the Nuclear Translocation of ERK1/2 Rather Than the Direct Phosphorylation of ERK1/2

In a study investigating the effect of DJ-1 on increasing superoxide dismutase (SOD) expression and its subsequent ability to decrease ROS generation caused by either 1-methyl-4-phenylpyridinium (MPP^{+}) or paraquat, the transcription factor Elk1 was found to bind to the *SOD1* promoter to increase transcription. Additionally, Elk1 activation following the administration of 1-methyl-4-phenyl-1,2,3,6-tetrahydropyridine (MPTP) was found to be decreased in the substantia nigra of DJ-1 KO mice compared to WT mice (Wang et al. 2011). Elk1 is phosphorylated and activated by MAPK kinases such as ERK1/2 (Mut et al. 2012), suggesting that one of the mechanisms through which DJ-1 protects cells against ROS is via ERK1/2 – Elk1 activation leading to SOD induction.

However, in contrast to in vitro data showing that knocking down DJ-1 decreases ERK1/2 phosphorylation (Lu et al. 2012), ERK1/2 phosphorylation is unaffected in DJ-1 KO mice (Wang et al. 2011). This has led to the hypothesis that under oxidative insult, DJ-1 may act as a molecular chaperone, a function that is attributed to DJ-1 in previous studies (Shendelman et al. 2004), to affect the nuclear translocation of ERK1/2, rather than its phosphorylation. Evidence for direct interaction between DJ-1 and ERK2 has been presented in HEK293T (human embryonic kidney) cells as well as in mouse brain lysates using co-immunoprecipitation (Wang et al. 2011), although not yet replicated in human brain lysates (Meulener et al. 2005). Residues 1–100 of DJ-1 are necessary for this interaction, while mutation of its cysteine 106, which is necessary for its role as a redox sensor and peroxide scavenger (Taira et al. 2004a; Canet-Aviles et al. 2004; Martinat et al. 2004), does not affect its ability to interact with ERK1/2 (Wang et al. 2011). Additionally, the nuclear translocation of ERK1/2 is reduced in DJ-1 knockdown SH-SY5Y cells and in DJ-1 KO primary mouse neurons but is rescued by reconstituting DJ-1 expression (Wang et al. 2011). This suggests that DJ-1 may promote the translocation of ERK1/2 to the nucleus upon oxidative stress, allowing ERK1/2 to phosphorylate Elk1, leading to increased SOD expression, and allowing the cell to mount a defense response and suppress the production of superoxide induced by insults such as MPP^+ or paraquat.

8.2.8 DJ-1 Can Bind Directly to and Phosphorylate c-Raf, Which Can Then Activate MEK and ERK1/2

When epidermal growth factor (EGF) binds to its receptor (EGFR), it can trigger the activation of Ras, which can then activate c-Raf, a serine/threonine kinase that can phosphorylate MEK, allowing it to activate ERK1/2 (McCubrey et al. 2007; Cargnello and Roux 2011). Pulldown assays using GST-DJ-1 mixed with ^{35}S-labeled c-Raf, and co-immunoprecipitation assays in HeLa (human cervical epithelial adenocarcinoma) and HEK293T cells transfected with FLAG-c-Raf and DJ-1-HA, have shown that DJ-1 binds directly to the kinase domain of c-Raf, the MAPKKK in the ERK signaling pathway (Takahashi-Niki et al. 2015). This interaction is enhanced when cells are treated with EGF (Takahashi-Niki et al. 2015), which augments c-Raf activation.

Phosphorylation of c-Raf at several residues regulates its activity. This kinase is normally inactive and phosphorylated at serines 43, 259, and 621. However, when Ser-259 and Ser-621 are dephosphorylated by protein phosphatase 2A (PP2A), and when Ser-338 is phosphorylated following EGF stimulation, c-Raf becomes activated (Dhillon et al. 2002; Abraham et al. 2000; Oehrl et al. 2003). When DJ-1 knockdown NIH3T3 (mouse embryonic fibroblast) and HeLa cells or DJ-1 knockout (KO) mouse primary fibroblasts are treated with EGF, the loss of DJ-1 is associated with decreased levels of phospho-Ser-338 c-Raf compared to control. However, loss of DJ-1 does not impact the levels of phospho-Ser-259 c-Raf. The reintroduction of DJ-1 into cells derived from DJ-1 KO mice rescues the level of phospho-Ser-338 c-Raf to control levels (Takahashi-Niki et al. 2015). Additionally, when a constitutively active c-Raf fragment (consisting of the kinase domain that binds DJ-1) is incubated with purified DJ-1 and ATP, the fragment exhibits increased autophosphorylation activity in a DJ-1-dose-dependent manner (Takahashi-Niki et al. 2015). This suggests that by binding to c-Raf, DJ-1 stimulates the kinase activity and autophosphorylation of c-Raf at Ser-338, which aids in the activation of the protein and the ERK pathway.

Interestingly, expression of C106S mutant DJ-1 into DJ-1 null cells reportedly does not rescue the level of phospho-Ser-338 c-Raf to control levels, and the binding activity of C106S mutant DJ-1 to c-Raf in pulldown assays is weaker than that of wild-type DJ-1 (Takahashi-Niki et al. 2015). The cysteine 106 (Cys-106) residue is highly sensitive to oxidative stress and is essential for the role of DJ-1 as a cytoprotective redox sensor (Taira et al. 2004a, b; Canet-Aviles et al. 2004; Martinat et al. 2004). The level of Cys-106 oxidation dictates the level of DJ-1 activation, and when Cys-106 is excessively oxidized, it renders DJ-1 inactive (Wilson 2011). To investigate whether oxidized forms of DJ-1 are necessary to activate c-Raf, the level of DJ-1 oxidized at Cys-106 was examined in cells treated with EGF, H_2O_2, or both. As expected, H_2O_2 treatment increases highly oxidized forms of DJ-1 at Cys-106 (SO_2H and SO_3H forms), but does not increase the level of phospho-Ser-338 c-Raf, indicating that oxidized forms of DJ-1 do not aid in the autophosphorylation of c-Raf. Also, EGF treatment, which increases the level of phospho-Ser-338, does not induce DJ-1 oxidation, and subsequent H_2O_2 treatment in EGF-treated cells decreases the level of phospho-Ser-338 c-Raf (Takahashi-Niki et al. 2015). This suggests that C106 is important for the interaction of DJ-1 with c-Raf and for subsequent c-Raf activation and that the oxidation of Cys-106 to SO_2H and SO_3H forms is not necessary for this interaction.

8.2.9 DJ-1 Suppresses the Expression of PP2A, An Inhibitor of MEK1/2 and ERK1/2 Family Kinases

Members of the ERK1/2 pathway are regulated by a variety of kinases and phosphatases, one of which is protein phosphatase 2A (PP2A). PP2A is a serine/threonine phosphatase, a holoenzyme whose well-conserved catalytic (C) subunit is tightly regulated by a regulatory (B) subunit and held together by a scaffolding (A) subunit (Janssens and Goris 2001). PP2A can negatively regulate the ERK pathway by dephosphorylating MEK1/2 and ERK1/2 kinases (Letourneux et al. 2006; Zhou et al. 2002; Alessi et al. 1995; Sonoda et al. 1997; Silverstein et al. 2002) and also by dephosphorylating and inhibiting cellular SRC kinase (c-SRC), an activator of the ERK pathway (Eichhorn et al. 2007).

Overexpression of WT DJ-1 in MN9D and COS-7 cells reduces PP2A levels, while the overexpression of L166P mutant DJ-1 has negligible effect on PP2A (Gu et al. 2009). Additionally, overexpression of neither WT DJ-1 nor L166P mutant had an effect on protein kinase A (PKA) levels (Gu et al. 2009), a protein that has been shown to activate the ERK pathway (Gu et al. 2009; Zanassi et al. 2001). This suggests that DJ-1 may positively regulate ERK1/2 signaling by decreasing the expression of PP2A.

It should be noted, however, that PP2A has also been shown to activate the ERK pathway through its interactions with c-Raf and Ras (Abraham et al. 2000; Adams et al. 2005), and the effect of DJ-1 on the ERK pathway through PP2A regulation may be context-dependent (Junttila et al. 2008).

8.2.10 *DJ-1 May Sequester p53 Away from Promoters in a DNA-Binding Affinity-Dependent Manner, Resulting in the Downregulation of ERK1/2 Inhibitor, DUSP1*

p53 is a crucial tumor-suppressor protein and a transcription factor that is activated by cellular stress to induce transcription of genes involved in DNA repair, apoptosis, cell cycle arrest, or autophagy (Vousden and Prives 2009; Menendez et al. 2009). p53 is closely associated with DJ-1, with experimental evidence showing that DJ-1 may repress p53 transcriptional activity to prevent apoptosis (Fan et al. 2008), that p53 may inhibit DJ-1 activation through phosphorylation (Rahman-Roblick et al. 2008), and that p53 may decrease DJ-1 protein levels through a posttranscriptional route (Vasseur et al. 2012). In a study using co-immunoprecipitation experiments, p53 was shown to bind strongly to WT DJ-1 in an oxidative stress-dependent manner, with their interaction enhanced by H_2O_2 treatment (Kato et al. 2013). On the other hand, C106S mutant DJ-1 failed to co-immunoprecipitate with p53 under oxidative stress, suggesting that the Cys-106 residue is necessary for this interaction. Further examination of their interaction using pulldown assays of tagged DJ-1 with p53 deletion mutants showed that DJ-1 binds p53 at its DNA-binding domain, suggesting that it may interfere with the transcriptional activity of p53 (Kato et al. 2013). As expected, this interaction also impacts the transcriptional expression of p53 target genes such as DUSP1, which is a mitogen-activated protein kinase phosphatase that can dephosphorylate ERK and regulate apoptosis by inhibiting downstream effectors of the ERK pathway (Jeffrey et al. 2007). In mouse primary fibroblasts, DUSP1 mRNA and protein expression is increased under oxidative stress, and this induction is even greater in fibroblasts obtained from DJ-1 KO mice, suggesting that DJ-1 may modulate DUSP1 levels (Kato et al. 2013). These findings provide support to the notion that under conditions of oxidative stress, DJ-1 forms a complex with p53, sequestering it away from the DUSP1 promoter. Decreased DUSP1 transcription would prevent the dephosphorylation of ERK, allowing for the promotion of cell survival.

8.3 DJ-1 Activates the PI3K/Akt Pathway

The PI3K/Akt pathway is another classic signaling cascade that regulates cell growth, proliferation, and survival by affecting a multitude of complementary downstream pathways. The cascade is initiated when various stimuli induce the lipid kinase phoaphatidylinositol-3-kinase (PI3K) to generate phosphatidylinositol-3,4,5-triphosphate (PIP3) from phosphatidylinositol-4,5 bisphosphate (PIP2). This allows for the translocation of the serine/threonine kinase Akt (also known as protein kinase B/PKB) to the plasma membrane. At the membrane, Akt is phosphorylated at residue Thr-308 by phosphoinositide-dependent kinase 1 (PDK1) to induce partial activation. Subsequent phosphorylation of Akt at residue Ser-473 by mTORC2 or members of the PI3K-related kinase (PIKK) family induces full enzymatic Akt activity. Phosphatase and tensin homologue (PTEN) negatively regulates the PI3K/Akt pathway by dephosphorylating PIP3 back to PIP2. And Akt is dephosphorylated by protein phosphatase 2A (PP2A) and the PH-domain leucine-rich-repeat-containing protein phosphatases (PHLPP1/2) (Maehama and Dixon 1998; Andjelkovic et al. 1996; Brognard et al. 2007; Gao et al. 2005).

One of the major downstream effects of Akt is the mammalian target of rapamycin complex 1 (mTORC1) which promotes the translation of proteins necessary for cell growth and protein synthesis. Akt can directly phosphorylate and activate mTOR and also inhibit mTOR inhibitor proteins proline-rich Akt substrate of 40 kDa (PRAS40) and tuberin (TSC2). Phosphorylation of mTORC1 activates the mTORC1/S6 kinase axis and its downstream effectors such as eukaryote translation initiation factor 4E-binding protein 1(4E-BP1) and 40S ribosomal protein S6 (RPS6) (Manning and Cantley 2007).

Another important target of Akt is glycogen synthase kinase 3 (GSK-3), which is inhibited by Akt (Cross et al. 1995). This results in cell cycle progression through the inhibition of GSK3-mediated phosphorylation and degradation of cyclin D and cyclin E as well as the transcription factors c-Jun and c-Myc. Inhibition of GSK3 can also promote glycogen metabolism, regulate wnt signaling, and affect the formation of neurofibrillary tangles in Alzheimer's disease. Akt also positively regulates cell proliferation through inhibition of cyclin-dependent kinase inhibitors p21 (CDKN1A/CIP1/WAF1) and p27 (CDKN1B/KIP1).

Akt can mediate cell survival by inhibiting pro-apoptotic proteins such as Bcl-2-associated death promoter (Bad) or by inhibiting forkhead transcription factors (FoxO1/3a) from generating pro-apoptotic proteins such as Bim and cytokine Fas ligand (FasL) (Zhang et al. 2011). Akt also phosphorylates and activates the E3 ubiquitin ligase murine double minute 2 (MDM2), which triggers p53 degradation. This prevents the transcription of pro-apoptotic BH3-only proteins such as Puma and Noxa (Mayo and Donner 2002).

Additionally, there is significant cross talk between Akt and other major signaling pathways, often under specific conditions. Under stimuli such as tumor necrosis factor (TNFα) or platelet-derived growth factor (PDGF), Akt activates NF-kB/p65 signaling by phosphorylating pathway activators – IκB kinase α (IκKα). In certain cell types, or under stresses such as ischemia, Akt has also been reported to block ERK signaling through direct phosphorylation and inhibition of c-Raf (Zhou et al. 2015).

DJ-1 has been shown to activate the PI3k/Akt pathway, a process that may mediate its transforming effects during oncogenesis, as well as its cytoprotective effects against oxidative and nitrosative stress as discussed below (Fig. 8.3) (Kim et al. 2005; Yang et al. 2005).

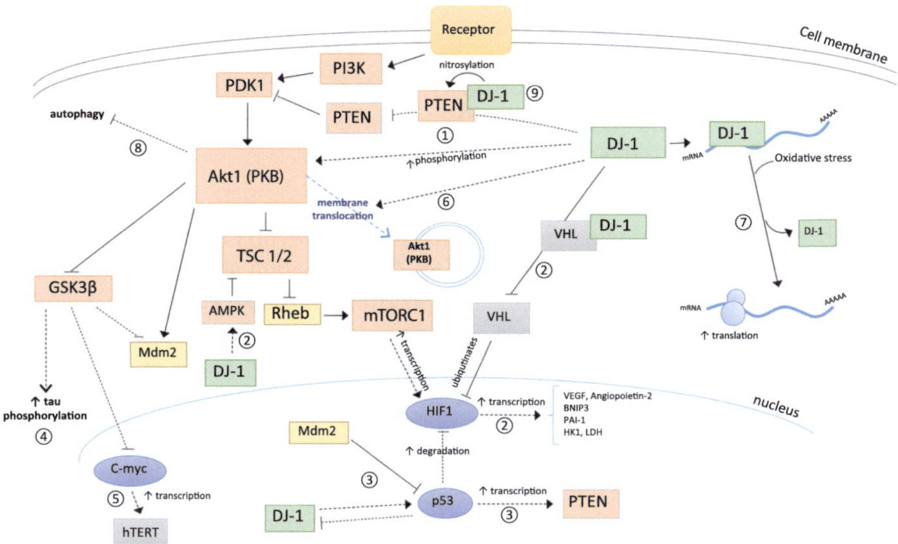

Fig. 8.3 DJ-1 activates PI3K/AKT1 signaling pathway
1. DJ-1 increases phosphorylated Akt, and activates the PI3k/Akt pathway, which contributes to oncogenesis and neuronal survival (Kim et al. 2005; Yang et al. 2005; Aleyasin et al. 2010; Zhang et al. 2016)
2. Under hypoxic conditions, DJ-1 mediates the activation of HIF1 that regulates the expression of many genes needed for cells to adapt to hypoxic conditions. DJ-1 does this by increasing mTORC activity through PI3k/Akt pathway modulation, decreasing p53 activation, and increasing AMPK activation of mTORC (Vasseur et al. 2009), as well as by inhibiting HIF-VHL interaction through direct binding with VHL (Parsanejad et al. 2014)
3. There may be significant cross talk and feedback regulation between p53, DJ-1, and the Akt pathway. p53 may inhibit DJ-1 activation through phosphorylation (Rahman-Roblick et al. 2008) and may also decrease DJ-1 protein levels through a posttranscriptional route (Vasseur et al. 2012). p53 negatively regulates the IGF-1/PI3k/Akt pathway by increasing the transcription of target genes such as IGF-BP3 or PTEN (Feng 2010), and is also conversely regulated by Akt, which activates the E3 ubiquitin ligase MDM2, a trigger for p53 degradation (Mayo and Donner 2002). Oxidative stress may induce DJ-1-mediated phosphorylation/activation of p53 (Vasseur et al. 2012), a process that may be mediated by active Akt (Zhan et al. 2010). Active p53 may then respond appropriately to stress by activating transcriptional programs (Kruiswijk et al. 2015). In a negative feedback loop, p53 can also decrease DJ-1 levels and attenuate Akt activation, preventing abnormal cellular transformation (Vasseur et al. 2012)
4. PD pathogenic mutants (L166P, D146A) of DJ-1 lose the ability to inhibit GSK-3β through the regulation of the Akt pathway and may contribute to increased tau phosphorylation and PD pathogenesis (Wang et al. 2013)
5. By activating the PI3K/Akt pathway, DJ-1 may increase the activity of downstream transcription factor c-myc to modulate the expression of human telomerase reverse transcriptase (hTERT), which is implicated in cellular differentiation and neoplastic transformation (Sitaram et al. 2009)
6. DJ-1 may play a role in modulating recruitment of Akt to the membrane in an ROS-dependent manner (Aleyasin et al. 2010)
7. DJ-1 binds mRNAs of PI3K/Akt pathway (Akt1, IGF2) and may release them to be translated under oxidative stress (van der Brug et al. 2008)
8. DJ-1 may decrease ceramide-induced autophagy and cell death through its effects on the PI3k/AKT pathway (Jaramillo-Gomez et al. 2015)
9. DJ-1 may bind PTEN directly in a manner dependent on the redox status of its Cys-106 residue (Kim et al. 2009). And under nitrosative stress, DJ-1 may directly bind and inhibit PTEN activity via transnitrosylation
[*positive regulation is indicated by pointed arrows, and negative regulation is indicated by blunted arrows. Direct or known regulation is indicated by solid lines; indirect or unknown regulation is indicated by dotted lines*]

8.3.1 DJ-1 Negatively Regulates PTEN to Activate the PI3K/ Akt Pathway Mediating Oncogenic and Cytoprotective Properties of DJ-1

A study using *Drosophila* genetic screen for gain-of-function mutants was the first to report in 2005 that DJ-1 may antagonize PTEN function and activate the PI3K cell survival pathway to induce oncogenesis (Kim et al. 2005). The negative regulation of PTEN by DJ-1 has since been corroborated in various other model systems (Yang et al. 2005; Klawitter et al. 2013; Sitaram et al. 2009; Yao et al. 2011; Fang et al. 2010; Davidson et al. 2008; Liu et al. 2015). DJ-1 overexpression in PTEN overexpressing NIH-3T3 fibroblasts and in PTEN+/− mouse embryonic fibroblasts (MEFs) rescues cell survival under apoptotic stress, suggesting that DJ-1 may protect cells from apoptosis in a PTEN-dependent manner. In COS-7 cells, A597 cells, NIH-3T3, and PTEN+/− MEFs, DJ-1 increases phosphorylation of Akt and its downstream effectors in a PI3K-dependent manner, whereas knockdown of DJ-1 decreases the phosphorylation of Akt. In PTEN−/− MEFs, however, DJ-1 knockdown does not affect Akt phosphorylation, suggesting that PTEN is necessary for DJ-1 effects on Akt phosphorylation and that DJ-1 may suppress PTEN functionality to aid in PI3K/Akt pathway activation (Kim et al. 2005).

The connection between DJ-1, PTEN, and human carcinogenesis has also been examined in carcinomas (Kim et al. 2005). In primary breast cancer tissue samples, mutations in PTEN are not a major factor in the development of sporadic breast cancers (Feilotter et al. 1999; Freihoff et al. 1999), and alterations in PTEN levels in primary ductal adenocarcinomas are thought to be primarily epigenetic (Perren et al. 1999; Shi et al. 2003). Examination of the expression of DJ-1, phospho-Akt, and PTEN in serial histologic sections of breast cancer from 73 patients with lymph node-negative disease suggested an inverse relationship between PTEN and DJ-1 expression and a positive relationship between DJ-1 and phosphorylated Akt (Kim et al. 2005). Further, an examination of DJ-1 and phosphorylated Akt levels in primary lung cancer samples from 40 patients showed positive correlation between DJ-1 and phosphorylated Akt levels (Kim et al. 2005). Taken together, this suggests that DJ-1 may affect oncogenesis through epigenetic modulation of PTEN functionality in multiple cancer types, allowing for Akt to become hyperphosphorylated.

This assertion was further supported by a report showing that inhibition of DJ-1A in *Drosophila* leads to impaired PI3K/Akt signaling (Yang et al. 2005). DJ-1A RNAi induces phenotypes such as photoreceptor loss in the eye, dopaminergic neuron reduction in aging brains, and hypersensitivity to oxidative stress. When wild-type (WT) PTEN or a dominant-negative form of the PI3K catalytic subunit is co-expressed with DJ-1A RNAi transgene, the pathologic eye phenotype is exacerbated. Conversely, co-expression of the WT form of the PI3K catalytic subunit or Akt overexpressing transgene with DJ-1A RNAi suppresses the dysfunctional phenotype (Yang et al. 2005). This suggests that the dysfunctional phenotypes induced by DJ-1 knockdown are mediated through the PI3K/Akt pathway. Finally, phosphorylated Akt levels are significantly reduced in DJ-1A RNAi expressing fly head extracts, suggesting that DJ-1A is necessary for maintaining normal phosphorylation of Akt and the activation of PI3K/Akt signaling in the fly brain that may be needed to mediate dopaminergic neuron survival and responses to oxidative stress (Yang et al. 2005; Aleyasin et al. 2010). The protective effect of DJ-1 against oxidative stress is also seen in cardiomyocytes from mice (Billia et al. 2013; Dongworth et al. 2014; Mukherjee et al. 2011) and humans (Klawitter et al. 2013). DJ-1 is upregulated in cardiac tissue samples taken from patients with chronic ischemia dilated cardiomyopathy (Klawitter et al. 2013), a condition which has been linked to endothelial dysfunction and ROS (Tentolouris et al. 2004; Sorescu et al. 2002). Increased levels of DJ-1 in these tissues are associated with decreased PTEN protein expression and increased Akt phosphorylation (Klawitter et al. 2013), which may serve as a compensatory mechanism to protect from ischemic injury.

8.3.2 The Effect of DJ-1 on the PI3K/Akt Pathway Is Implicated in Multiple Disease States

The role of DJ-1 in the PI3k/Akt pathway is implicated in many disease models. For example, DJ-1 may affect the phosphorylation of tau, a protein that when hyperphosphorylated is found as a main component of neurofibrillary tangles in Alzheimer's disease and progressive supranuclear palsy (Avila et al. 2004). Overexpression of the Parkinson's disease-linked L166P mutant or D149A DJ-1 in cells induces less phosphorylation of Akt compared to wild-type DJ-1, leading to increased activity of GSK-3β and increased tau phosphorylation (Wang et al. 2013). This suggests that abnormal function of mutant DJ-1 in Akt activation may play a role in tauopathies.

In human clear cell renal cell carcinoma (ccRCC), DJ-1 may regulate the PTEN/PI3K/Akt pathway to modulate hTERT levels and contribute to disease progression (Sitaram et al. 2009). Human telomerase reverse transcriptase (hTERT) is the catalytic subunit of telomerase, a ribonucleoprotein reverse transcriptase, which has been implicated in cellular differentiation and neoplastic transformation (Kim et al. 1994; Takakura et al. 1999). *hTERT* gene may be regulated by the transcription factor c-*myc* which is a downstream target of the PI3K/Akt pathway (Wu et al. 1999; Asano et al. 2004). In human ccRCC tissue samples, DJ-1 mRNA levels are positively correlated with c-*myc* and hTERT mRNA levels (Sitaram et al. 2009). Follow-up experiments in human kidney carcinoma A498 cells showed that knockdown of DJ-1 reduces c-*myc* and hTERT RNA expression and also decreases protein levels of p-PTEN and p-Akt, as well as that of Akt's downstream effectors, p-GSK-3β and c-Myc (Sitaram et al. 2009). As such, the effect of DJ-1 on the activation of PI3k/Akt pathway appears to have far-reaching consequences on downstream effectors that can contribute to disease states.

Indeed, the effect of DJ-1 on the PI3k/Akt pathway may also be involved in several pathologic states including renal tubular epithelial-mesenchymal transition

(EMT), which can lead to renal interstitial fibrosis and end-stage renal failure (Yao et al. 2011), in the peritoneal metastasis of gastric carcinoma by modulating levels of matrix metallopeptidases MMP-2 and MMP-9 (Zhu et al. 2014), in the migration and invasion of human glioma SWO-38 cells by affecting focal adhesion kinase (FAK) phosphorylation (Fang et al. 2010), in the cell survival and aggressiveness of ovarian carcinoma (Davidson et al. 2008), in the progression of uterine cervical neoplasia (Choi et al. 2015), in the tumorigenesis of medulloblastomas (Lin et al. 2014), and in the malignant properties of the hepatocellular carcinoma (HCC) cell line, HepG2 (Liu et al. 2015).

8.3.3 DJ-1 Mediates Stress-Induced Cellular Responses Through PI3K/Akt Pathway, Possibly Through PTEN-Independent Mechanisms

Under cellular stress, many of the protective effects of DJ-1 appear to be mediated through the PI3k/Akt pathway. For example, under hypoxic stress, DJ-1 may regulate transcription factor hypoxia-inducible factor-1 (HIF1) through the PI3K/Akt/mTOR survival pathway as well as the metabolic sensor AMPK to protect cells against hypoxia-induced cell death (Vasseur et al. 2009). HIF1 regulates the expression of many genes known to be affected by Akt and may operate in parallel with AMPK to create a concerted response to hypoxia (Laderoute et al. 2006). mTOR activity has been reported to stabilize HIF1 protein (Zhong et al. 2000), while excess p53 may promote HIF1 degradation (Ravi et al. 2000). Under hypoxia, loss of DJ-1 was shown to significantly reduce mTOR activity (as measured by phosphorylation of p70-S6K and 4E-BP1) and increase p53 induction (as measured by p53 phosphorylation) in human osteosarcoma U2OS cells and in MEFs derived from DJ-1 KO mice. Congruently, DJ-1 is required for full HIF1 induction and activation under hypoxic stress, as DJ-1 knockdown in U2OS cells and MEFs exhibits decreased HIF1 protein levels and HIF1 target gene expression compared to control (Vasseur et al. 2009). DJ-1 has also been shown to interact with von Hippel-Lindau (VHL) protein, an E3 ubiquitin ligase, which can form a complex with HIF1 to induce its degradation (Parsanejad et al. 2014). DJ-1 inhibits HIF-VHL interaction and protects cells from hypoxia-induced apoptosis. Additionally, loss of DJ-1 alters AMPK activity and expression, where DJ-1 knockdown cells exhibit total decreased AMPK levels as well as decreased phosphorylation of AMPK under hypoxic stress (Vasseur et al. 2009). Taken together, these findings suggest that the ability of DJ-1 to modulate AMPK activity and regulate the PI3k/Akt/mTOR pathway or VHL-HIF1 interaction to increase HIF1 transcriptional activity may be crucial in coordinating the induction of genes necessary to adapt to hypoxia.

DJ-1 may also regulate Akt through its association with p53. p53 negatively regulates the IGF-1/PI3k/Akt pathway by increasing the transcription of target genes such as IGF-BP3 or PTEN (Feng 2010), and is also conversely regulated by Akt, which activates the MDM2, a trigger for p53 degradation (Mayo and Donner 2002). Accordingly, a study investigating the link between DJ-1, p53, and cell transformation showed that when p53 is deleted, DJ-1 is able to exert oncogenic effects through the phosphorylation and activation of Akt (Vasseur et al. 2012). In wild-type MEFs with normal p53 levels, DJ-1 knockdown results in inhibition of Akt phosphorylation. And in p53 null MEFs, which exhibit a higher level of Akt phosphorylation, DJ-1 is required for full activation of Akt. Interestingly, under oxidative stress, DJ-1 is necessary for the full phosphorylation and activation of p53 at serine 15 (Vasseur et al. 2012), which may be mediated by active Akt protein in certain cellular contexts (Vasseur et al. 2012; Zhan et al. 2010). Thus, there may be cross talk and feedback regulation between p53, DJ-1, and the Akt pathway, whereby oxidative stress induces DJ-1-mediated phosphorylation of p53, which may be mediated by active Akt. Active p53 may then be able to respond appropriately to stress by activating transcriptional programs (Kruiswijk et al. 2015), and also by consequently decreasing DJ-1 expression in a negative feedback loop, leading to attenuated Akt activation and preventing abnormal cellular transformation.

Under oxidative stress, DJ-1 may regulate Akt signaling by affecting Akt localization, and this may be crucial for neuronal protection in the context of Parkinson's disease (PD). Upon hydrogen peroxide (H_2O_2) treatment, primary neuronal cultures from DJ-1 KO mice show a reduction in Akt phosphorylation (Aleyasin et al. 2010). Similarly, dopaminergic neurons in the substantia nigra from DJ-1 KO mice exhibit reduced Akt phosphorylation in response to in vivo administration of MPTP (Aleyasin et al. 2010). Further, when lymphoblasts isolated from PD patients harboring the Parkinson's disease pathogenic mutation L166P are treated with H_2O_2, Akt phosphorylation is found to be reduced compared to that of healthy control lymphoblasts (Aleyasin et al. 2010). Suppression of Akt phosphorylation using a pharmacological inhibitor of AKT, LY294002 (LY), diminishes the neuroprotective function of DJ-1 in primary mouse neurons. This suggests that DJ-1 exerts some of the aforementioned protective effects through the Akt pathway (Aleyasin et al. 2010). Yet, exogenous WT Akt is unable to protect DJ-1-deficient neurons from oxidative stress both in vivo and in vitro (Aleyasin et al. 2010). On the other hand, myristoylated Akt (Myr-Akt), a membrane-anchored, constitutively active form of Akt, is able to protect DJ-1-deficient neurons (Aleyasin et al. 2010). Early studies have shown that Akt localization to the membrane occurs prior to its phosphorylation and activation (James et al. 1996; Franke et al. 1997) and that membrane-bound Myr-Akt is sufficient to provide cytoprotection (Ries et al. 2006). As loss of DJ-1 reduces Akt phosphorylation, and only membrane-anchored constitutively active form of Akt can provide protection to neurons lacking DJ-1, these findings suggest that DJ-1 may be an upstream activator of WT Akt. And consistent with this notion, DJ-1 has been shown to be necessary for Akt to translocate from the cytoplasmic compartment to membranous fractions following H_2O_2 treatment (Aleyasin et al. 2010). This suggests that DJ-1 may play a role in modulating recruitment of Akt to the membrane in an ROS-dependent manner, further supporting the notion that DJ-1 is necessary for Akt pathway activation and Akt-mediated neuroprotection from ROS.

It is also worthwhile to note that DJ-1 has been shown to associate with mRNA that encode members of the PTEN/PI3k/Akt pathway, raising the possibility that DJ-1 may also regulate this pathway posttranscriptionally. mRNA interacting with DJ-1 includes Akt1, IGF2, JUND, RPS6KB2, PPP2R2C, BCL2L1, RASL10B, MAPK8IP1, and EIF3EPI1 (van der Brug et al. 2008).

8.3.4 DJ-1 Inhibits Autophagy Through Activation of the PI3K/AKT Pathway

While the exact role of DJ-1 in regulating autophagy is under debate, it is clear that it can affect autophagy in a variety of models (Vasseur et al. 2009; McCoy and Cookson 2014). Previously, it was shown that DJ-1 activates autophagy in a neuronal model through the activation of the non-canonical MEK-ERK-mTOR pathway (Gao et al. 2012). In contrast, DJ-1 has been reported to activate autophagy through the PI3K/AKT pathway, independent of downstream mTOR effects.

The effect of DJ-1 on autophagy has also been studied in relation to C2-ceramide, a neurotoxic lipid that is associated with early inhibition of the PI3K/Akt pathway. Overexpression of DJ-1 in CAD cells (mouse catecholaminergic neuronal tumor cells) reportedly prevents C2-ceramide-induced inhibition of the PI3K/Akt pathway by keeping PTEN phosphorylated (inhibited). DJ-1 overexpression also decreases C2-ceramide-induced autophagy in a manner independent of changes in mTOR, exhibiting decreased autophagosome formation and autophagic flux. This suggests that DJ-1 may decrease ceramide-induced autophagy and cell death through its effects on the PI3k/AKT pathway (Jaramillo-Gomez et al. 2015).

8.3.5 DJ-1 May Inhibit PTEN Through Direct Binding and Transnitrosylation

Evidence has been reported suggesting that DJ-1 can interact with PTEN directly (Kim et al. 2009; Choi et al. 2014). Both wild-type and C106S mutant DJ-1 bind PTEN to inhibit its phosphatase activity in NIH3T3 cells, but the non-oxidizable C106S mutant binds and inhibits PTEN activity to a greater extent than WT DJ-1 does (Kim et al. 2009). Early in oxidative stress, WT DJ-1 strongly inhibits PTEN activity. But this is not sustained as stress continues and highly oxidized forms of DJ-1 at Cys-106 (SO_2H and SO_3H forms) accumulate. On the other hand, C106S mutant DJ-1, which cannot be oxidized at the Cys-106, strongly inhibits PTEN activity in a sustained manner (Kim et al. 2009). MALDI-TOF/TOF-MS (matrix-assisted laser desorption/ionization/time-of-flight mass spectrometry) analysis of the oxidation state of Cys-106 has revealed that in order for DJ-1 to inhibit PTEN activity and increase phosphorylation of Akt, over 50% of the DJ-1 population needs to be in the reduced/non-oxidized (SH) form (Kim et al. 2009). This suggests that under acute oxidative stress, the ability of DJ-1 to inhibit PTEN and consequently increase protective PI3k/Akt signaling is initially increased. However, with prolonged oxidation, when highly oxidized forms of DJ-1 make up more than 50% of total DJ-1, DJ-1 loses its ability to inhibit PTEN, indicating that DJ-1's modulation of the PI3k/Akt pathway is dependent on its redox status.

Under mild nitrosative stress, DJ-1 directly binds and may inhibit PTEN activity via transnitrosylation (Choi et al. 2014). In HEK cells stably expressing neuronal nitric oxide synthase (nNOS) that can be made to generate cellular nitric oxide (NO) using calcium ionophore A23187 treatment, both DJ-1 and PTEN are nitrosylated at cysteine residues when endogenous NO production is induced: DJ-1 at its Cys106 residue forming SNO-DJ-1 and PTEN at Cys83 residue forming SNO-PTEN (Choi et al. 2014). As the transfer of an NO group from one protein thiol to another according to their respective Nernstian redox potentials is a common mechanism in mammalian systems (Kornberg et al. 2010; Nakamura and Lipton 2013), it was hypothesized that DJ-1 may interact with and S-nitrosylated PTEN. In in vitro pull-down assays using GST-tagged DJ-1 and PTEN, as well as in co-immunoprecipitation experiments in HEK293A cells using antibodies directed against endogenous PTEN or DJ-1, it was found that PTEN and DJ-1 form a complex in cells (Choi et al. 2014). When purified recombinant SNO-DJ-1 is incubated with PTEN, SNO-DJ-1 but not unmodified DJ-1 is able to act as an NO donor to PTEN. In the converse experiment using SNO-PTEN and DJ-1, nitrosylated PTEN is unable to transfer its NO group to DJ-1 (Choi et al. 2014). This transnitrosylation from DJ-1 to PTEN is abrogated in DJ-1 knockdown SH-SY5Y cells, while DJ-1 overexpression increases the level of SNO-PTEN. Additionally, while the co-transfection of WT DJ-1 protects SH-SY5Y cells from cell death induced by exogenous PTEN overexpression, co-transfection of the system with either non-nitrosylatable Cys-106 DJ-1 mutant or non-nitrosylatable C83A PTEN mutant cannot confer the same cytoprotection (Choi et al. 2014). This suggests that the neuroprotective activity of DJ-1 may require at least in part the transnitrosylation and inhibition of PTEN. Notably, in human PD brains, SNO-PTEN levels have been found to be significantly elevated and SNO-DJ-1 slightly decreased compared to controls, suggesting that DJ-1 may transnitrosylate PTEN in PD brains to detoxify neurotoxic levels of NO until this adaptive protective system is overwhelmed. Dysfunctional DJ-1 may lose this protective transnitrosylation activity due to mutations or through oxidation of its Cys106 residue (Choi et al. 2014).

8.4 DJ-1 Inhibits the ASK1 Pathway

Apoptosis signal-regulating kinase 1 (ASK1), also known as mitogen-activated protein kinase kinase kinase 5 (MAP3K5), is a MAPKKK that plays a key role in stress-induced apoptosis as well as cell survival and differentiation. Downstream of ASK1 are its MAPKKs, mitogen-activated protein kinase kinases (MKK4/MKK7/

SEK1 and MKK3/MKK6), which in turn activate the MAPK proteins, c-Jun N-terminal kinases (JNKs), and p38 mitogen-activated protein kinases. This results in activation of the mitochondrial cell death pathway to induce apoptosis. These separate pathways are also referred to as the MKK4/MKK7-JNK pathway and MKK3/MKK6-p38 pathway (Cargnello and Roux 2011; Lawler et al. 1998; Keshet and Seger 2010; Tobiume et al. 2001).

Under normal conditions, ASK1 is oligomerized through its C-terminal coiled-coil domain but rendered inactive by various inhibitors such as thioredoxin (Trx) (Saitoh et al. 1998) and 14-3-3 proteins (Goldman et al. 2004). Reduced thioredoxin (Trx) binds ASK1 at its N-terminal coiled-coil domain, and 14-3-3 protein binds ASK1 at its phosphorylated Ser967 residue (Saitoh et al. 1998; Goldman et al. 2004; Yoon et al. 2009). These inhibitors regulate ASK1 activation in a redox sensitive manner and compete with ASK1 activators – such as TNF-alpha receptor-associated factors (TRAFs) or death-domain-associated protein 6 (Daxx).

Under cellular stresses such as oxidative stress, ultraviolet (UV) light, endoplasmic reticulum stress, tumor necrosis factor (TNF), and withdrawal of growth factor or serum, phosphorylation of ASK1 at Ser-967 is lost. Daxx then helps relieve inhibitory intramolecular interactions between the N- and C- termini of the ASK1 kinase, and TRAF2 and TRAF6 are recruited to ASK1 to form a larger molecular mass complex dubbed the ASK1 signalosome. ASK1 is then able to form homo-oligomeric interactions through both its C-terminal and N-terminal coiled-coil domains, undergo autophosphorylation at threonine 845, and become fully activated (Matsuzawa et al. 2005; Chang et al. 1998; Nishitoh et al. 1998; Ichijo et al. 1997; Leisner et al. 2016).

DJ-1 has been shown to inhibit ASK1 activation through a variety of mechanisms: (1) through sequestration or inhibition of the ASK1 activator, Daxx; (2) through direct binding with ASK1; by (3) increasing the expression of the ASK1 inhibitor, Trx1; and by (4) stabilizing Trx1-ASK1 interaction. This ultimately enhances cell survival by inhibiting cell death signaling and the mitochondrial apoptotic cascade (Fig. 8.4) (Junn et al. 2005; Waak et al. 2009; Hwang et al. 2013; Tang et al. 2014; Mo et al. 2010; Karunakaran et al. 2007).

8.4.1 DJ-1 Inhibits Daxx/ASK1 Activity Through Nuclear Sequestration of Daxx

The death protein Daxx has been identified as a DJ-1-interacting protein through yeast two-hybrid screen and subsequent immunoprecipitation experiments (Junn et al. 2005). In SH-SY5Y and COS-7 cells, overexpression of Daxx and ASK1 increases ASK1 activation (detected by in vitro kinase assay) and cell death. Co-expression of wild-type (WT) DJ-1 in this system is able to repress ASK1 activation and reduce cell death, whereas the Parkinson's disease-associated pathogenic L166P DJ-1 mutant cannot protect cells from the effects of Daxx and ASK1

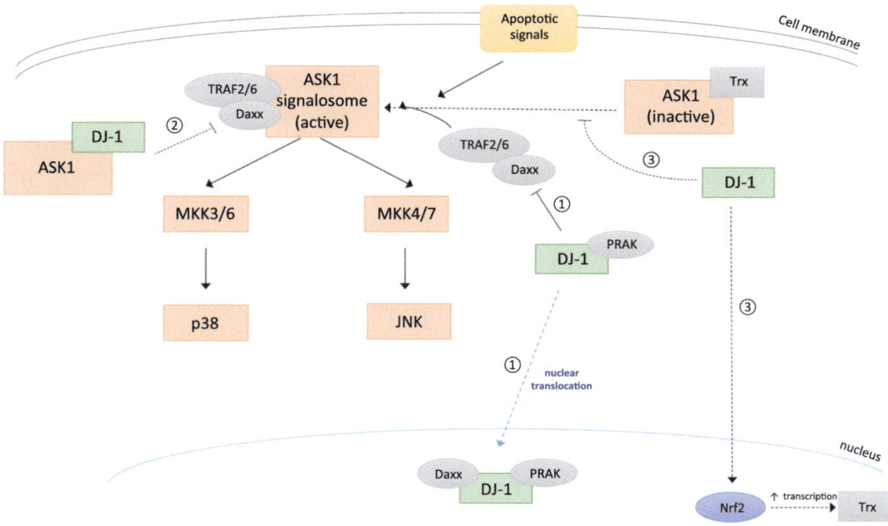

Fig. 8.4 DJ-1 inhibits the ASK1 pathway
1. DJ-1 inhibits ASK1 activity through nuclear sequestration of Daxx, an ASK1 activator and a part of the active ASK1 signalosome (Junn et al. 2005; Hwang et al. 2013; Karunakaran et al. 2007). Nuclear localization of DJ-1 may occur through its interaction with PRAK/MK5 which contains a nuclear localization sequence (NLS) (Tang et al. 2014)
2. DJ-1 may regulate ASK1 activity through direct binding (Waak et al. 2009; Mo et al. 2010; Cao et al. 2014)
3. DJ-1 inhibits ASK1 activation by preventing the dissociation of Trx1 from ASK1 and also by increasing the transcription of Trx1 (Im et al. 2010; Im et al. 2012)
[*positive regulation is indicated by pointed arrows, and negative regulation is indicated by blunted arrows. Direct or known regulation is indicated by solid lines; indirect or unknown regulation is indicated by dotted lines*]

overexpression (Junn et al. 2005). In an unstressed cell, Daxx is localized mainly in the nucleus, while ASK1 is localized in the cytoplasm. Under conditions of cell death signaling, Daxx translocates to the cytoplasm where it can interact with ASK1 to activate apoptosis (Chang et al. 1998; Ko et al. 2001). Fluorescent immunocytochemical experiments showed that WT DJ-1 blocks the translocation of Daxx to the cytoplasm by binding Daxx and sequestering it in the nucleus, whereas L166P mutant DJ-1 fails to do the same. This trend is magnified when cells are treated with H_2O_2 (Junn et al. 2005). M26I mutant DJ-1 is also unable to suppress the export of Daxx from the nucleus to the cytoplasm (Waak et al. 2009), indicating that WT DJ-1 but not its pathogenic mutants can protect cells against oxidative stress-induced Daxx/ASK1 cell death signaling.

The stress of MPTP administration in mice also results in the activation of ASK1 signaling as well as the translocation of Daxx from the nucleus to cytosol (Karunakaran et al. 2007; Lee et al. 2012). As expected in ASK1 KO mice, MPTP-induced motor impairments are not as pronounced, and striatonigral dopaminergic neurons are relatively preserved compared to wild-type littermates (Lee et al. 2012).

MPTP administration also lowers DJ-1 levels in the ventral midbrain of the animals, particularly in the nuclear fraction, which is consistent with the increase in the nuclear export of Daxx and its association with and activation of ASK1 in the cytosol (Karunakaran et al. 2007). Together, this indicates that ASK1 is a crucial effector of MPTP and oxidative stress-induced toxicity and apoptosis in the brain and that the cytoprotective function of DJ-1 may be mediated in part by its ability to regulate Daxx and ASK1 signaling.

DJ-1 may also bind to p38 regulated/activated kinase (PRAK/MK5) under cellular stress to help localize DJ-1 into the nucleus, allowing it to sequester Daxx in the nucleus and prevent cell death (Tang et al. 2014). PRAK is a downstream effector of the ASK1-MKK3/MKK6-p38 pathway and can be activated by p38 MAPK (Cargnello and Roux 2011). PRAK, which contains a putative nuclear localization sequence (NLS) and a nuclear export sequence (NES), is localized primarily in the cytoplasm under normal conditions (New et al. 2003). DJ-1 has been identified as a PRAK interactor in a yeast two-hybrid screen and found to bind directly to PRAK in immunoprecipitation assays and fluorescence resonance energy transfer (FRET) assays (Tang et al. 2014). DJ-1 also co-localizes with PRAK in the nuclei of NIH3T3 cells under oxidative stress. Following H_2O_2 treatment, PRAK increases phosphorylation of DJ-1 and its nuclear localization, as well as the nuclear sequestration of Daxx. Conversely, cells lacking PRAK exhibit impaired nuclear localization of DJ-1 and Daxx as well as increased cell death under oxidative stress (Tang et al. 2014). As DJ-1 lacks both a NLS and a NES, PRAK may be the crucial partner that assists DJ-1 in regulating the cellular localization of Daxx and ASK1 signaling.

Finally, DJ-1 may not only modulate Daxx localization but also its expression through the PI3k/Akt/dFOXO axis. In *Drosophila*, *DJ-1β* loss-of-function mutants are acutely sensitive to oxidative stress conditions and exhibit neuronal death after H_2O_2 treatment (Meulener et al. 2005; Menzies et al. 2005). As Daxx has been shown in mammalian systems to induce apoptosis by activating the JNK/FOXO cell death signaling pathway (Chang et al. 1998), one study examined the relationship of *DJ-1β* with the *Drosophila* homologue of Daxx, *Daxx-like protein* (*DLP*) (Hwang et al. 2013). *DLP*, seen to be elevated by H_2O_2 and UV exposure in wild-type *Drosophila*, has been shown to render flies more sensitive to oxidative stress when overexpressed in neurons (Hwang et al. 2013). In *DJ-1β* mutant flies, DLP expression as well as the translocation of DLP from the nucleus to the cytoplasm is increased, whereas overexpression of WT *DJ-1β* reduces the level of endogenous DLP. Additionally, *DLP* deficiency rescues the phenotypes of *DJ-1β Drosophila* mutants (Hwang et al. 2013). These findings suggest that DJ-1β may regulate the activity of DLP at least in part by limiting its expression and cytosolic localization. As *DLP* promoter harbors a consensus forkhead box subgroup O (FoxO) response element (FRE), it was also investigated whether DJ-1β could downregulate *DLP* expression transcriptionally under oxidative stress. Indeed, in *DJ-1β* mutants, the transcriptional activity of dFOXO is increased, inducing DLP transcription and apoptosis (Hwang et al. 2013). Interestingly, DLP overexpression also activates the JNK/dFOXO axis in *Drosophila*, meaning that *DLP* activation may further increase *DLP* expression in a feed-forward loop of DLP-JNK-dFOXO (Hwang et al. 2013).

However, the PI3k/Akt pathway, which is activated by DJ-1, inhibits dFOXO (Greer and Brunet 2005). Thus, *DJ-1β* may play a complex role in the regulation of DLP, where it cannot only modulate the cellular localization of the protein, but also inhibit the activation and overexpression of *DLP* and the subsequent propagation of apoptotic signaling through modulating the PI3k/Akt/dFOXO axis.

8.4.2 DJ-1 Inhibits ASK1 Activity Through Direct Interaction

In addition to DJ-1 binding to Daxx leading to ASK1 regulation (Junn et al. 2005), evidence has also been reported that DJ-1 may regulate ASK1 activity through direct binding as well (Waak et al. 2009; Mo et al. 2010; Cao et al. 2014). Following overexpression of both tagged DJ-1 and ASK1, the two proteins were shown to co-immunoprecipitate. By binding to ASK1, DJ-1 may disrupt ASK1 homo-oligomerization and activation, thus inhibiting H_2O_2-induced ASK1 activation of MKK3 and p38 (Mo et al. 2010). Although DJ-1 is able to inhibit ASK1 activity, it has little effect on the enzyme activity of either MKK3 or p38 (Mo et al. 2010), indicating that DJ-1 targets primarily ASK1 in the ASK1/MKK2/p38 cascade.

While one study observed DJ-1/ASK1 co-immunoprecipitation both in the presence and absence of oxidative stress (Mo et al. 2010), others have reported that oxidative stress and the resulting oxidized DJ-1 forms are necessary for DJ-1/ASK1 interaction (Waak et al. 2009; Cao et al. 2014). In support of the latter point, substitution of Cys-106 residue of DJ-1 to non-oxidizable alanine reportedly abrogates DJ-1/ASK1 interaction, whereas mutations of peripheral conserved cysteine residues, such as C53A and C46A, still allow the association of DJ-1 with ASK1. This suggests that oxidation of DJ-1 at Cys-106 may be crucial for its binding to ASK1 (Waak et al. 2009).

Size-exclusion chromatography has also shown that oxidized DJ-1 is incorporated into native ASK1 complexes. These complexes are dissolved upon reducing SDS-PAGE, implying that DJ-1 is incorporated into ASK1 signalosome by mixed disulfide formation that is dependent on the central Cys-106 residue (Waak et al. 2009). This mixed disulfide formation is similar to the interaction between thioredoxin 1 (Trx1) and ASK1 at its N-terminal Trx1 binding site. And, overexpression of DJ-1 or Trx1 is able to repress ASK1 homo-oligomerization (Mo et al. 2010), suggesting that DJ-1 acts similarly to Trx1 in suppressing ASK1 activation.

Interestingly, WT DJ-1 is unable to bind ASK1 that lacks this N-terminal Trx1 binding site, but M26I mutant DJ-1 is able to constitutively bind both WT ASK1 and the N-terminal deleted ASK1, presumably at a dysfunctional site (Waak et al. 2009). Additionally, while L166P mutant DJ-1 also associates with ASK1, the interaction is much weaker than that of wild-type DJ-1 (Mo et al. 2010). Unlike WT DJ-1, its C106A, L166P, and M26I mutants fail to provide cytoprotection against oxidative stress (Wilson 2011; Malgieri and Eliezer 2008). This shows that proper interaction of WT DJ-1 with ASK1, presumably at the N-terminal Trx1 binding site, is necessary to provide protection. Parkinson-associated pathogenic mutant DJ-1, which differs in protein stability and lacks normal functionality compared to WT DJ-1, may be unable to properly interact with ASK1, thus contributing to pathogenic consequences.

8.4.3 DJ-1 Inhibits ASK1 Activation by Preventing the Dissociation of Trx1 from ASK1 and by Upregulating Trx1 Expression

An additional mechanism by which DJ-1 may inhibit ASK1 signaling is through affecting the inhibitory complex of Trx1-ASK1 (Im et al. 2010). In unstressed HEK293T cells, Trx1 co-immunoprecipitates with ASK1 equally in the presence or absence of exogenous DJ-1. However, under oxidative stress, the interaction between ASK1 and Trx1 decreases dramatically. This dissociation of Trx1 from ASK1 upon oxidative stress is prevented by co-expression of WT DJ-1, whereas co-expression of L166P or C106S mutant DJ-1 fails to do the same. Additionally, in DJ-1 KO mouse brain homogenates, the Trx1-ASK1 complex is found to dissociate more readily under oxidative challenge compared to brains from WT mice, suggesting that DJ-1 plays an important role in the maintenance of the Trx1-ASK1 inhibitory complex (Im et al. 2010).

A follow-up study reported that WT DJ-1, but not its L166P or M26I mutants, can also upregulate Trx1 mRNA and protein expression by upregulating the levels of the transcription factor Nrf2 and stimulating its translocation into the nucleus (Im et al. 2012). This enhances Nrf2 recruitment to the antioxidant response element (ARE) of the Trx1 gene promoter, increasing the level of Trx1 within the cell and blocking ASK1 activation (Im et al. 2012). Interestingly, Trx1 has also been shown to bind PTEN to activate the PI3K/Akt pathway (Meuillet et al. 2004). Knocking down Trx1 impairs the ability of DJ-1 to induce AKT phosphorylation and activation (Im et al. 2012), suggesting that DJ-1-mediated modulation of Trx1 may impact both the ASK1 pathway and the PI3K/Akt pathway, dually regulating major cell death and cell survival signaling pathways.

8.5 DJ-1 in Other Signaling Pathways

8.5.1 DJ-1 Can Interact with MEKK1 to Suppress MEKK1-MKK4-JNK1 Cell Death Signaling

Apoptotic signaling via MKK4-JNK can be activated not only by ASK1 but also by other MAPKKKs such as MAP-ERK kinase kinase 1 (MEKK1), MEKK4, TAK1, and MLKs (Whitmarsh and Davis 1998). It has been demonstrated that wild-type (WT) DJ-1, but not its L166P mutant form, can interact physically with MEKK1 to inhibit its activity, protecting cells from UV-induced MEKK1-MKK4-JNK1 apoptotic signaling (Mo et al. 2008). By binding to MEKK1, DJ-1 also sequesters the kinase in the cytoplasm, preventing it from translocating to the nucleus and regulating gene expression through its downstream transcription factor effectors such as NF-kB and c-Jun. On the other hand, L166P mutant DJ-1 enhances the nuclear accumulation of MEKK1. DJ-1 does not appear to inhibit other MAPKKKs such as MLK3 and TAK1 (Mo et al. 2008).

8.5.2 DJ-1 Suppresses the JNK/Beclin 1 Pathway to Regulate Autophagy

In addition to its effects on autophagy through the ERK and AKT pathway, DJ-1 can inhibit autophagy in a cancer cell model through the JNK/Beclin1 pathway (Ren et al. 2010). In H1299 cells, a p53 null lung cancer cell line, overexpression of DJ-1 decreases autophagy, whereas knocking down of DJ-1 increases autophagy. DJ-1 knockdown also activates JNK1/2 in H1299 cells without affecting ERK or p38 activation and increases Beclin1 expression (Ren et al. 2010). JNK1/2 phosphorylation, Beclin1 upregulation, and autophagy activation that occur with DJ-1 knockdown are abrogated in the presence of the JNK-specific inhibitor, SP600125, suggesting that regulation of autophagy by DJ-1 is JNK dependent. Under conditions of starvation, DJ-1 knockdown in H1299 cells also exhibits increased autophagy as well as increased cell death (Ren et al. 2010). Thus, in a cancer model, the upregulation of DJ-1 may enhance oncogenesis by inhibiting autophagy in a JNK-dependent manner. This would lead to downregulation of the tumor-suppressor Beclin1 as well as accumulation of p62, which has been shown to contribute to tumorigenesis (Moscat and Diaz-Meco 2009; Mathew et al. 2009).

8.5.3 DJ-1 Positively Regulates Androgen Receptor Signaling

The nuclear receptor androgen receptor (AR) relays androgen signaling from the cell surface to the nucleus and activates the transcription of genes essential for male reproductive function such as spermatogenesis (O'Hara and Smith 2015). DJ-1 has been shown to be necessary for normal AR function and to positively regulate AR signaling in a variety of ways (Taira et al. 2004b). DJ-1 directly binds to the protein PIASxa/ARIP3, an inhibitor of AR, and prevents PIASxa/ARIP3 from forming a complex with AR (Takahashi et al. 2001). In addition, DJ-1 binds to AR to stimulate its transcriptional activity in hormonally treated prostate cancer cells, potentially contributing to cancer progression and androgen independence (Pitkanen-Arsiola et al. 2006; Tillman et al. 2007).

DJ-1 also binds and sequesters a novel DJ-1-binding protein (DJBP), which was identified by yeast two-hybrid screen (Niki et al. 2003). DJBP can bind to the DNA-binding domain of AR and repress its transcriptional activity through recruitment of a histone deacetylase (HDAC) co-repressor complex. Normally, when hormone receptors bind their putative ligand, they assume a configuration that leads to transcriptional activation (Xu et al. 1999). However, an HDAC-co-repressor complex may change the active form of AR into an inactive form. DJ-1 can partially restore AR function by abrogating the DJBP-HDAC complex.

Interestingly, Daxx has been shown to sumoylate and repress the DNA binding activity of AR, leading to inhibition of its transcriptional function (Shih et al. 2007). Daxx also binds DJ-1 in the nucleus (Junn et al. 2005). Additionally, sumoylation of DJ-1 at lysine residue K130 has been found to be crucial for its normal activity (Shinbo et al. 2005). The latter finding raises the possibility that Daxx might play a part in the sumoylation of DJ-1 and that DJ-1 binding to Daxx may impact the ability of Daxx to repress AR through a competitive mechanism as well. However, any direct link between Daxx and sumoylated DJ-1 as well as a link between sumoylated DJ-1 and AR remain to be investigated.

8.5.4 Mutant DJ-1 Interacts with TTRAP and TRAF6, Affecting Cell Death Signaling, Protein Aggregation, and rRNA Biogenesis

Several pathogenic missense mutations in DJ-1, such as L166P, M26I, and E64D, have been linked to recessively inherited PD (Bonifati et al. 2003; Bonifati et al. 2004; Abou-Sleiman et al. 2003). L166P and M26I mutations affect DJ-1 stability (although with mixed results for M26I), and its ability to form homodimers, and result in reduced cellular DJ-1 levels (Moore et al. 2003; Takahashi-Niki et al. 2004; Milkovic et al. 2015). E64D mutant DJ-1 may retain the ability to form homodimers, but this mutation may potentiate DJ-1 aggresome formation (Repici et al. 2013).

As there is still no unifying consensus on how these various mutants abrogate normal DJ-1 function, one study sought to investigate whether these mutants may be involved in a pathological "gain-of-function" protein-protein interaction (Zucchelli et al. 2009). To this end, a yeast two-hybrid screen of a human fetal brain cDNA library was employed to identify proteins that interact with both wild-type and mutant DJ-1. The TNFR-associated factor (TRAF) and tumor necrosis factor (TNF) receptor-associated protein (TTRAP/EAPII) was one such protein identified as a novel DJ-1 interactor (Zucchelli et al. 2009). TTRAP is a member of the non-canonical TGFβ-TRAF6-TAK1 apoptotic pathway and was originally isolated for its ability to bind TNF receptor and TRAFs and consequently inhibit NF-kβ activation (Pype et al. 2000; Varady et al. 2011).

TTRAP is found in the nucleus of SH-SY5Y cells and throughout the adult mouse brain, with strong expression in the dentate gyrus of the hippocampus. In the adult mouse mesencephalon, TTRAP is expressed both in non-dopaminergic and dopaminergic neurons of the substantia nigra. Interestingly, TTRAP solubility and subcellular localization are modified by proteasomal impairment. Treatment with the proteasomal inhibitor MG132 shifts TTRAP from the soluble fraction into the insoluble fraction, as TTRAP moves from the nucleus to the cytoplasm to form a single large juxta-nuclear aggresome-like structure. In co-immunoprecipitation experiments, TTRAP binding is stronger to M26I and L166P mutant DJ-1 than to WT DJ-1. When proteasomal function is impaired with MG132 treatment, the binding of all DJ-1 isoforms to TTRAP is enhanced. TTRAP overexpression protects SH-SY5Y cells from MG132-induced apoptosis, reducing JNK phosphorylation and poly-(ADP-ribose) polymerase (PARP) activation. However, co-transfection of M26I or L166P mutant DJ-1 with TTRAP results in a "gain-of-function" phenomenon, in which TTRAP assumes a novel signaling property that not only abrogates its protective effects against proteasomal impairment, but induces the activation of JNK- and p38 MAPK-mediated apoptosis. Cells that express only M26I or L166P DJ-1 without TTRAP do not activate the JNK- and p38 MAPK-mediated apoptotic pathway (Zucchelli et al. 2009). Thus L166P and M26I mutant DJ-1 may be involved in a pathological interaction with TTRAP, in a process that can render cells more sensitive to proteasomal stress and activate apoptotic pathways.

In a follow-up investigation, TRAF6, which interacts with TTRAP, was also shown to interact with L166P mutant but not with WT DJ-1 (Zucchelli et al. 2010). TRAF6 is an E3 ubiquitin ligase that is involved in activating the non-canonical TGFβ-TRAF6-TAK1 apoptotic pathway and can help relay neuronal cell signaling from the neurotrophin receptor p75 and TrkA (Geetha et al. 2005). It is also part of the active ASK1 signalosome, which can induce apoptosis (Fujino et al. 2007). Notably, TRAF6 promotes an atypical mode of polyubiquitination of DJ-1, stimulating the accumulation of mutant DJ-1 into insoluble aggregates (Zucchelli et al. 2010). Normally, polyubiquitin chains form through covalent conjugations using any one of seven lysine residues present in the ubiquitin protein (K6, K11, K27, K29, K33, K48, and K63), each contributing distinct but sometimes intersecting roles. For example, to target a protein for degradation by the proteasome, polyubiquitin chains are mainly formed through linkages of the K48 residue (Chen and Sun 2009). K11 linkages appear to function in endoplasmic reticulum-associated degradation (Xu et al. 2009). TRAF6 promotes K63-specific chain assembly that contributes to the non-canonical TGFβ-TRAF6-TAK1 apoptotic pathway, ultimately leading to NF-kβ activation (Chen 2005). While TRAF6 does not bind or affect the ubiquitination of WT DJ-1, it does bind and enhance the ubiquitination of L166P mutant DJ-1. Furthermore, instead of using the typical K63-specific chain assembly, TRAF6 promotes atypical polyubiquitination of L166P DJ-1 by using K6, K27, K29, and K33 isotype linkages. And rather than triggering degradation, these atypical polyubiquitin chains on L166P DJ-1 mediate its accumulation into insoluble aggregates (Zucchelli et al. 2010). In addition, an examination of postmortem brains from sporadic PD patients in a subsequent study revealed that TTRAP is associated with cytoplasmic Lewy bodies and localized to the nucleolus of surviving dopamine neurons. In a cell model, L166P mutant but not WT DJ-1 promotes the accumulation of TTRAP into insoluble cytoplasmic aggresomes, which in turn impair rRNA biogenesis by inhibiting TTRAP localization into nucleolar cavities (Vilotti et al. 2012).

Thus L166P mutant DJ-1 appears to gain dominant-negative functions over WT DJ-1 through its binding interactions with TTRAP and TRAF6. Under proteasomal impairment, L166P DJ-1 binds strongly to TTRAP, which localizes as an aggregate around the nucleus, and triggers p38-/JNK-mediated apoptosis (Zucchelli et al. 2009). In addition, L166P DJ-1 can bind TRAF6, an E3 ubiquitin ligase, and become poly-ubiquitinated in an atypical fashion. The poly-ubiquitinated L166P DJ-1 is able to form insoluble aggregates and sequester TTRAP in the cytoplasm, preventing TTRAP from localizing to the nucleolus and negatively affecting rRNA biogenesis (Zucchelli et al. 2010; Vilotti et al. 2012).

8.6 Conclusion

In a mere 10 years, DJ-1 has emerged as a significant player in many major signaling pathways, including ERK, Akt/PI3k, and ASK1, with distinct effects on different cancers and neuronal models. However, a bulk of the studies utilize in vitro reductionist methods, studying how cultured cells respond to various stresses and stimuli. Much of the data remains to be confirmed in in vivo models, where complex cross talk between signaling pathways and resulting biological compensation may attenuate or strengthen effects by DJ-1.

Nevertheless, many studies have shown that DJ-1 undoubtedly impacts the signaling processes that maintain cellular homeostasis, as well as the careful balance between uncontrolled oncogenesis and premature cell death. This indicates the importance of pursuing the role of DJ-1 in regulating signal transduction and of elucidating ways to target it for therapeutic intervention in cancer and neurodegeneration.

References

Abou-Sleiman PM et al (2003) The role of pathogenic DJ-1 mutations in Parkinson's disease. Ann Neurol 54(3):283–286

Abraham D et al (2000) Raf-1-associated protein phosphatase 2A as a positive regulator of kinase activation. J Biol Chem 275(29):22300–22304

Adams DG et al (2005) Positive regulation of Raf1-MEK1/2-ERK1/2 signaling by protein serine/threonine phosphatase 2A holoenzymes. J Biol Chem 280(52):42644–42654

Alessi DR et al (1995) Inactivation of p42 MAP kinase by protein phosphatase 2A and a protein tyrosine phosphatase, but not CL100, in various cell lines. Current Biology 5(3):283–295

Aleyasin H et al (2010) DJ-1 protects the nigrostriatal axis from the neurotoxin MPTP by modulation of the AKT pathway. Proc Natl Acad Sci U S A 107(7):3186–3191

Andjelkovic M et al (1996) Activation and phosphorylation of a pleckstrin homology domain containing protein kinase (RAC-PK/PKB) promoted by serum and protein phosphatase inhibitors. Proc Natl Acad Sci U S A 93(12):5699–5704
Aron L et al (2010) Pro-survival role for Parkinson's associated gene DJ-1 revealed in trophically impaired dopaminergic neurons. PLoS Biol 8(4):e1000349
Asano T et al (2004) The PI 3-kinase/Akt signaling pathway is activated due to aberrant Pten expression and targets transcription factors NF-kappaB and c-Myc in pancreatic cancer cells. Oncogene 23(53):8571–8580
Avila J et al (2004) Role of tau protein in both physiological and pathological conditions. Physiol Rev 84(2):361–384
Billia F et al (2013) Parkinson-susceptibility gene DJ-1/PARK7 protects the murine heart from oxidative damage in vivo. Proc Natl Acad Sci U S A 110(15):6085–6090
Blasi F, Carmeliet P (2002) uPAR: a versatile signalling orchestrator. Nat Rev Mol Cell Biol 3(12):932–943
Bonifati V et al (2003) Mutations in the DJ-1 gene associated with autosomal recessive early-onset parkinsonism. Science 299(5604):256–259
Bonifati V, Oostra BA, Heutink P (2004) Linking DJ-1 to neurodegeneration offers novel insights for understanding the pathogenesis of Parkinson's disease. J Mol Med (Berl) 82(3):163–174
Brognard J et al (2007) PHLPP and a second isoform, PHLPP2, differentially attenuate the amplitude of Akt signaling by regulating distinct Akt isoforms. Mol Cell 25(6):917–931
Burotto M et al (2014) The MAPK pathway across different malignancies: a new perspective. Cancer 120(22):3446–3456
Canet-Aviles RM et al (2004) The Parkinson's disease protein DJ-1 is neuroprotective due to cysteine-sulfinic acid-driven mitochondrial localization. Proc Natl Acad Sci U S A 101(24):9103–9108
Cao J et al (2014) The oxidation states of DJ-1 dictate the cell fate in response to oxidative stress triggered by 4-HPR: autophagy or apoptosis? Antioxid Redox Signal 21(10):1443–1459
Cao J et al (2015) DJ-1 as a human oncogene and potential therapeutic target. Biochem Pharmacol 93(3):241–250
Cargnello M, Roux PP (2011) Activation and function of the MAPKs and their substrates, the MAPK-activated protein kinases. Microbiol Mol Biol Rev 75(1):50–83
Caunt CJ et al (2015) MEK1 and MEK2 inhibitors and cancer therapy: the long and winding road. Nat Rev Cancer 15(10):577–592
Chang HY et al (1998) Activation of apoptosis signal-regulating kinase 1 (ASK1) by the adapter protein Daxx. Science 281(5384):1860–1863
Chang F et al (2003) Signal transduction mediated by the Ras/Raf/MEK/ERK pathway from cytokine receptors to transcription factors: potential targeting for therapeutic intervention. Leukemia 17(7):1263–1293
Chen ZJ (2005) Ubiquitin signalling in the NF-kappaB pathway. Nat Cell Biol 7(8):758–765
Chen ZJ, Sun LJ (2009) Nonproteolytic functions of ubiquitin in cell signaling. Mol Cell 33(3):275–286
Choi MS et al (2014) Transnitrosylation from DJ-1 to PTEN attenuates neuronal cell death in parkinson's disease models. J Neurosci 34(45):15123–15131
Choi SK et al (2015) Overexpression of PI3K-p110alpha in the progression of uterine cervical neoplasia and its correlation with pAkt and DJ-1. Eur J Gynaecol Oncol 36(4):389–393
Cross DA et al (1995) Inhibition of glycogen synthase kinase-3 by insulin mediated by protein kinase B. Nature 378(6559):785–789
Davidson B et al (2008) Expression and clinical role of DJ-1, a negative regulator of PTEN, in ovarian carcinoma. Hum Pathol 39(1):87–95
Devine MJ, Plun-Favreau H, Wood NW (2011) Parkinson's disease and cancer: two wars, one front. Nat Rev Cancer 11(11):812–823
Dhillon AS et al (2002) Cyclic AMP-dependent kinase regulates Raf-1 kinase mainly by phosphorylation of serine 259. Mol Cell Biol 22(10):3237–3246

Dongworth RK et al (2014) DJ-1 protects against cell death following acute cardiac ischemia-reperfusion injury. Cell Death Dis 5:e1082
Eichhorn PJ et al (2007) A RNA interference screen identifies the protein phosphatase 2A subunit PR55gamma as a stress-sensitive inhibitor of c-SRC. PLoS Genet 3(12):e218
Fan J et al (2008) DJ-1 decreases Bax expression through repressing p53 transcriptional activity. J Biol Chem 283(7):4022–4030
Fang M et al (2010) Role of DJ-1-induced PTEN down-regulation in migration and invasion of human glioma cells. Chin J Cancer 29(12):988–994
Feilotter HE et al (1999) Analysis of the 10q23 chromosomal region and the PTEN gene in human sporadic breast carcinoma. Br J Cancer 79(5–6):718–723
Feng Z (2010) p53 Regulation of the IGF-1/AKT/mTOR pathways and the endosomal compartment. Cold Spring Harb Perspect Biol 2(2):a001057
Franke TF et al (1997) Direct regulation of the Akt proto-oncogene product by phosphatidylinositol-3,4-bisphosphate. Science 275(5300):665–668
Freihoff D et al (1999) Exclusion of a major role for the PTEN tumour-suppressor gene in breast carcinomas. Br J Cancer 79(5–6):754–758
Fujino G et al (2007) Thioredoxin and TRAF family proteins regulate reactive oxygen species-dependent activation of ASK1 through reciprocal modulation of the N-terminal homophilic interaction of ASK1. Mol Cell Biol 27(23):8152–8163
Gao T, Furnari F, Newton AC (2005) PHLPP: a phosphatase that directly dephosphorylates Akt, promotes apoptosis, and suppresses tumor growth. Mol Cell 18(1):13–24
Gao H et al (2012) DJ-1 protects dopaminergic neurons against rotenone-induced apoptosis by enhancing ERK-dependent mitophagy. J Mol Biol 423(2):232–248
Geetha T, Jiang J, Wooten MW (2005) Lysine 63 polyubiquitination of the nerve growth factor receptor TrkA directs internalization and signaling. Mol Cell 20(2):301–312
Goldman EH, Chen L, Fu H (2004) Activation of apoptosis signal-regulating kinase 1 by reactive oxygen species through dephosphorylation at serine 967 and 14-3-3 dissociation. J Biol Chem 279(11):10442–10449
Green DR, Llambi F (2015) Cell death signaling. Cold Spring Harb Perspect Biol 7(12):a006080
Greer EL, Brunet A (2005) FOXO transcription factors at the interface between longevity and tumor suppression. Oncogene 24(50):7410–7425
Gu L et al (2009) Involvement of ERK1/2 signaling pathway in DJ-1-induced neuroprotection against oxidative stress. Biochem Biophys Res Commun 383(4):469–474
He X et al (2012) DJ-1 promotes invasion and metastasis of pancreatic cancer cells by activating SRC/ERK/uPA. Carcinogenesis 33(3):555–562
Hermanson E et al (2003) Nurr1 regulates dopamine synthesis and storage in MN9D dopamine cells. Exp Cell Res 288(2):324–334
Hwang S et al (2013) Drosophila DJ-1 decreases neural sensitivity to stress by negatively regulating Daxx-like protein through dFOXO. PLoS Genet 9(4):e1003412
Ichijo H et al (1997) Induction of apoptosis by ASK1, a mammalian MAPKKK that activates SAPK/JNK and p38 signaling pathways. Science 275(5296):90–94
Im JY et al (2010) DJ-1 protects against oxidative damage by regulating the thioredoxin/ASK1 complex. Neurosci Res 67(3):203–208
Im JY et al (2012) DJ-1 induces thioredoxin 1 expression through the Nrf2 pathway. Hum Mol Genet 21(13):3013–3024
Ishikawa S et al (2009) Oxidative status of DJ-1-dependent activation of dopamine synthesis through interaction of tyrosine hydroxylase and 4-dihydroxy-L-phenylalanine (L-DOPA) decarboxylase with DJ-1. J Biol Chem 284(42):28832–28844
Iwawaki T, Kohno K, Kobayashi K (2000) Identification of a potential nurr1 response element that activates the tyrosine hydroxylase gene promoter in cultured cells. Biochem Biophys Res Commun 274(3):590–595
James SR et al (1996) Specific binding of the Akt-1 protein kinase to phosphatidylinositol 3,4,5-trisphosphate without subsequent activation. Biochem J 315(3):709–713

Janssens V, Goris J (2001) Protein phosphatase 2A: a highly regulated family of serine/threonine phosphatases implicated in cell growth and signalling. Biochem J 353(3):417–439
Jaramillo-Gomez J et al (2015) Overexpression of DJ-1 protects against C2-ceramide-induced neuronal death through activation of the PI3K/AKT pathway and inhibition of autophagy. Neurosci Lett 603:71–76
Jeffrey KL et al (2007) Targeting dual-specificity phosphatases: manipulating MAP kinase signalling and immune responses. Nat Rev Drug Discov 6(5):391–403
Junn E et al (2005) Interaction of DJ-1 with Daxx inhibits apoptosis signal-regulating kinase 1 activity and cell death. Proc Natl Acad Sci U S A 102(27):9691–9696
Junttila MR, Li SP, Westermarck J (2008) Phosphatase-mediated crosstalk between MAPK signaling pathways in the regulation of cell survival. FASEB J 22(4):954–965
Kahle PJ, Waak J, Gasser T (2009) DJ-1 and prevention of oxidative stress in Parkinson's disease and other age-related disorders. Free Radic Biol Med 47(10):1354–1361
Karunakaran S et al (2007) Activation of apoptosis signal regulating kinase 1 (ASK1) and translocation of death-associated protein, Daxx, in substantia nigra pars compacta in a mouse model of Parkinson's disease: protection by alpha-lipoic acid. FASEB J 21(9):2226–2236
Kato I et al (2013) Oxidized DJ-1 inhibits p53 by sequestering p53 from promoters in a DNA-binding affinity-dependent manner. Mol Cell Biol 33(2):340–359
Keshet Y, Seger R (2010) The MAP kinase signaling cascades: a system of hundreds of components regulates a diverse array of physiological functions. Methods Mol Biol 661:3–38
Kim NW et al (1994) Specific association of human telomerase activity with immortal cells and cancer. Science 266(5193):2011–2015
Kim RH et al (2005) DJ-1, a novel regulator of the tumor suppressor PTEN. Cancer Cell 7(3):263–273
Kim YC et al (2009) Oxidation of DJ-1-dependent cell transformation through direct binding of DJ-1 to PTEN. Int J Oncol 35(6):1331–1341
Klawitter J et al (2013) Association of DJ-1/PTEN/AKT- and ASK1/p38-mediated cell signalling with ischaemic cardiomyopathy. Cardiovasc Res 97(1):66–76
Ko YG et al (2001) Apoptosis signal-regulating kinase 1 controls the proapoptotic function of death-associated protein (Daxx) in the cytoplasm. J Biol Chem 276(42):39103–39106
Kornberg MD et al (2010) GAPDH mediates nitrosylation of nuclear proteins. Nat Cell Biol 12(11):1094–1100
Kramer ER et al (2007) Absence of ret signaling in mice causes progressive and late degeneration of the nigrostriatal system. PLoS Biol 5(3):e39
Krebiehl G et al (2010) Reduced basal autophagy and impaired mitochondrial dynamics due to loss of Parkinson's disease-associated protein DJ-1. PLoS One 5(2):e9367
Kruiswijk F, Labuschagne CF, Vousden KH (2015) p53 In survival, death and metabolic health: a lifeguard with a licence to kill. Nat Rev Mol Cell Biol 16(7):393–405
Laderoute KR et al (2006) 5′-AMP-activated protein kinase (AMPK) is induced by low-oxygen and glucose deprivation conditions found in solid-tumor microenvironments. Mol Cell Biol 26(14):5336–5347
Lawler S et al (1998) Synergistic activation of SAPK1/JNK1 by two MAP kinase kinases in vitro. Curr Biol 8(25):1387–1391
Lee KW et al (2012) Apoptosis signal-regulating kinase 1 mediates MPTP toxicity and regulates glial activation. PLoS One 7(1):e29935
Leisner TM et al (2016) CIB1: a small protein with big ambitions. FASEB J 30(8):2640–2650
Letourneux C, Rocher G, Porteu F (2006) B56-containing PP2A dephosphorylate ERK and their activity is controlled by the early gene IEX-1 and ERK. EMBO J 25(4):727–738
Lev N et al (2008) Oxidative insults induce DJ-1 upregulation and redistribution: implications for neuroprotection. Neurotoxicology 29(3):397–405
Lev N et al (2009) DJ-1 protects against dopamine toxicity. J Neural Transm (Vienna) 116(2):151–160

Lin E et al (2008) Rapid activation of ERK by 6-hydroxydopamine promotes survival of dopaminergic cells. J Neurosci Res 86(1):108–117
Lin J et al (2014) DJ-1 is activated in medulloblastoma and is associated with cell proliferation and differentiation. World J Surg Oncol 12:373
Liu S et al (2015) DJ-1 knockdown inhibits growth and xenograft induced tumor generation of human hepatocellular carcinoma cells. Oncol Rep 33(1):201–206
Lu L et al (2012) DJ-1 upregulates tyrosine hydroxylase gene expression by activating its transcriptional factor Nurr1 via the ERK1/2 pathway. Int J Biochem Cell Biol 44(1):65–71
Lu L et al (2016) DJ-1/PARK7, but not its L166P mutant linked to autosomal recessive parkinsonism, modulates the transcriptional activity of the orphan nuclear receptor Nurr1 In vitro and in vivo. Mol Neurobiol 53:7363
Maehama T, Dixon JE (1998) The tumor suppressor, PTEN/MMAC1, dephosphorylates the lipid second messenger, phosphatidylinositol 3,4,5-trisphosphate. J Biol Chem 273(22):13375–13378
Malgieri G, Eliezer D (2008) Structural effects of Parkinson's disease linked DJ-1 mutations. Protein Sci 17(5):855–868
Manning BD, Cantley LC (2007) AKT/PKB signaling: navigating downstream. Cell 129(7):1261–1274
Martinat C et al (2004) Sensitivity to oxidative stress in DJ-1-deficient dopamine neurons: an ES-derived cell model of primary parkinsonism. PLoS Biol 2(11):e327
Mathew R et al (2009) Autophagy suppresses tumorigenesis through elimination of p62. Cell 137(6):1062–1075
Matsuzawa A et al (2005) ROS-dependent activation of the TRAF6-ASK1-p38 pathway is selectively required for TLR4-mediated innate immunity. Nat Immunol 6(6):587–592
Mayo LD, Donner DB (2002) The PTEN, Mdm2, p53 tumor suppressor-oncoprotein network. Trends Biochem Sci 27(9):462–467
McCoy MK, Cookson MR (2014) DJ-1 regulation of mitochondrial function and autophagy through oxidative stress. Autophagy 7(5):531–532
McCubrey JA et al (2007) Roles of the Raf/MEK/ERK pathway in cell growth, malignant transformation and drug resistance. Biochim Biophys Acta 1773(8):1263–1284
Menendez D, Inga A, Resnick MA (2009) The expanding universe of p53 targets. Nat Rev Cancer 9(10):724–737
Menzies FM, Yenisetti SC, Min KT (2005) Roles of drosophila DJ-1 in survival of dopaminergic neurons and oxidative stress. Curr Biol 15(17):1578–1582
Meuillet EJ et al (2004) Thioredoxin-1 binds to the C2 domain of PTEN inhibiting PTEN's lipid phosphatase activity and membrane binding: a mechanism for the functional loss of PTEN's tumor suppressor activity. Arch Biochem Biophys 429(2):123–133
Meulener M et al (2005) Drosophila DJ-1 mutants are selectively sensitive to environmental toxins associated with Parkinson's disease. Curr Biol 15(17):1572–1577
Milkovic NM et al (2015) Transient sampling of aggregation-prone conformations causes pathogenic instability of a parkinsonian mutant of DJ-1 at physiological temperature. Protein Sci 24(10):1671–1685
Mo JS et al (2008) DJ-1 modulates UV-induced oxidative stress signaling through the suppression of MEKK1 and cell death. Cell Death Differ 15(6):1030–1041
Mo JS et al (2010) DJ-1 modulates the p38 mitogen-activated protein kinase pathway through physical interaction with apoptosis signal-regulating kinase 1. J Cell Biochem 110(1):229–237
Moore DJ et al (2003) A missense mutation (L166P) in DJ-1, linked to familial Parkinson's disease, confers reduced protein stability and impairs homo-oligomerization. J Neurochem 87(6):1558–1567
Moscat J, Diaz-Meco MT (2009) p62 at the crossroads of autophagy, apoptosis, and cancer. Cell 137(6):1001–1004
Mukherjee UA et al (2011) 37 A novel role for DJ-1 in cardioprotection. Heart 97(24):e8
Mut M et al (2012) Both mitogen-activated protein kinase (MAPK)/extracellular-signal-regulated kinases (ERK) 1/2 and phosphatidylinositide-3-OH kinase (PI3K)/Akt pathways regulate acti-

vation of E-twenty-six (ETS)-like transcription factor 1 (Elk-1) in U138 glioblastoma cells. Int J Biochem Cell Biol 44(2):302–310
Nagakubo D et al (1997) DJ-1, a novel oncogene which transforms mouse NIH3T3 cells in cooperation with ras. Biochem Biophys Res Commun 231(2):509–513
Nakamura T, Lipton SA (2013) Emerging role of protein-protein transnitrosylation in cell signaling pathways. Antioxid Redox Signal 18(3):239–249
New L, Jiang Y, Han J (2003) Regulation of PRAK subcellular location by p38 MAP kinases. Mol Biol Cell 14(6):2603–2616
Niki T et al (2003) DJBP: a novel DJ-1-binding protein, negatively regulates the androgen receptor by recruiting histone deacetylase complex, and DJ-1 antagonizes this inhibition by abrogation of this complex. Mol Cancer Res 1(4):247–261
Nishitoh H et al (1998) ASK1 is essential for JNK/SAPK activation by TRAF2. Mol Cell 2(3):389–395
Nordzell M et al (2004) Defining an N-terminal activation domain of the orphan nuclear receptor Nurr1. Biochem Biophys Res Commun 313(1):205–211
O'Hara L, Smith LB (2015) Androgen receptor roles in spermatogenesis and infertility. Best Pract Res Clin Endocrinol Metab 29(4):595–605
Oehrl W, Rubio I, Wetzker R (2003) Serine 338 phosphorylation is dispensable for activation of c-Raf1. J Biol Chem 278(20):17819–17826
Parsanejad M et al (2014) Regulation of the VHL/HIF-1 pathway by DJ-1. J Neurosci 34(23):8043–8050
Perren A et al (1999) Immunohistochemical evidence of loss of PTEN expression in primary ductal adenocarcinomas of the breast. Am J Pathol 155(4):1253–1260
Pitkanen-Arsiola T et al (2006) Androgen and anti-androgen treatment modulates androgen receptor activity and DJ-1 stability. Prostate 66(11):1177–1193
Plotnikov A et al (2011) The MAPK cascades: signaling components, nuclear roles and mechanisms of nuclear translocation. Biochim Biophys Acta 1813(9):1619–1633
Pype S et al (2000) TTRAP, a novel protein that associates with CD40, tumor necrosis factor (TNF) receptor-75 and TNF receptor-associated factors (TRAFs), and that inhibits nuclear factor-kappa B activation. J Biol Chem 275(24):18586–18593
Rahman-Roblick R et al (2008) Proteomic identification of p53-dependent protein phosphorylation. Oncogene 27(35):4854–4859
Ravi R et al (2000) Regulation of tumor angiogenesis by p53-induced degradation of hypoxia-inducible factor 1alpha. Genes Dev 14(1):34–44
Ren H et al (2010) DJ-1, a cancer and Parkinson's disease associated protein, regulates autophagy through JNK pathway in cancer cells. Cancer Lett 297(1):101–108
Repici M et al (2013) Parkinson's Disease-associated mutations in DJ-1 modulate its dimerization in living cells. J Mol Med (Berl) 91(5):599–611
Ries V et al (2006) Oncoprotein Akt/PKB induces trophic effects in murine models of Parkinson's disease. Proc Natl Acad Sci U S A 103(49):18757–18762
Saitoh M et al (1998) Mammalian thioredoxin is a direct inhibitor of apoptosis signal-regulating kinase (ASK) 1. EMBO J 17(9):2596–2606
Sekito A et al (2006) DJ-1 interacts with HIPK1 and affects H2O2-induced cell death. Free Radic Res 40(2):155–165
Shamoto-Nagai M et al (2003) An inhibitor of mitochondrial complex I, rotenone, inactivates proteasome by oxidative modification and induces aggregation of oxidized proteins in SH-SY5Y cells. J Neurosci Res 74(4):589–597
Shendelman S et al (2004) DJ-1 is a redox-dependent molecular chaperone that inhibits alpha-synuclein aggregate formation. PLoS Biol 2(11):e362
Shi W et al (2003) Dysregulated PTEN-PKB and negative receptor status in human breast cancer. Int J Cancer 104(2):195–203
Shih HM et al (2007) Daxx mediates SUMO-dependent transcriptional control and subnuclear compartmentalization. Biochem Soc Trans 35(Pt 6):1397–1400

Shinbo Y et al (2005) Proper SUMO-1 conjugation is essential to DJ-1 to exert its full activities. Cell Death Differ 13(1):96–108
Silverstein AM et al (2002) Actions of PP2A on the MAP kinase pathway and apoptosis are mediated by distinct regulatory subunits. Proc Natl Acad Sci 99(7):4221–4226
Sitaram RT et al (2009) The PTEN regulator DJ-1 is associated with hTERT expression in clear cell renal cell carcinoma. Int J Cancer 125(4):783–790
Sonoda Y et al (1997) Stimulation of Interleukin-8 production by Okadaic acid and vanadate in a human Promyelocyte cell line, an HL-60 subline: possible role of mitogen-activated protein kinase on the Okadaic acid-induced nf-κb activation. J Biol Chem 272(24):15366–15372
Sorescu D et al (2002) Superoxide production and expression of Nox family proteins in human atherosclerosis. Circulation 105(12):1429
Taira T et al (2004a) DJ-1 has a role in antioxidative stress to prevent cell death. EMBO Rep 5(2):213–218
Taira T, Iguchi-Ariga SM, Ariga H (2004b) Co-localization with DJ-1 is essential for the androgen receptor to exert its transcription activity that has been impaired by androgen antagonists. Biol Pharm Bull 27(4):574–577
Takahashi K et al (2001) DJ-1 positively regulates the androgen receptor by impairing the binding of PIASx alpha to the receptor. J Biol Chem 276(40):37556–37563
Takahashi-Niki K et al (2004) Reduced anti-oxidative stress activities of DJ-1 mutants found in Parkinson's disease patients. Biochem Biophys Res Commun 320(2):389–397
Takahashi-Niki K et al (2015) Epidermal growth factor-dependent activation of the extracellular signal-regulated kinase pathway by DJ-1 protein through its direct binding to c-Raf protein. J Biol Chem 290(29):17838–17847
Takakura M et al (1999) Cloning of human telomerase catalytic subunit (hTERT) gene promoter and identification of proximal core promoter sequences essential for transcriptional activation in immortalized and cancer cells. Cancer Res 59(3):551–557
Tang J et al (2014) PRAK interacts with DJ-1 and prevents oxidative stress-induced cell death. Oxidative Med Cell Longev 2014:735618
Tentolouris C et al (2004) Endothelial function and proinflammatory cytokines in patients with ischemic heart disease and dilated cardiomyopathy. Int J Cardiol 94(2):301–305
Tillman JE et al (2007) DJ-1 binds androgen receptor directly and mediates its activity in hormonally treated prostate cancer cells. Cancer Res 67(10):4630–4637
Tobiume K et al (2001) ASK1 is required for sustained activations of JNK/p38 MAP kinases and apoptosis. EMBO Rep 2(3):222–228
van der Brug MP et al (2008) RNA binding activity of the recessive parkinsonism protein DJ-1 supports involvement in multiple cellular pathways. Proc Natl Acad Sci U S A 105(29):10244–10249
Varady G, Sarkadi B, Fatyol K (2011) TTRAP is a novel component of the non-canonical TRAF6-TAK1 TGF-beta signaling pathway. PLoS One 6(9):e25548
Vasseur S et al (2009) DJ-1/PARK7 is an important mediator of hypoxia-induced cellular responses. Proc Natl Acad Sci U S A 106(4):1111–1116
Vasseur S et al (2012) Consequences of DJ-1 upregulation following p53 loss and cell transformation. Oncogene 31(5):664–670
Vilotti S et al (2012) Parkinson's disease DJ-1 L166P alters rRNA biogenesis by exclusion of TTRAP from the nucleolus and sequestration into cytoplasmic aggregates via TRAF6. PLoS One 7(4):e35051
Vousden KH, Prives C (2009) Blinded by the light: the growing complexity of p53. Cell 137(3):413–431
Waak J et al (2009) Oxidizable residues mediating protein stability and cytoprotective interaction of DJ-1 with apoptosis signal-regulating kinase 1. J Biol Chem 284(21):14245–14257
Wang J et al (2009) A non-canonical MEK/ERK signaling pathway regulates autophagy via regulating Beclin 1. J Biol Chem 284(32):21412–21424

Wang Z et al (2011) DJ-1 modulates the expression of cu/Zn-superoxide dismutase-1 through the Erk1/2-Elk1 pathway in neuroprotection. Ann Neurol 70(4):591–599

Wang Y et al (2013) Parkinson's Disease-associated DJ-1 mutations increase abnormal phosphorylation of tau protein through Akt/GSK-3beta pathways. J Mol Neurosci 51(3):911–918

West AB, Dawson VL, Dawson TM (2005) To die or grow: Parkinson's disease and cancer. Trends Neurosci 28(7):348–352

Whitmarsh AJ, Davis RJ (1998) Structural organization of MAP-kinase signaling modules by scaffold proteins in yeast and mammals. Trends Biochem Sci 23(12):481–485

Wilson MA (2011) The role of cysteine oxidation in DJ-1 function and dysfunction. Antioxid Redox Signal 15(1):111–122

Wu KJ et al (1999) Direct activation of TERT transcription by c-MYC. Nat Genet 21(2):220–224

Xu L, Glass CK, Rosenfeld MG (1999) Coactivator and corepressor complexes in nuclear receptor function. Curr Opin Genet Dev 9(2):140–147

Xu P et al (2009) Quantitative proteomics reveals the function of unconventional ubiquitin chains in proteasomal degradation. Cell 137(1):133–145

Yang Y et al (2005) Inactivation of drosophila DJ-1 leads to impairments of oxidative stress response and phosphatidylinositol 3-kinase/Akt signaling. Proc Natl Acad Sci U S A 102(38):13670–13675

Yao Y et al (2011) Upregulated DJ-1 promotes renal tubular EMT by suppressing cytoplasmic PTEN expression and Akt activation. J Huazhong Univ Sci Technolog Med Sci 31(4):469–475

Yoon KW et al (2009) CIB1 Functions as a ca(2+)-sensitive modulator of stress-induced signaling by targeting ASK1. Proc Natl Acad Sci U S A 106(41):17389–17394

Yoshioka K (2004) Scaffold proteins in mammalian MAP kinase cascades. J Biochem 135(6):657–661

Zanassi P et al (2001) cAMP-dependent protein kinase induces cAMP-response element-binding protein phosphorylation via an intracellular calcium release/ERK-dependent pathway in striatal neurons. J Biol Chem 276(15):11487–11495

Zhan H et al (2010) Ataxia telangiectasia mutated (ATM)-mediated DNA damage response in oxidative stress-induced vascular endothelial cell senescence. J Biol Chem 285(38):29662–29670

Zhang X et al (2011) Akt, FoxO and regulation of apoptosis. Biochim Biophys Acta (BBA) Mol Cell Res 1813(11):1978–1986

Zhang Y et al (2016) Protective effects of DJ-1 medicated Akt phosphorylation on mitochondrial function are promoted by da-Bu-yin-wan in 1-methyl-4-phenylpyridinium-treated human neuroblastoma SH-SY5Y cells. J Ethnopharmacol 187:83–93

Zhong H et al (2000) Modulation of hypoxia-inducible factor 1alpha expression by the epidermal growth factor/phosphatidylinositol 3-kinase/PTEN/AKT/FRAP pathway in human prostate cancer cells: implications for tumor angiogenesis and therapeutics. Cancer Res 60(6):1541–1545

Zhou B et al (2002) The specificity of extracellular signal-regulated kinase 2 dephosphorylation by protein phosphatases. J Biol Chem 277(35):31818–31825

Zhou J et al (2015) Crosstalk between MAPK/ERK and PI3K/AKT signal pathways during brain ischemia/reperfusion. ASN Neuro 7(5):175909141560246

Zhu ZM et al (2014) DJ-1 is involved in the peritoneal metastasis of gastric cancer through activation of the Akt signaling pathway. Oncol Rep 31(3):1489–1497

Zucchelli S et al (2009) Aggresome-forming TTRAP mediates pro-apoptotic properties of Parkinson's disease-associated DJ-1 missense mutations. Cell Death Differ 16(3):428–438

Zucchelli S et al (2010) TRAF6 promotes atypical ubiquitination of mutant DJ-1 and alpha-synuclein and is localized to Lewy bodies in sporadic Parkinson's disease brains. Hum Mol Genet 19(19):3759–3770

Chapter 9
Protein Repair from Glycation by Glyoxals by the DJ-1 Family Maillard Deglycases

Mouadh Mihoub, Jad Abdallah, and Gilbert Richarme

Abstract DJ-1 and its prokaryotic homologs, Hsp31, YhbO and YajL from *Escherichia coli* and PfpI from *Pyrococcus furiosus,* repair proteins from glycation by glyoxals (R-CO-CHO), which constitute their major glycating agents. Glycation is a non-enzymatic covalent reaction discovered by Louis Camille Maillard in 1912, between reactive carbonyls (reducing sugars and glyoxals) and amino acids (cysteine, arginine and lysine), which inactivates proteins. By degrading Maillard adducts formed between carbonyls and thiols or amino groups, the DJ-1 family Maillard deglycases prevent the formation of the so-called advanced glycation end products (AGEs) that arise from Maillard adducts after dehydrations, oxidations and rearrangements. Since glycation is involved in ageing, cancer, atherosclerosis and cataracts, as well as post-diabetic, neurovegetatives and renal and autoimmune diseases, the DJ-1 deglycases are likely to play an important role in preventing these diseases. These deglycases, especially those from thermophilic organisms, may also be used to prevent the formation of dietary AGEs during food processing, sterilization and storage. They also prevent acrylamide formation in food, likely by degrading the asparagine/glyoxal Maillard adducts responsible for its formation. Since Maillard adducts are the substrates of the DJ-1 family deglycases, we propose renaming them Maillard deglycases.

Keywords Glycation • Glyoxal • Methylglyoxal • Advanced glycation end products • Maillard adducts • Carbonyl stress • Parkinson • Diabetes • Protein repair • Acrylamide

---

M. Mihoub • G. Richarme (✉)
Stress Molecules, Institut Jacques Monod, Université Paris 7, UMR 7592,
15 rue Hélène Brion, 75013 Paris, France
e-mail: mouadhmihoub@yahoo.fr; richarme@paris7.jussieu.fr

J. Abdallah
Stress Molecules, Institut Jacques Monod, Université Paris 7, UMR 7592,
15 rue Hélène Brion, 75013 Paris, France

School of Pharmacy, Lebanese American University, Byblos, Lebanon
e-mail: jabdallah@lau.edu.lb

H. Ariga, S.M.M. Iguchi-Ariga (eds.), *DJ-1/PARK7 Protein*,
Advances in Experimental Medicine and Biology 1037,
https://doi.org/10.1007/978-981-10-6583-5_9

9.1 Introduction

The Parkinsonism-associated protein DJ-1/Park7 functions as a chaperone for synuclein, covalent chaperone for the thiol proteome, protease, glyoxalase, Nrf2 stabilizer, apoptosis inhibitor and translational regulator (reviewed in Wilson 2011). Our recent work shows that DJ-1 and its prokaryotic homologs function as deglycases that repair amino acids and proteins that have been glycated by glyoxals (methylglyoxal [CH_3-CO-CHO] and glyoxal [CHO-CHO]).

Glycation is initiated by a non-enzymatic reaction between carbonyl and amino groups, a reaction discovered by the French chemist Louis Camille Maillard in 1912 (Maillard 1912), and an increase in the steady-state level of reactive carbonyls results in the phenomenon called carbonyl stress. The major glycating agents are reducing sugars (glucose, fructose, ribose and their phosphorylated derivatives), and glyoxals, which arise as glycolytic by-products and are responsible for approximately 65% of cellular glycation events (Thornalley 2008). The condensation reaction between carbonyl groups and amino acids begins with the rapid formation of a hemithioacetal with cysteine and formation of aminocarbinols with arginine or lysine (Fig. 9.1), after which a series of dehydrations, oxidations and rearrangements result in a myriad of products, including Schiff bases, Amadori products, advanced glycation end products (AGEs) and protein cross-links (Thornalley 2008). Because glycation results in protein damage, it is responsible for ageing, cancer, atherosclerosis, cataracts and many post-diabetic, neurovegetatives and renal and autoimmune diseases.

Similar glycation events, activated by temperature, occur during food processing, cooking, sterilization and storage. Dietary AGEs are detrimental because they accelerate oxidative stress and inflammation, especially in patients with diabetes or renal failure (Poulsen et al. 2013, Tessier and Birlouez-Aragon 2012). Moreover, a Maillard reaction that occurs between free asparagine and sugars or glyoxals is responsible for acrylamide formation in heated food, which is a matter of concern because acrylamide is carcinogenic, neurotoxic and reprotoxic.

Carbonyl scavengers and deglycases protect cells against glycation. Carbonyl scavengers include aldo-keto reductases which reduce carbonyls into alcohols, glyoxalases, which degrade glyoxals into acid alcohols, and detoxification systems, which export electrophile-glutathione conjugates (Thornalley 2008). Deglycases include fructosamine-3-kinases (FN3Ks) which detoxify lysine-fructosamine, the Amadori product resulting from the glycation of lysine by glucose (van Schaftingen et al. 2012), and the DJ-1 family deglycases, which repair methylglyoxal- and glyoxal-glycated amino acids and proteins (Richarme et al. 2015).

In addition to performing protein repair, DJ-1 family deglycases prevent acrylamide formation in asparagine/sugar and asparagine/glyoxal mixtures, whose components are responsible for most acrylamide formation in food (Richarme et al. 2016).

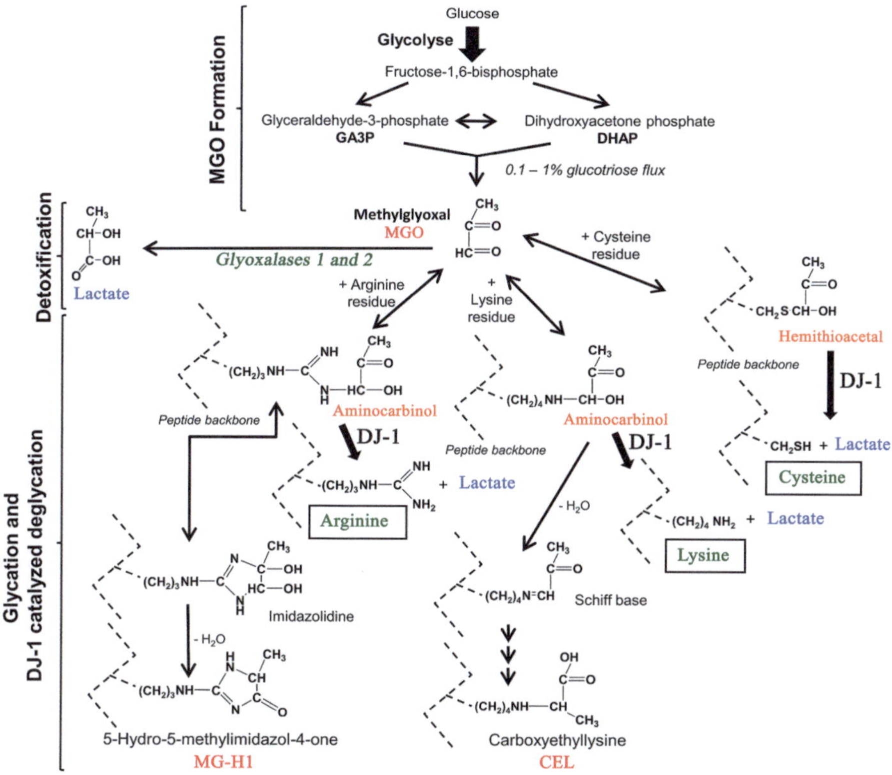

Fig. 9.1 Methylglyoxal metabolism, protein glycation by methylglyoxal and deglycation by DJ-1. Methylglyoxal (MGO) forms spontaneously as a glycolytic by-product and is detoxified by the sequential action of glyoxalases 1 and 2. It forms covalent Maillard adducts with cysteines, arginines and lysines (hemithioacetals with cysteines and aminocarbinols with arginines and lysines) and also forms imidazolidines with the arginine guanidino group. Hemithioacetals, aminocarbinols and imidazolidines are transformed into intermediate glycation products (such as Schiff bases) and advanced glycation end products (AGEs), including MG-H1 and CEL. DJ-1 transforms hemithioacetal (via a thioester) into cysteine and lactate and aminocarbinols (via an amide) into lysine/arginine and lactate and in this way prevents the formation of Schiff bases and AGEs. The deglycation-dependent formation of lactate is responsible for the pseudo-glyoxalase activity of DJ-1 (see below). Similar reactions occur with glyoxal, which result in the formation of the advanced glycation end products G-H1 and CML (carboxymethyllysine) in the absence of DJ-1 (From Richarme et al. 2015)

9.2 Amino Acid and Protein Deglycation

9.2.1 Cysteine, Arginine and Lysine Deglycation

Cysteine, arginine and lysine are the three main glycated amino acids in proteins (we used their αN-acetyl forms to restrict glycation to the lateral chains). Deglycation of amino acids occurs as follows (Richarme et al. 2015):

Cysteine

NacCys-SH + CHO-CO-CH_3 → NacCys-S-CHOH-CO-CH_3 (spontaneous hemithioacetal formation)

NacCys-S-CHOH-CO-CH_3 → NacCys-S-CO-CHOH-CH_3 (H migration catalysed by DJ-1)

NacCys-S-CO-CHOH-CH_3 → NacCys-SH + COOH-CHOH-CH_3 (thioester hydrolysis catalysed by DJ-1)

DJ-1 degrades the hemithioacetal formed between N-acetylcysteine and methylglyoxal into N-acetylcysteine and lactate (Figs. 9.1 and 9.2).

Arginine and Lysine

NacArg/Lys-NH_2 + CHO-CO-CH_3 → NacArg/Lys-NH-CHOH-CO-CH_3 (spontaneous aminocarbinol formation)

NacArg/Lys-NH-CHOH-CO-CH_3 → NacArg/Lys-NH-CO-CHOH-CH_3 (H migration catalysed by DJ-1)

NacArg/Lys-NH-CO-CHOH-CH_3 → NacArg/Lys-NH_2 + COOH-CHOH-CH_3 (amidolysis catalysed by DJ-1)

DJ-1 degrades the aminocarbinol formed between N-acetylarginine or N-acetyllysine and methylglyoxal into N-acetylarginine or N-acetyllysine and lactate (Fig. 9.1). In the absence of deglycase, the aminocarbinol formed upon lysine glycation (Lys-NH-CHOH-CO-CH_3) dehydrates into a Schiff base (Lys-N = CH-CO-CH_3), which is finally converted into the AGE carboxyethyllysine (Lys-NH-CH(CH_3)-COOH; CEL) (Fig. 9.1). If the carbonyl of the aminocarbinol reacts with another lysine, a methylglyoxal lysine dimer is formed (MOLD), which may constitute a protease-resistant covalent cross-link (not shown). In the absence of deglycase, the aminocarbinol formed upon arginine glycation reacts via its

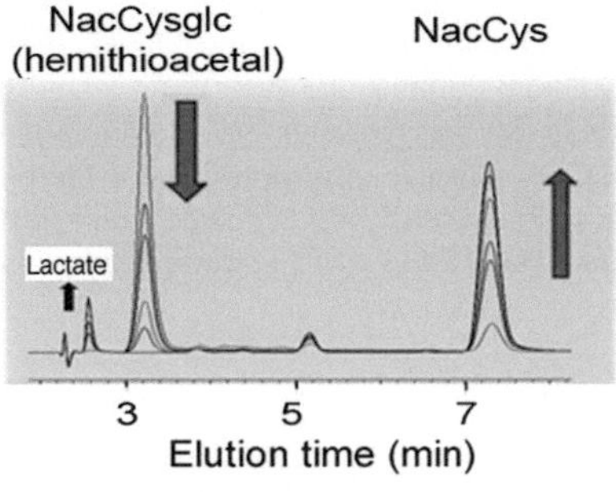

Fig. 9.2 Deglycation of glycated NacCys by DJ-1. NacCys and MGO (2 mM each) were incubated for 2 min, leading to the formation of glycated NacCys (NacCysglc), after which 4 μM DJ-1 was added to the glycation mixture. The amounts of NacCys and glycated NacCys were analysed by RP-HPLC at various time points after the addition of DJ-1. Arrows indicate the decrease in glycated NacCys and increase in NacCys after addition of DJ-1. Degradation of glycated NacCys results in lactate formation (From Richarme et al. 2015)

residual carbonyl with the imine function of arginine to form an imidazolidine, which then dehydrates into the advanced glycation end product MG-H1(Fig. 9.1). Thus, by deglycating early Maillard adducts (hemithioacetals and aminocarbinols), DJ-1 prevents formation of advanced glycation end products (AGEs).

9.2.2 *Protein Deglycation*

DJ-1 deglycated all proteins tested, including bovine serum albumin (BSA), glyceraldehyde-3-phosphate dehydrogenase (GAPDH), fructose-bisphosphate aldolase and aspartate transaminase (Richarme et al. 2015).

DJ-1 repaired cysteines, arginines and lysines in bovine serum albumin (BSA). After treatment of BSA with methylglyoxal (MGO), the number of titratable SH groups decreased from 0.62 to 0.28, and DJ-1 restored SH group levels to a value of 0.53 in less than 5 min (Fig. 9.3a). In native BSA or BSA incubated for 2 h with MGO and DJ-1, 19 arginines were titrated by phenanthrenequinone versus 12 arginines in BSA incubated with MGO alone (Fig. 9.3b). Glycation of BSA lysines was followed by measurement of absorbance at 330 nm, which reflects Schiff base formation. The absorption spectrum of BSA incubated with MGO for 3 h displayed a shoulder between 300 and 370 nm, but the shoulder was considerably lower when DJ-1 was present in the BSA/MGO mixture, reflecting lysine repair (Fig. 9.3c).

Glycation of fructose-bisphosphate aldolase by glyoxal (GO) was analysed by mass spectrometry, AGE detection and enzyme activity. When aldolase was glycated by GO in the absence or presence of DJ-1, DJ-1 prevented the glycation of the residues R22, K28, K42, R43, K111, C202, K208, K230, K243, R259, C290, K318, K330, R331, C339 and K342 (Fig. 9.4a, d). Glycation ratios in the absence or presence of DJ-1 ranged from 1.9 for C290 and 2.6 for K230 (two buried residues) to 330 for K243 and 980 for K42. The median glycation ratio was 25. K42, R43 and K230 are located in the active site, with K42 and R43 being involved in substrate binding (R43 mutation results in 14-fold decrease in activity) and K230 forming a Schiff base with the substrate (K230 mutation results in complete loss of activity). DJ-1 also abolished the reactivity of aldolase towards anti-AGE antibodies (Fig. 9.4b) and prevented its inactivation by glyoxal (Fig. 9.4c). Thus DJ-1 efficiently prevents the glycation of cysteines, arginines and lysines in proteins.

9.2.3 *DJ-1 and Its Prokaryotic Homologs Prevent Protein Glycation in Cell Culture and In Vivo*

Glycation is responsible for skin dysfunction and ageing. Accumulation of AGEs has been detected in fibroblasts and keratinocytes and is increased in aged skin, UV-irradiated skin and diabetes (Pageon et al. 2015). In HaCaT keratinocytes, DJ-1

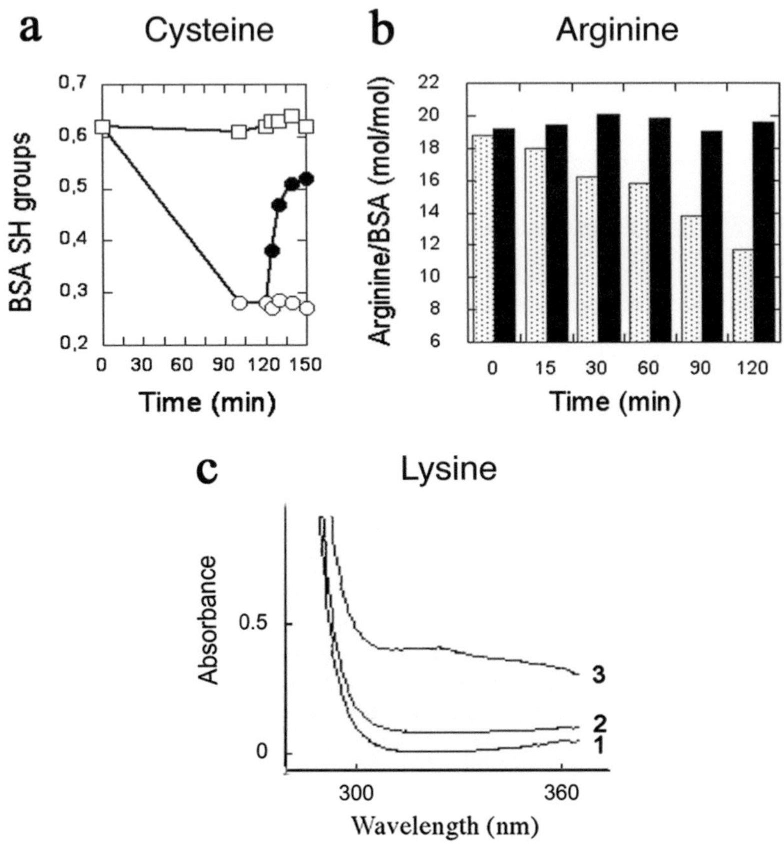

Fig. 9.3 Serum albumin deglycation. (**a**) Cysteine deglycation. BSA was incubated with MGO over a period of 120 min, separated from MGO by gel permeation chromatography and incubated in the absence (empty circles) or presence (filled circles) of 3 μM DJ-1. Squares represent BSA SH groups without MGO treatment. (**b**) Arginine deglycation. BSA was incubated with MGO in the absence (grey columns) or presence of DJ-1 (*black* columns), and free arginines were titrated at various times. (**c**) Lysine deglycation. BSA was incubated with MGO in the absence (curve 3) or presence of DJ-1 (curve 2), and absorption spectra were recorded; the spectrum of BSA without MGO treatment is displayed in curve 1 (From Richarme et al. 2015)

depletion by siRNA treatment led to a significant increase in protein glycation (up to 3.2-fold), suggesting that DJ-1 is important for preventing protein glycation in eukaryotic cells (Fig. 9.5a; Advedissian et al. 2016).

Escherichia coli contains three DJ-1 homologs, Hsp31 (the *hchA* gene product), YhbO and YajL (deglycases 1, 2 and 3, respectively) (Mihoub et al. 2015; Abdallah et al. 2016), which were previously characterized as stress-resistance proteins (Malki et al. 2005; Abdallah et al. 2007; Kthiri et al. 2010). The glycation levels of *hchA*, *yhbO*, *yajL*, *yhbOyajL* and *yhbOyajLhchA* mutants were 2-, 1.5-, 3-, 10- and tenfold higher, respectively, than that of the wild-type strain (Fig. 9.5b; not shown).

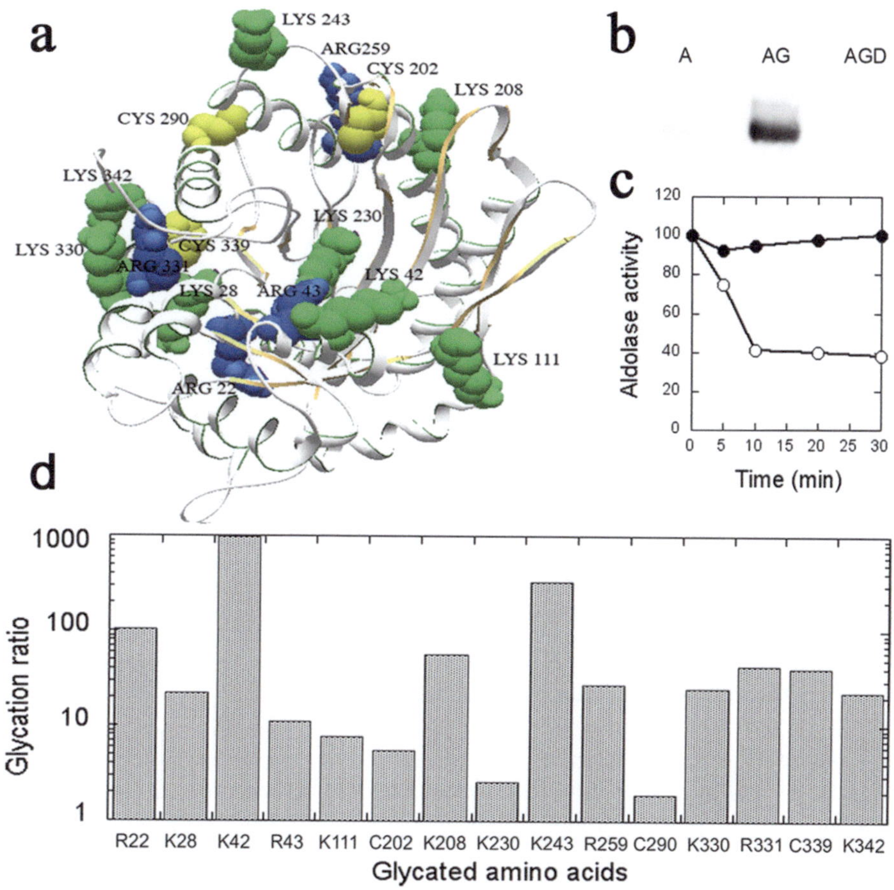

Fig. 9.4 Aldolase deglycation. (**a**) 3D structure of fructose bisphosphate aldolase, displaying cysteines (yellow), lysines (green) and arginines (blue) repaired by DJ-1 from their glycation by 5 mM glyoxal (**b**) Glycation of aldolase assayed by immunoblotting with anti-AGE antibodies: aldolase (A), aldolase glycated by GO in the absence (AG) or presence (AGD) of DJ-1. (**c**) Aldolase activity after incubation with glyoxal, in the absence (empty circles) or presence of DJ-1 in the glycation mixture (filled circles). (**d**) Mass spectrometry analysis of aldolase deglycation by DJ-1. Aldolase was glycated for 1 h by 5 mM glyoxal in the absence or presence of 5 μM DJ-1, separated from glyoxal by gel permeation chromatography and analysed by mass spectrometry. The glycation ratio is the ratio between glycation levels in the absence or presence of DJ-1 (From Richarme et al. 2015)

Thus, despite the presence of active glyoxal scavengers (glyoxalases, aldo-keto reductases, glutathione S-transferases and multidrug exporters), deglycases clearly play a major role in preventing protein glycation in vivo. This refutes the assertion that deglycases play only a minor role in vivo compared to glyoxalases (Rabbani and Thornalley 2015).

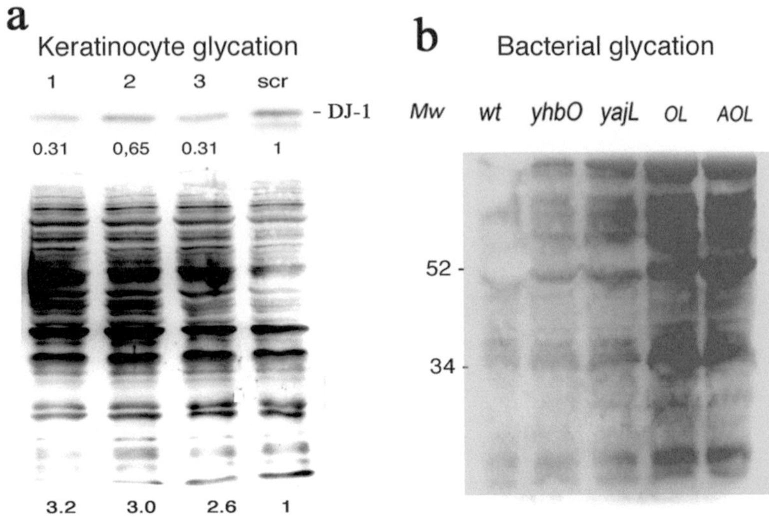

Fig. 9.5 Increased protein glycation levels in deglycase-depleted cells. (**a**) Protein extracts from keratinocytes treated with DJ-1 siRNA1, siRNA2, siRNA3 or scrambled RNA for 48 h were probed with anti-DJ-1 and anti-AGEs antibodies. DJ-1 and protein glycation levels are indicated below each lane. (**b**) Lysates from the wild type, *yajL*, *yhbO*, *yhbOyajL (OL)* and *hchAyhbOyajL (AOL)* strains, grown overnight in LB medium containing 0.6% glucose, were probed with anti-AGE antibodies (**a** from Advedissian et al. 2016; **b** from Abdallah et al. 2016)

Conversely, overexpression of DJ-1 in keratinocytes decreased protein glycation (Fig. 9.6a), and overexpression of YhbO in bacterial cells decreased the glycation level of an overexpressed protein (the kinase YeaG) by 2.5-fold (Fig. 9.6b).

9.2.4 The Apparent Glyoxalase Activity of the DJ-1 Family Deglycases Reflects Their Deglycase Activity

DJ-1 was reported to function as a glutathione-independent glyoxalase displaying a 1000-fold lower activity than the bona fide glutathione-dependent glyoxalases Glo1 and Glo2 (Lee et al. 2012). The apparent glyoxalase activity of DJ-1, however, reflects its deglycase activity. First, the kinetics of MGO degradation at micromolar DJ-1 concentrations displayed a lag, which is required for the spontaneous formation of its substrate, glycated DJ-1 (Richarme et al. 2015). Second, the apparent glyoxalase activity of DJ-1 increased with the square of DJ-1 concentration, in accordance with DJ-1 being both an enzyme (deglycase) and a substrate (glycated DJ-1). Third, the negligible level of apparent glyoxalase activity of DJ-1 at low concentrations was stimulated by BSA up to 40-fold (BSA is a substrate for spontaneous glycation and DJ-1-catalysed deglycation) with an apparent Ka of approximately 5 μM BSA (Richarme et al. 2015). These results are consistent with DJ-1

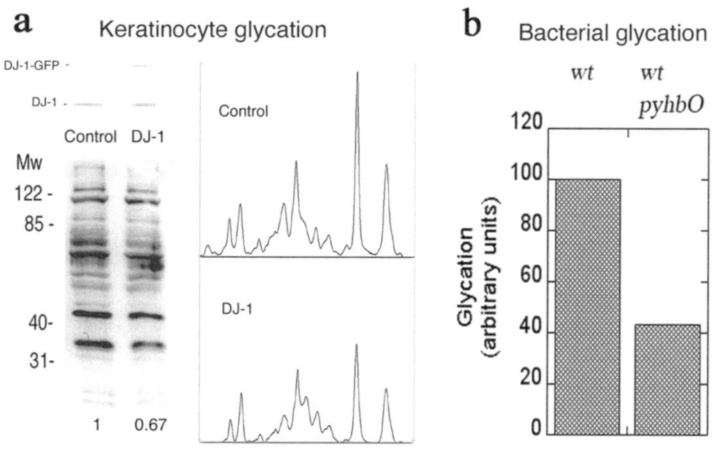

Fig. 9.6 Decreased protein glycation levels in deglycase overproducing cells. (**a**) Exponentially growing HaCaT keratinocytes were transfected with the DJ-1/Park7 plasmid or with the empty vector. Protein extracts were probed with anti-DJ-1 and anti-AGE antibodies at 48 h after transfection. Protein glycation levels are indicated below each lane. The densitometric scan of the distribution of glycated proteins is shown in the right panel. Since the transfection efficiency was 50%, the 33% decrease of glycation represents a 66% decrease in transfected cells. (**b**) Protein kinase YeaG was overexpressed in the *E. coli* BL21 strain containing the pET-21a-*yeaG* plasmid alone or with the co-expressed plasmid pBAD33-*yhbO*. Protein extracts were analysed for AGE content by recording fluorescence at 440 nm ($\lambda_{ex} = 340$ nm) (**a**, from Advedissian et al. 2016; **b** from Abdallah et al. 2016)

substrates being glycated proteins instead of glyoxals and with DJ-1 being a deglycase rather than a glyoxalase. We suggest that, like DJ-1, Hsp31, YhbO and YajL, glutathione-independent glyoxalases from yeast (reviewed in Wilson 2014) also function as deglycases.

9.3 Prevention of Acrylamide Formation in Food

Glycation reactions similar to those observed in living cells also occur during food processing, cooking, sterilization and storage, and are responsible for the rapid formation of Maillard adducts and dietary AGEs which are detrimental because they accelerate oxidative stress and inflammation, especially in patients with diabetes or renal failure (Poulsen et al. 2013; Tessier and Birlouez-Aragon 2012).

In addition to these dietary AGEs, acrylamide has been detected in many foods (Stadler et al. 2002; Mottram et al. 2002), including bakery products, chips, grilled cereals, instant coffees, evaporated milk and baby food (Erkekoglu and Baydar 2010; Tessier and Birlouez-Aragon 2012). Mottram and Stadler reported that acrylamide was formed by a Maillard reaction between reducing sugars or glyoxals and free asparagine, followed by the Strecker degradation of asparagine into acrylamide

(Stadler et al. 2002; Mottram et al. 2002). DJ-1 family deglycases prevent acrylamide formation in mixtures containing asparagine and glucose, fructose or glyoxals, likely by degrading the Maillard adducts responsible for its formation (Richarme et al. 2016). Moreover, PfpI from *Pyrococcus furiosus*, which belongs to the PfpI/Hsp31/DJ-1 superfamily and was characterized as a peptidase (Halio et al. 1997), functions as a deglycase that prevents acrylamide formation at 100 °C (Richarme et al. 2016).

9.3.1 Mechanism of Acrylamide Formation and Prevention of Its Formation by DJ-1 Family Deglycases

The proposed mechanism of acrylamide formation is as follows (Stadler 2002):

NH_2-CO-CH_2-CH-(COOH)-NH_2 + CHO-CO-CH_3 (asparagine + methylglyoxal)
NH_2-CO-CH_2-CH-(COOH)-NH-CHOH-CO-CH_3 (condensation into aminocarbinol)
NH_2-CO-CH_2-CH-(COOH)-N = CH-CO-CH_3 (Schiff base formation)
NH_2-CO-CH_2-CH_2-N = CH-CO-CH_3 (decarboxylation of the Schiff base)
NH_2-CO-CH_2-CH_2-NH_2 + CHO-CO-CH_3 (hydrolysis into 3-aminopropionamide and methylglyoxal)
NH_2-CO-CH = CH_2 + CHO-CO-CH_3 (3-aminopropionamide deamination into acrylamide)

The spontaneous condensation of asparagine and methylglyoxal leads to the formation of an aminocarbinol, which then dehydrates to a Schiff base. The Schiff base activates decarboxylation of its asparagine moiety, and its hydrolysis produces methylglyoxal and 3-aminopropionamide, which is deaminated to form acrylamide.

The proposed pathways for the prevention of acrylamide formation by DJ-1 family deglycases are as follows (Richarme et al. 2016):

NH_2-CO-CH_2-CH-(COOH)-NH_2 + CHO-CO-CH_3 (asparagine + methylglyoxal)
NH_2-CO-CH_2-CH-(COOH)-NH-CHOH-CO-CH_3 (condensation into aminocarbinol)
NH_2-CO-CH_2-CH-(COOH)-NH-CO-CHOH-CH_3 (amide formation by catalysed H migration)
NH_2-CO-CH_2-CH-(COOH)-NH_2 + COOH-CHOH-CH_3 (catalysed amidolysis to asparagine + lactate)

We propose a sequence of reactions similar to those involved in lysine/arginine deglycation by DJ-1 (Richarme et al. 2015). The spontaneous condensation of asparagine with methylglyoxal leads to the formation of an aminocarbinol. The deglycase then converts the aminocarbinol into an amide and subsequently cleaves the amide into asparagine and lactate in a reaction reminiscent of its peptidase activity. Similar reactions would occur with glyoxal (CHO-CHO), leading to the formation of asparagine and glycolate (COOH-CH_2OH).

9.3.2 Deglycases Prevent Acrylamide Formation in Asparagine/Glyoxal Mixtures

Methylglyoxal and glyoxal form quickly from fructose and glucose upon heating, especially in the presence of additives such as the baking agent NH_4HCO_3, and they react much more efficiently than fructose or glucose with asparagine (Amrein et al. 2006).

We incubated asparagine and glyoxal for 30 min at 55 °C in the presence of DJ-1 (5 μM), at 70 °C in the presence of YhbO (5 μM) or at 95 °C in the presence of PfpI (5 μM) and measured acrylamide formation by analysing the mixtures on a C18 reverse-phase HPLC column or on a Hypercarb hydrophilic interaction liquid chromatography (HILIC) column. Acrylamide formation from the asparagine/glyoxal mixture was reduced by 78%, 82% and 98% by YhbO, PfpI and DJ-1, respectively (Fig. 9.7, not shown).

9.3.3 Deglycases Prevent Acrylamide Formation in Asparagine/Fructose and Asparagine/Glucose Mixtures

Glucose and fructose by themselves react poorly with asparagine: 370-fold less acrylamide was formed in asparagine/glucose mixtures than in asparagine/glyoxal mixtures (Amrein et al. 2006). However, if asparagine/glucose mixtures contained 1% NH_4HCO_3, acrylamide formation increased by 45-fold because glyoxal formation from glucose increased by a similar factor. Moreover, acrylamide formation in asparagine/glucose mixtures was reduced by 75% by orthophenylenediamine, a complexant of glyoxals, suggesting that 75% of acrylamide formed in asparagine/

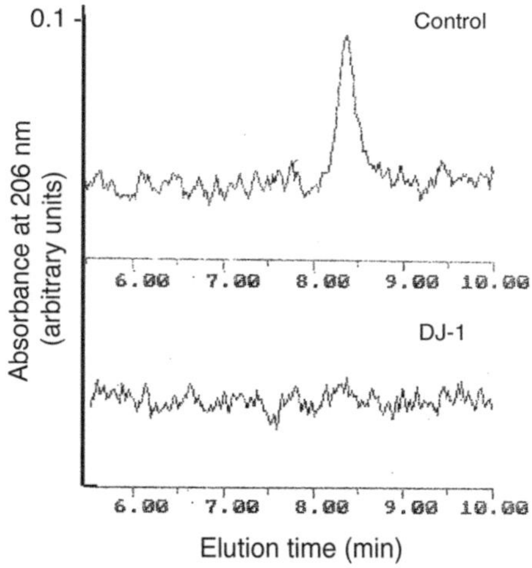

Fig. 9.7 Deglycases prevent acrylamide formation. Asparagine and glyoxal (6 mM each) were incubated for 30 min at 55 °C in the absence or presence of 5 μM DJ-1 as indicated, and the mixtures were analysed by hydrophilic interaction chromatography on a Hypercarb column (From Richarme et al. 2016)

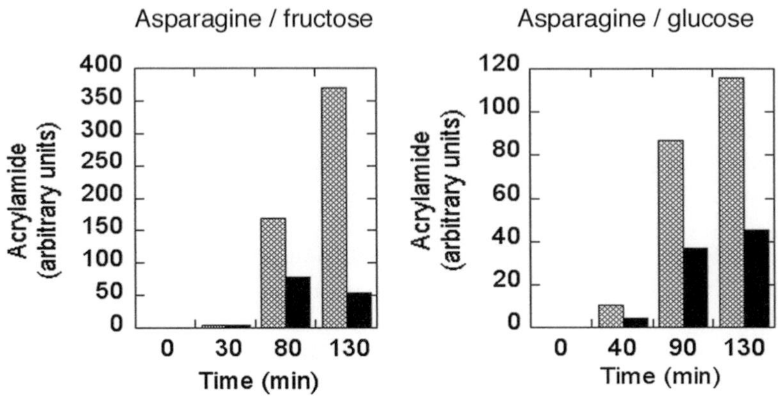

Fig. 9.8 Deglycases prevent acrylamide formation. Asparagine and fructose or glucose (25 mM each) were incubated for 30 min at 95 °C in the absence (grey columns) or presence (black columns) of 5 μM PfpI, and the mixtures were analysed by hydrophilic interaction chromatography on a Hypercarb column (135 μM acrylamide was formed in 130 min in the control asparagine/fructose solution) (From Richarme et al. 2016)

glucose mixtures resulted from the reaction of asparagine with glyoxals (Yuan et al. 2008). For this reason, the DJ-1 family deglycases whose substrates are glyoxal-containing Maillard adducts prevent acrylamide formation in asparagine/sugar mixtures (Fig. 9.8).

A scheme describing acrylamide formation from asparagine, and the role of Maillard deglycases in its prevention, is presented (Fig. 9.9).

9.4 Conclusion

The Maillard reaction, discovered by Louis Camille Maillard in 1912, is responsible for the formation of covalent adducts between reactive carbonyls and thiol and amino groups of amino acids and proteins. By degrading Maillard adducts formed by glyoxals, which are responsible for approximately 65% of glycation damage, DJ-1 deglycases prevent the formation of AGEs in proteins. Since glycation is responsible for ageing and for glycation-related diseases, these deglycases are likely to play an important role in their prevention.

The deglycase activity of DJ-1 suggests that DJ-1-associated Parkinsonism results from excessive protein glycation, which affects dopaminergic transmission in the *substantia nigra*, and strengthens the role of glycation in the etiology of neurovegetative diseases.

Since diabetes and post-diabetic diseases are characterized by increased levels of glucose and glyoxals, DJ-1 deglycases should be important players in alleviating atherosclerosis and hypertension (glycation is responsible for arterial stiffening), cardiomyopathy (glycation results in contractile dysfunction), nephropathy

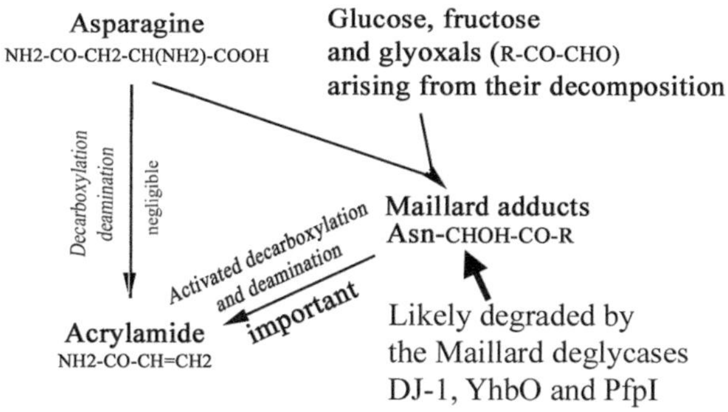

Fig. 9.9 Acrylamide formation from asparagine and the role of Maillard deglycases in its prevention. Asparagine alone could in principle afford acrylamide by direct decarboxylation and deamination, but the reaction is inefficient with extremely low yields. However, the Maillard adducts between asparagine and reducing sugars or glyoxals furnish acrylamide in the range of 0.1–1 mol% in model systems. The Maillard deglycases of the DJ-1 family prevent acrylamide formation, likely by degrading Maillard adducts responsible for its formation (From Richarme et al. 2016)

(glycation is responsible for glomerulosis), cataracts (glycation results in lens protein aggregation) and retinopathy (glycation is a cause of vascular endothelial growth factor upregulation and proliferative retinopathy) (reviewed in Singh et al. 2014).

The ability of DJ-1 deglycases to repair proteins is important for their role in cancer. Since glycolytic flux and glyoxals levels are increased in cancer cells (Warburg effect), DJ-1 deglycases are likely involved in preventing increased protein glycation damage in cancer cells. Moreover, the high levels of glyoxals in cancer cells (Thornalley 2008) are likely responsible for the overexpression of DJ-1 in most of these cells (Ariga 2015).

Glycation is also responsible for the formation of AGEs in food products. Because glycation is accelerated at high temperatures (60–200 °C), the use of thermoresistant deglycases, such as PfpI from *P. furiosus*, has potential utility for preventing glycation in food. Moreover, the formation of acrylamide in food is dependent on a Maillard reaction between free asparagine and reducing sugars or glyoxals, which can be prevented by the activity of DJ-1 deglycases. Consequently, many foods whose heating processes occur at around 100 °C, such as baby foods, evaporated milk, dry soups and prune juices, could be clients of the deglycase method to prevent acrylamide formation, and it should be possible to raise its operating temperature by protein engineering or immobilization on heterofunctional supports (Mateo et al. 2010).

Thus, more than 100 years after Maillard's discovery (Maillard 1912), the DJ-1 family deglycases (Richarme et al. 2015; Mihoub et al. 2015; Abdallah et al. 2016; Richarme et al. 2016) constitute a major biological solution to the glycation problem and the fight against carbonyl stress. Since Maillard adducts are the substrate of these deglycases, we propose renaming them Maillard deglycases.

References

Abdallah J, Caldas T, Kthiri F, Kern R, Richarme G (2007) YhbO protects cells against multiple stresses. J Bacteriol 189:9140–9144

Abdallah J, Mihoub M, Gautier V, Richarme G (2016) The DJ-1 superfamily members YhbO and YajL from *Escherichia coli* repair proteins from glycation by methylglyoxal and glyoxal. Biochem Biophys Res Commun 470:282–286

Advedissian T, Deshayes F, Poirier F, Viguier M, Richarme G (2016) The Parkinsonism-associated protein DJ-1/Park7 prevents glycation damage in human keratinocyte. Biochem Biophys Res Commun 473:87–91

Amrein TM, Andres L, Manzardo GG, Amado RJ (2006) Investigations on the promoting effect of ammonium hydrogen carbonate on the formation of acrylamide in model systems. Agric Food Chem 54:10253–10261

Ariga H (2015) Common mechanisms of onset of cancer and neurodegenarative diseases. Biol Pharm Bull 38:795–808

Erkekoğlu P, Baydar T (2010) Toxicity of acrylamide and evaluation of its exposure in baby foods. Nutrition Res Rev 23:323–333

Halio SB, Bauer MW, Mukund S, Adams M, Kelly RM (1997) Purification and characterization of two functional forms of intracellular protease PfpI from the hyperthermophilic archaeon *Pyrococcus furiosus*. Appl Environ Micr 63:289–295

Kthiri F, Le HT, Gautier V, Caldas T, Malki A, Landoulsi A, Bohn C, Bouloc P, Richarme G (2010) Protein aggregation in a mutant deficient in yajL, the bacterial homolog of the Parkinsonism-associated protein DJ-1. J Biol Chem 285:10328–10336

Lee JY, Song J, Kwon K, Jang S, Kim C, Baek K, Kim J, Park JC (2012) Human DJ-1 and its homologs are novel glyoxalases. Hum Mol Genetics 21:3215–3225

Maillard LC (1912) Action des acides aminées sur les sucres: formation des mélanoïdines par voie métabolique. CR Hebd Séances. Acad Sci 2:66–68

Malki A, Caldas T, Abdallah J, Kern R, Eckey V, Kim SJ, Cha SS, Mori H, Richarme G (2005) Peptidase activity of the Escherichia coli Hsp31 chaperone. J Biol Chem 280:14420–14426

Mateo C, Bolivar JM, Godoy CA, Rocha-Martin J, Pessela BC, Curiel JA, Muñoz R, Guisan JM, Fernández-Lorente G (2010) Improvement of enzyme properties with a two-step immobilization process on novel heterofunctional supports. Biomacromolecules 11:3112–3117

Mihoub M, Abdallah J, Gontero B, Dairou J, Richarme G (2015) The DJ-1 superfamily member Hsp31 repairs proteins from glycation by methylglyoxal and glyoxal. Biochem Biophys Res Commun 463:1305–1310

Mottram DS, Wedzicha BL, Dodson AT (2002) Acrylamide is formed in the Maillard reaction. Nature 419:448–449

Pageon H, Zucchi H, Rousset F, Monnier VM, Asselineau D (2015) The reconstructed skin model as a new tool for investigating in vitro dermal fillers: increased fibroblast activity by hyaluronic acid. Eur J Dermatol 25:312–322

Poulsen MW, Hedegaard RV, Andersen JM, de Courten B, Bügel S, Nielsen J, Skibsted LH, Dragsted LO (2013) Advanced glycation endproducts in food and their effects on health. Food Chem Toxicol 60:10–37

Rabbani N, Thornalley PJ (2015) Dicarbonyl stress in cell and tissue dysfunction. Biochem Biophys Res Commun 458:221–226

Richarme G, Mihoub M, Dairou J, Bui LC, Leger T, Lamouri A (2015) Parkinsonism-associated protein DJ-1/Park7 is a major protein deglycase that repairs methylglyoxal- and glyoxal-glycated cysteine, arginine, and lysine residues. J Biol Chem 290:1885–1897

Richarme G, Marguet E, Forterre P, Ishino S, Ishino Y (2016) DJ-1 family Maillard deglycases prevent acrylamide formation. Biochem Biophys Res Commun 478:1111–1116

Singh VP, Bali A, Singh N, Jaggi AS (2014) Advanced glycation end products and diabetic complications. Korean J Physiol Pharmacol 18(1):14

Stadler RH, Blank I, Varga N, Robert F, Hau J, Guy PA, Robert MC, Riediker S (2002) Acrylamide from Maillard reaction products. Nature 419:449–450

Tessier FJ, Birlouez-Aragon I (2012) Health effects of dietary Maillard reaction products: the results of ICARE and other studies. Amino Acids 42:1119–1131

Thornalley PJ (2008) Protein and nucleotide damage by glyoxal and methylglyoxal in physiological systems – role in ageing and disease. Drug Metabol Drug Interact 23:125–150

Van Schaftingen E, Collard F, Wiame E, Veiga-da-Cunha M (2012) Enzymatic repair of Amadori products. Amino Acids 42:1143–1150

Wilson MA (2011) The role of cysteine oxidation in DJ-1 function and dysfunction. Antioxid Redox Signal 15:111–122

Wilson MA (2014) Metabolic role for yeast DJ-1 superfamily proteins. Proc Natl Acad Sci U S A 111:6858–6859

Yuan Y, Zhao G, Chen F, Liu J, Wu J, Hu X (2008) Correlation of methylglyoxal with acrylamide formation in fructose/asparagine Maillard reaction model system. Food Chem 108:885–890

Chapter 10
DJ-1 as a Biomarker of Parkinson's Disease

Yoshiro Saito

Abstract Parkinson's disease is a progressive, age-related, neurodegenerative disorder, and oxidative stress is an important mediator in its pathogenesis. *DJ-1* has been identified as a causative gene of a familial form of Parkinson's disease, *PARK7*, and plays a significant role in antioxidative defense, protecting cells from oxidative stress. A cysteine residue of DJ-1 at position 106 (Cys-106) is preferentially oxidized under oxidative stress. This reactive Cys-106 plays a critical role in the biological function of DJ-1, which could act as a sensor of oxidative stress by regulating antioxidative defense depending on Cys-106 oxidation. Thus, the levels of Cys-106-oxidized DJ-1 (oxDJ-1) could be a possible biomarker of oxidative stress. This chapter focuses on the properties of DJ-1 and oxDJ-1 levels as a biomarker of Parkinson's disease. In particular, the usability of these biomarkers to prevent and treat this neurodegenerative disease is discussed. Further, this section deals with the importance of identifying a biomarker of early-phase Parkinson's disease. Finally, this chapter summarizes the features of oxDJ-1 levels in the brain and blood as a biomarker candidate for early-phase Parkinson's disease based on our results using oxDJ-1-specific antibodies.

Keywords Oxidized DJ-1 • Parkinson's disease • Biomarker • Oxidative stress • Reactive oxygen species • Dopamine • Cysteine oxidation • Antioxidative defense

10.1 Introduction

Parkinson's disease (PD) is a progressive, age-related, neurodegenerative disorder with worldwide prevalence. PD is characterized by bradykinesia, rigidity, tremors, and gait dysfunction with postural instability (Hughes et al. 1992). The pathological hallmark of PD is the degeneration of dopamine neurons in the substantia nigra (SN) of the midbrain and the subsequent depletion of striatal dopamine (Forno 1996). Although the etiology of PD is not fully elucidated, increasing evidence

Y. Saito (✉)
Department of Medical Life Systems, Faculty of Life and Medical Sciences, Doshisha University, 1-3 Miyakodani, Tatara, Kyotanabe, Kyoto 610-0394, Japan
e-mail: ysaito@mail.doshisha.ac.jp

H. Ariga, S.M.M. Iguchi-Ariga (eds.), *DJ-1/PARK7 Protein*,
Advances in Experimental Medicine and Biology 1037,
https://doi.org/10.1007/978-981-10-6583-5_10

indicates that oxidative stress—defined as an imbalance between oxidants and antioxidants resulting in increased levels of oxidants—is an important mediator in its pathogenesis (Jenner 2003; Sayre et al. 2008; Dias et al. 2013). Because reactive oxygen species (ROS) are remarkably generated in the enzymatic and nonenzymatic metabolism of dopamine, nigral dopaminergic neurons are rich in ROS and are vulnerable to oxidative stress compared with other neurons (Lotharius and Brundin 2002).

DJ-1 has been identified as the product of a causative gene of a familial form of PD, *PARK7*, and plays a significant role in antioxidative defense to protect the cells from oxidative stress (Bonifati et al. 2003; Taira et al. 2004). DJ-1 undergoes preferential oxidation at the cysteine residue at position 106 (Cys-106) under oxidative stress (Kinumi et al. 2004). A critical role of reactive Cys-106 in its biological function has been demonstrated, and DJ-1 might act as a sensor of oxidative stress by regulating the gene expression of antioxidative defense depending on Cys-106 oxidation (Kahle et al. 2009; Ariga et al. 2013). Several studies have reported that DJ-1 and Cys-106-oxidized DJ-1 (oxDJ-1) levels may be used as a biomarker of oxidative stress-related diseases including PD (Kahle et al. 2009; Shi et al. 2010; Parnetti et al. 2013; Saito 2014). This chapter focuses on the properties of DJ-1 and oxDJ-1 levels as a biomarker of PD and, in particular, the usability of oxDJ-1 in the care and prevention of this neurodegenerative disease. This chapter further discusses the features of oxDJ-1 levels in the brain and blood as a biomarker candidate for early-phase PD based on our results using oxDJ-1-specific antibodies.

10.2 Parkinson's Disease and Oxidative Stress

The degradation of dopamine neurons in the SN of the midbrain and the subsequent depletion of striatal dopamine are primary causes of PD symptoms (Fig. 10.1a). The midbrain is located in the upper brainstem and plays a regulatory role in motor function. The SN of the midbrain is rich in dopaminergic neurons containing neuromelanin, which can be observed as a dark pigment in brain sections (Zecca et al. 2003). A neuropathological hallmark of PD is the presence of clumps of insoluble protein, called Lewy bodies (LBs), in the SN of the midbrain (Fig. 10.1b). LBs are composed mainly of α-synuclein and are found where ubiquitinated proteins accumulate (Braak et al. 2003). There is a close relationship between the formation of LBs in the SN of the midbrain, degeneration of dopamine neurons, and movement disorder in PD (Braak et al. 2004).

Dopamine neurons are vulnerable to various stressors compared with other neurons (Lotharius and Brundin 2002; Sayre et al. 2008; Wang and Michaelis 2010). Nigral dopaminergic neurons are remarkably rich in ROS because the enzymatic and nonenzymatic metabolism of dopamine leads to the generation of ROS, including superoxide anions and hydrogen peroxide (H_2O_2) (Fig. 10.2). Monoamine oxidase, which is bound to the outer membrane of the mitochondria, catalyzes the oxidation of monoamines including dopamine and enzymatically generates H_2O_2

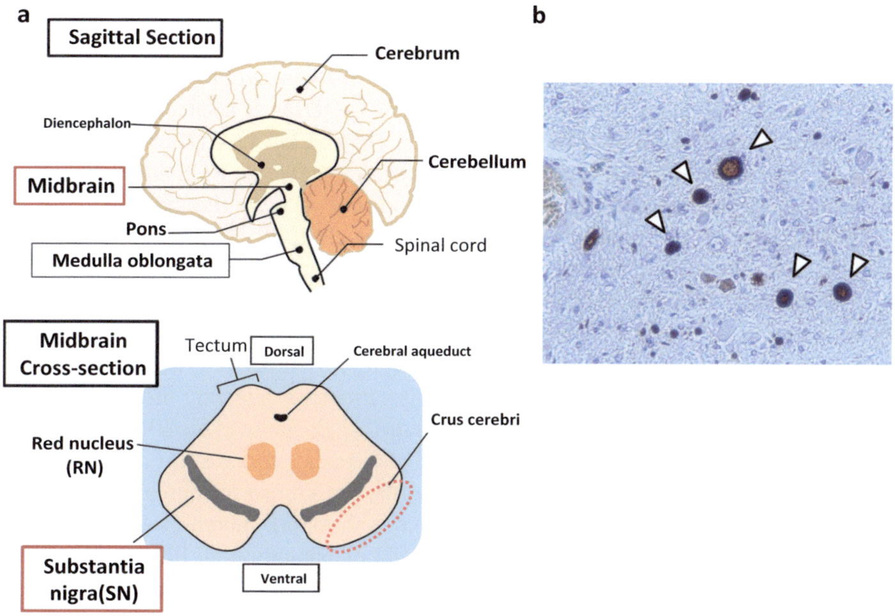

Fig. 10.1 Structure of the human brain (**a**) and Lewy bodies (LBs) in the substantia nigra (SN) of the human midbrain (**b**). (**a**) Sagittal section and cross-section of the midbrain are shown. (**b**) A section was immunostained with a polyclonal antibody against phosphorylated α-synuclein (indicative of LBs, white arrowheads)

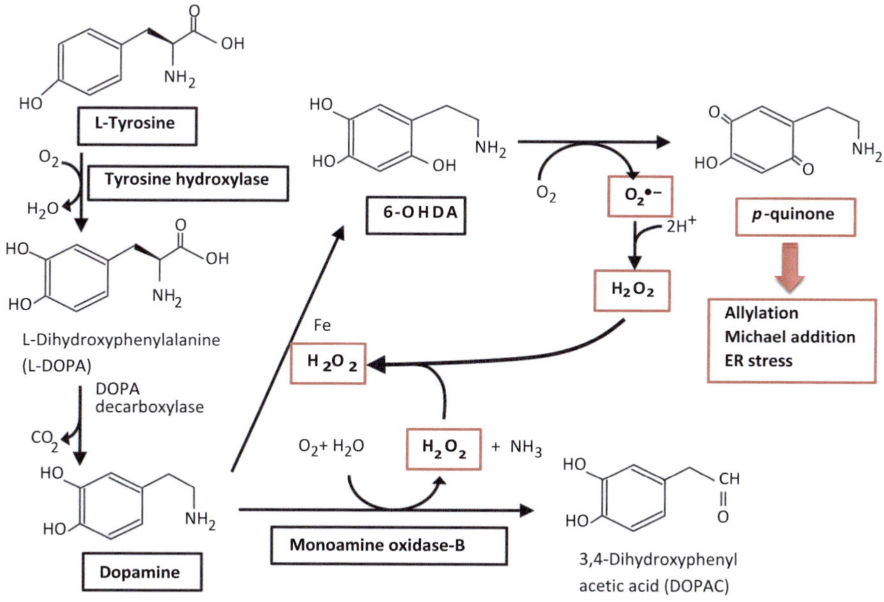

Fig. 10.2 Dopamine metabolism and generation of reactive oxygen species

(Maker et al. 1981; Lotharius and Brundin 2002). Furthermore, dopamine oxidation can occur spontaneously in the presence of iron, resulting in the generation of various dopamine oxidation products including 6-hydroxydopamine (6-OHDA) (Heikkila and Cohen 1973; Pezzella et al. 1997). 6-OHDA is a selective neurotoxin that has been used to produce models of PD in vitro and in vivo (Blum et al. 2001). 6-OHDA is oxidized rapidly by molecular oxygen to generate ROS and the reactive electrophile *p*-quinone, which plays an important role in the neurotoxicity of 6-OHDA (Fig. 10.2) (Izumi et al. 2005; Wang et al. 2006; Saito et al. 2007a). In addition, in nigral dopaminergic neurons, calcium entry through L-type channels occurs throughout the pacemaking cycle causing metabolic consumption of ATP, resulting in an increase of ROS generation (Guzman et al. 2010).

Increased levels of oxidation products of lipids, proteins, and nucleic acids are known in nigral dopaminergic neurons of PD patients, suggesting the involvement of oxidative stress in PD (Lotharius and Brundin 2002; Sayre et al. 2008; Dias et al. 2013). In addition, materials generating ROS and oxidative stress, for example, prolonged exposure to pesticides and heavy metals, are known to increase the risk of developing PD (Semchuk et al. 1992; Rybicki et al. 1993). Increased levels of transition metals, such as iron, are found in the SN of patients with PD (Dexter et al. 1992). Furthermore, decreased levels of factors that participate in antioxidative defense, such as glutathione (GSH) and GSH peroxidase 4 (GPx4), to remove ROS have been reported (Perry et al. 1982; Bellinger et al. 2011). The role of GSH in the protection of cells against oxidative stress and other xenobiotic compounds has been well established (Sies 1999). The cellular content of GSH is an important factor for cellular antioxidative defense and the cellular capacity to remove ROS (Saito et al. 2007b). GSH levels are a function of the balance between consumption, regeneration, and synthesis (Lu 2000). Although the reasons why GSH levels in the SN of the midbrain decrease have not been fully elucidated, it is well known that oxidative stress and reactive electrophiles such as *p*-quinone result in a decrease in GSH levels (Andersen 2004; Wang et al. 2006). GSH functions as a major reductant for GPxs including GPx4 to remove peroxides. GPx4, a selenium-containing protein, removes lipid hydroperoxide and H_2O_2 in the presence of GSH (Takebe et al. 2002). GPx4 partially colocalizes with LBs in the SN of brains affected by PD, and GPx4 levels are significantly decreased in the SN of patients with PD (Bellinger et al. 2011). Interestingly, higher expression of GPx4 is observed in the surviving nigral cells (Bellinger et al. 2011). GPx4 has been identified as a regulator of ferroptosis, nonapoptotic cell death depending on lipid peroxidation and iron (Yang et al. 2014). Decreased GSH levels have also been known to induce ferroptosis demonstrating the importance of the GSH-GPx4 axis (Dixon et al. 2012). The importance of GSH levels in the SN of PD patients has again come under the spotlight, and GSH repletion has been proposed as a therapeutic intervention (Sian et al. 1994; Sechi et al. 1996; Hauser et al. 2009; Mischley et al. 2016). It is also interesting to note that DJ-1 regulates the levels of expression of GPx4 and enzymes for GSH synthesis (van der Brug et al. 2008; Zhou and Freed 2005).

10.3 Significance of a Biomarker at Early Phases of Parkinson's Disease

Diagnosis of PD is greatly dependent on subjective symptoms, such as resting tremor, rigidity, and bradykinesia, and more than half of the dopaminergic neurons in the SN are lost by the time a patient is diagnosed as having PD (Hughes et al. 1992; Forno 1996). Recent studies have yielded new insights into the molecular mechanisms underlying the pathogenesis of PD, and various biological events that occur before the degradation of dopaminergic neurons in SN have been discovered (Bonifati 2012; Lesage and Brice 2012). Therefore, the movement dysfunction associated with PD is now considered to be "just the tip of the iceberg" (Langston 2006). Based on the presence of LBs in the brain of individuals including PD patients, Braak et al. have postulated a hypothesis regarding the distribution and progression of LBs in PD (Braak et al. 2004): LBs are first formed in the medulla oblongata and then progress into the upper layers of the brainstem to the midbrain and finally into the cerebral cortex (Fig. 10.3). The progressive formation of LBs and the sites of neurodegeneration coincide with symptoms (Braak et al. 2003; Braak et al. 2004). The formation of LBs and neurodegeneration in the cerebral cortex are related to dementia with PD (PDD), an advanced form of PD. By contrast, PD patients recognize disorders of smell and/or sleep before the onset of the movement disorder, which may reflect the formation of LBs and neurodegeneration in the olfactory bulb and/or the medulla oblongata (Fig. 10.3). These presymptomatic disorders have received much attention because these nonmovement disorders have the potential to serve as an early indicator of PD, such as used in the smell test (Jones 2010).

The identification of a biomarker of early-phase PD is vital for the treatment of PD (Jones 2010; Parnetti et al. 2013; Schapira 2013; Sharma et al. 2013). A bio-

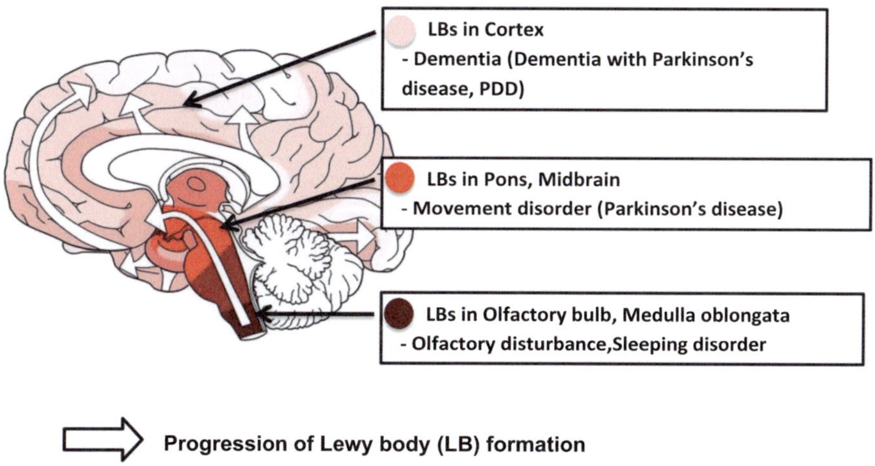

Fig. 10.3 Progression of Lewy body formation (Adapted from Braak et al. (2004) with permission)

marker indicating an early stage of the disease would serve not only to identify patients with preclinical PD for preventive treatment but also to facilitate the development of novel therapeutics to prevent the progression of PD. A number of avenues have been pursued to establish a biomarker of early-phase PD (Parnetti et al. 2013; Schapira 2013; Sharma et al. 2013). In vivo imaging markers have been sought, and specific probes for dopamine neurons and iron levels in the SN have been developed (Booij et al. 1997; Oakley et al. 2007). Changes in peripheral tissues have also been discovered and used for the diagnosis of PD. Postmortem samples showed degeneration of cardiac neurons in patients with PD (Orimo et al. 2007; Dickson et al. 2008). In the early phases of PD, cardiac sympathetic dysfunction and reduced cardiac ^{123}I-meta-iodobenzylguanidine (MIBG) uptake have been identified and used for the diagnosis of PD (Orimo et al. 1999). These imaging techniques provide reliable information for the diagnosis of PD; however, imaging biomarkers require specific instruments and the administration of the probe. Therefore, a number of issues need to be resolved before these techniques can be used to predict PD clinically. Like imaging biomarkers, biochemical biomarkers for early-phase PD have been the subject of extensive studies. For example, abnormal oligomeric forms of α-synuclein have been determined in the blood and cerebrospinal fluid (CSF) (El-Agnaf et al. 2006; Mollenhauer et al. 2010). Higher than normal levels of these oligomeric forms in PD patients have been reported (El-Agnaf et al. 2006; Mollenhauer et al. 2010).

However, the identification of a biomarker for early-phase PD is challenging (Parnetti et al. 2013). PD is complex and can be difficult to diagnose, particularly in its early phases; it is sometimes misdiagnosed as another disease, such as progressive supranuclear palsy (PSP). Therefore, developing a diagnostic system using a single biomarker is difficult. It is important to examine several biomarker candidates, such as imaging and biochemical markers, simultaneously (Schapira 2013). This strategy has been used in the research project known as the Parkinson's Progression Markers Initiative of the Michael J. Fox Foundation for PD (Parkinson's Progression Marker Initiative 2011). The identification of a biomarker of early-phase PD requires considerable funding, human resources, specialized instruments, and organized effort because investigators would have to evaluate several biomarker candidates in many patients over a long period.

10.4 Role of DJ-1 in Antioxidative Defense and Its Oxidation Under Oxidative Stress

The identification of *DJ-1* as *PARK7* opened the door to the relationship of DJ-1 to PD (Bonifati et al. 2003). The important role of DJ-1 in antioxidative defense implies the involvement of oxidative stress in the onset and progression of PD (Lev et al. 2006). Cells with high DJ-1 levels are resistant to both oxidative stress and PD-related neurotoxins, such as 6-OHDA, and can survive despite higher concentrations of these stressors, while decreased DJ-1 levels render cells vulnerable to oxidative stress (Taira et al. 2004; Inden et al., 2011). Thus, cellular DJ-1 levels are correlated with tolerance of oxidative stress. DJ-1 is induced under oxidative stress, and it is important for cells to survive by allowing adjustment to environmental change. From this point of view, cellular DJ-1 levels may be considered a biomarker of the ability of cells to protect themselves from oxidative stress (Ishiwatari et al. 2015).

An important property of DJ-1 is preferential oxidation at Cys-106 under oxidative stress (Kinumi et al. 2004). A proteomic study of cells cultured under oxidative stress identified several proteins, including peroxiredoxins and DJ-1 (Mitsumoto et al. 2001). This proteomic analysis using 2D-PAGE revealed that the isoelectric point of DJ-1 showed an acidic mobility shift under oxidative stress and to my knowledge is the first study to report the relationship of DJ-1 to oxidative stress (Mitsumoto et al. 2001). The structural characterization of an acidic isoform of DJ-1 using mass spectrometry revealed that the cysteines in DJ-1, particularly Cys-106, were oxidized to cysteine sulfonic acid (Cys-SO_3H) when cells were exposed to H_2O_2 (Kinumi et al. 2004). Cys-106 is now accepted to be the key residue involved in the antioxidative action of DJ-1 (Kahle et al. 2009; Ariga et al. 2013; Kato et al. 2013). Cysteine forms three different oxidized species, namely, cysteine sulfenic acid (Cys-SOH), cysteine sulfinic acid (Cys-SO_2H), and Cys-SO_3H through direct oxygen addition (Fig. 10.4). Acidic pI shifts in 2D-PAGE are the results of a post-translational process induced by the oxidation of the cysteine residue to Cys-SO_2H or Cys-SO_3H (Kinumi et al. 2004). Cys-SO_2H is chemically unstable and easily oxidized to Cys-SO_3H. However, Cys-SO_2H is stable within the oxDJ-1 protein because of the surrounding amino acid residues (Blackinton et al. 2009). The Cys-SO_2H form of DJ-1 is postulated to be the active form of DJ-1, based on studies that have shown a protective effect following an E18A point mutation, which depressed the pKa of Cys-106 and stabilized the Cys-SO_2H form of Cys-106 in DJ-1 (Witt et al. 2008; Blackinton et al. 2009). Further oxidation of Cys-106 to Cys-SO_3H leads to a loss of biological function (Zhou et al. 2006; Wilson 2011). The oxidation of DJ-1 results in the reduction of ROS such as H_2O_2; however, this reaction is not enzyme catalyzed and might be less efficient to remove ROS compared with other antioxidative enzymes, including GPx and catalase (Taira et al. 2004; Saito et al. 2006). Thus, DJ-1 might play an important role for antioxidative defense by acting as an oxidative stress sensor, detecting cellular redox status through the oxidation of Cys-106, and altering the activity of signal mediators and the expression levels of genes involved in antioxidative defense (Ariga et al. 2013; Cookson 2010; Kahle et al. 2009; Kato et al. 2013). It is thought that DJ-1 oxidation has an important biological meaning, and oxDJ-1 may be a promising biomarker candidate for oxidative stress, particularly oxidative stress related to PD (Wilson 2011; Saito 2014).

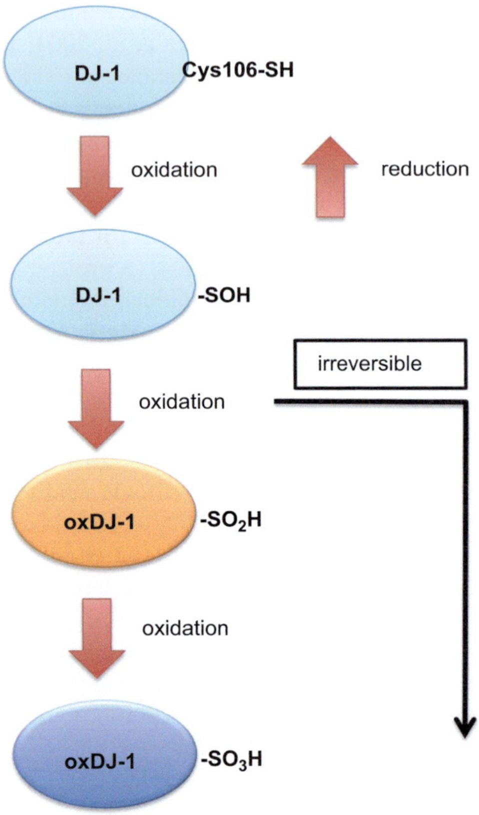

Fig. 10.4 Oxidation of a cysteine residue in DJ-1. Cys-106 of DJ-1 is sequentially oxidized to sulfenic acid (SOH), sulfinic acid (SO_2H), and sulfonic acid (SO_3H)

10.5 Preparation of Antibodies Against Oxidized DJ-1

Our research group has successfully prepared oxDJ-1-specific antibodies (Saito et al. 2009; Akazawa et al. 2010; Saito et al. 2014; Saito et al. 2016). In the first trial, a synthesized peptide (23 amino acids) including Cys-106 was treated with performic acid to oxidize Cys-SH to Cys-SO_3H, and the oxidized peptide was used as an immunogen. We could prepare specific antibodies against this oxidized peptide; however, these antibodies were not very reactive against oxDJ-1 protein. We considered the effects of tertiary protein structure of oxDJ-1, and we then examined purified recombinant virus expressing a surface glycoprotein gp64-DJ-1 (52 amino acids) fusion protein as an immunogen (Saito et al. 2009). This second trial was successful in preparing two clones of oxDJ-1-specific antibodies. We further examined full recombinant oxDJ-1 protein as an immunogen, and we have succeeded in preparing nine clones of oxDJ-1-specific antibodies (Saito et al. 2014). By using these oxDJ-1 antibodies, we developed immunological methods to detect oxDJ-1, such as Western blotting, enzyme-linked immunosorbent assay (ELISA), and immunostaining (Saito et al. 2014; Saito et al. 2016). The preparation of antibodies against Cys-SO_2H- and/or Cys-SO_3H-containing proteins has not been well established. The results might provide a good benchmark for the preparation of antibodies to Cys-oxidized proteins.

10.6 Oxidized DJ-1 Levels in the Brain of Patients with Parkinson's Disease

The distribution of DJ-1 and oxDJ-1 in the brain of patients with PD has been investigated by using immunohistochemistry of postmortem brains (Saito et al. 2014). The distribution of oxDJ-1 immunoreactivity (IR) in sections of mouse and human brains is summarized in Table 10.1. Oxidized DJ-1 IR is evident in PD-related sites such as the medulla oblongata, striatum, and the SN of the midbrain, suggesting preferential oxidation of DJ-1 and higher levels of oxidative stress in these sites (Table 10.1 and Fig. 10.5a). Furthermore, double immunostaining suggests that oxDJ-1 is a component of LBs (Fig. 10.5b), implying that oxDJ-1 is deeply involved in the pathogenesis of PD.

The dopaminergic neurons in the SN of the midbrain of mice and humans show oxDJ-1 IR; particularly obvious oxDJ-1 IR in cell bodies and neurites of dopaminergic neurons is observed in patients with LB stage (LS) III-PD (Fig. 10.6). The "criterion LS III-PD" means patients diagnosed as PD with attributable parkinsonism with abundant LB formation in the SN and macroscopic loss of dopaminergic neurons (Saito et al. 2004). LB pathology is classified into 6 LSs according to published criteria: LS 0 = no LBs, LS I = scattered LBs without cell loss, LS II = abundant LBs with macroscopic loss of neuromelanin in the SN but without attributable parkinsonism or dementia, LS III-PD = PD without dementia, and LS IV-PDD and LS V-PDD = dementia with PD (PDD). Oxidized DJ-1 IR has been observed in the SN of the midbrain from control subjects (LS 0). Further, oxDJ-1 IR of the SN is increased in LS II-PD and LS III-PD and then decreased in LS IV-PDD and LS V-PDD (Fig. 10.6), suggesting alteration of oxDJ-1 formation in dopaminergic nigral neuronal cells in the progression of PD, in particular, oxidation of DJ-1 before the onset of PD (Saito et al. 2014). Cellular GSH levels play a significant role in DJ-1 oxidation (Miyama et al. 2011), and there are decreased GSH levels in the SN of patients with PD (Perry et al. 1982; Sian et al. 1994). These findings strongly suggest the generation of oxidative stress in dopaminergic neurons of the SN of the midbrain before the onset of PD.

DJ-1 plays an important role in the synthesis of dopamine: DJ-1 upregulates the transcription of tyrosine hydroxylase, the rate-limiting enzyme of dopamine synthesis, by inhibiting the sumoylation of pyrimidine tract-binding protein-associated splicing factor (Zhong et al. 2006). Furthermore, DJ-1 activates dopamine synthesis through interaction with dopamine biosynthetic enzymes such as tyrosine hydroxylase and 4-dihydroxy-L-phenylalanine decarboxylase, in an oxidation-dependent manner (Ishikawa et al. 2009). Slight oxidative stress inducing DJ-1 oxidation to

Table 10.1 Distribution of oxidized DJ-1 in the human and mouse brain

Species	Area	Obvious staining site	Cell types	Others	Note
Human	Midbrain	SN	Dopaminergic neuronal cells	Neuropil	Increase in LS II-PD and LS III-PD,
			(Cell bodies, neurites)	Colocalization with LB	Decrease in LS IV-PDD and LS V-PDD
		RN	Glial cells (cell bodies, nucleus, neurites)	Neuropil	n.a.
	Striatum	Putamen	Astrocytes (cell bodies, nucleus, neurites)	Neuropil	Increase in LS II-PD and LS III-PD, decrease in LS IV-PDD and LS V-PDD
	Medulla oblongata	VNDN	Neuronal cells (cell bodies, neurites)	Neuropil	n.a.
				Colocalization with LB	
		ION	Neuronal cells (cell bodies, nucleus neurites), glial cells (cell bodies, nucleus, neurites	Neuropil	n.a.
Mouse	Midbrain	SN	Neuronal cells (cell bodies), glial cells (nucleus)	Neuropil	n.a.
	Striatum	n.a.	Small-medium spiny neurons (cell bodies), glial cells (nucleus)	Neuropil	n.a.
	Hippocamps	CA1	Pyramidal cells (cell bodies)	Stratum radiatum, mossy fibers, neuropil	n.a.
		DG	Granule cells	Molecular layer	n.a.
	Cortex	Laminae	Neuronal cells (cell bodies), glial cells (nucleus)	Neuropil	n.a.
	Cerebellum	n.a.	n.a.	n.a.	n.a.
	Olfactory bulb	n.a.	n.a.	n.a.	n.a.

SN substantia nigra, *n.a.* not assigned, *CA1* cornu ammonis 1, *DG* dentate gyrus, *LB* Lewy body, *LS* Lewy stage, *PD* Parkinson's disease, *PDD* dementia with Parkinson's disease, *RN* red nucleus, *VNDN* vagus nerve dorsum nucleus, and *ION* inferior olivary nucleus (Adapted from Saito et al. 2014 with permission)

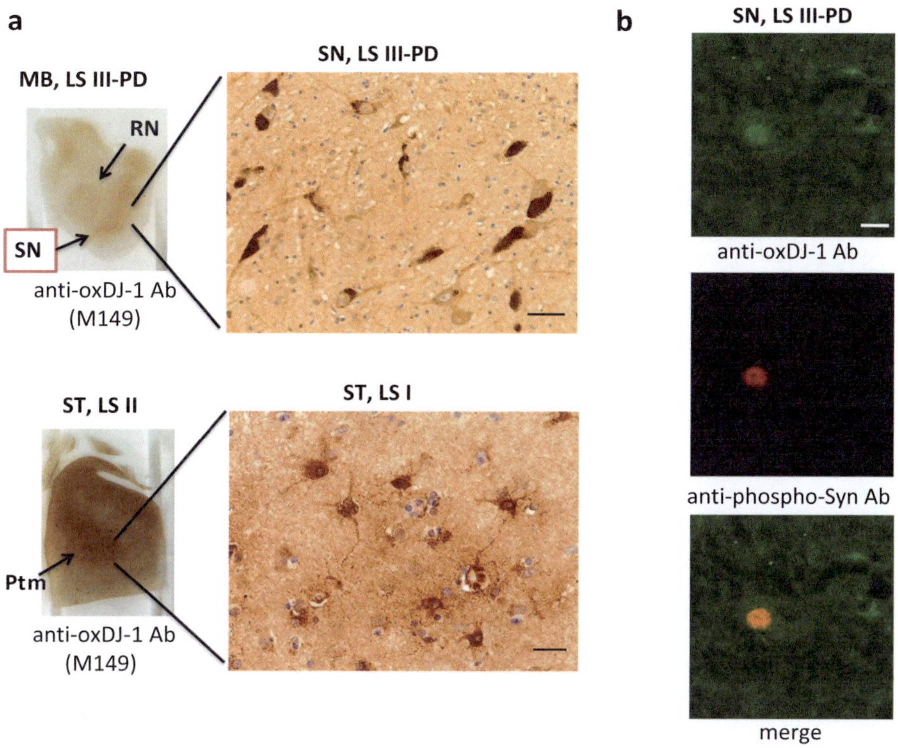

Fig. 10.5 Immunohistochemical distribution of oxDJ-1 in the human midbrain (MB) and striatum (ST). (**a**) Oxidized DJ-1 immunoreactivity (IR) was present throughout the MB (upper) and ST (lower). In the MB, particularly high levels were observed in the substantia nigra (SN) and red nucleus (RN). There was IR of cell bodies and processes of neuromelanin-containing neurons with the oxDJ-1 mAb at higher magnifications. In the ST, particularly high IR was observed in the putamen (Ptm). There was IR of cell bodies and proximal processes of glial cells at high magnification. Scale bars = 50 μm (upper) and 20 μm (lower). LS: Lewy body stage. (**b**) Confocal images of oxDJ-1 in dopaminergic nigral neuronal cells. A section containing the SN was immunostained with a mAb against oxDJ-1 and a polyclonal antibody against phosphorylated α-synuclein and then visualized using fluorescence confocal microscopy. The areas labeled with anti-oxDJ-1 and anti-phosphorylated α-synuclein (indicative of Lewy bodies) antibodies were colocalized. Scale bar = 20 μm (Adapted from Saito et al. 2014 with permission)

Cys-106-SOH upregulates dopamine synthesis, while extensive oxidative stress inducing DJ-1 oxidation to Cys-106-SO_2H or Cys-106-SO_3H does not (Ishikawa et al. 2009). Because DJ-1 Cys-106-SO_2H or Cys-106-SO_3H is detected in the normal SN of human and mouse midbrains, the regulation of dopamine synthesis via DJ-1 oxidation might occur under physiological conditions. The most intense oxDJ-1 IR in the SN was observed during LS II-PD and LS III-PD; therefore, oxDJ-1 might be extensively formed and thus decrease dopamine synthesis at an early stage of the disease.

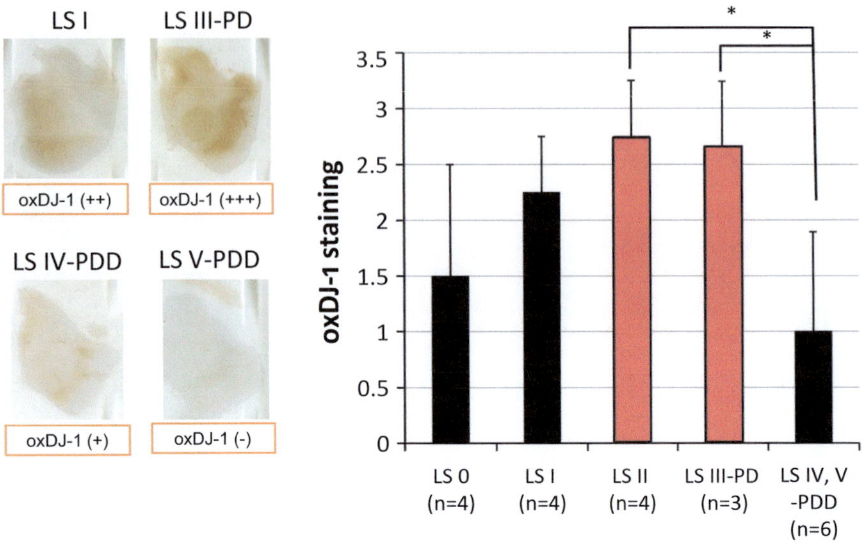

Fig. 10.6 Change of oxDJ-1 staining in the substantia nigra of the midbrain during the course of Parkinson's disease. Qualitative rating scales used for oxDJ-1 immunoreactivity of the substantia nigra. Brain sections were evaluated visually and immunoreactive staining classified as (−) none, (+) mild, (++) moderate, or (+++) high. The mean ± SD is shown. * $P < 0.05$, Tukey-Kramer test, ANOVA (Adapted from Saito et al. 2014 with permission)

Glial cells were immunostained with the oxDJ-1 mAb in several sites, including the cortex, striatum, SN of the midbrain, and medulla oblongata, suggesting the important role of DJ-1 in glial cells (Table 10.1). DJ-1 is abundantly expressed in the reactive astrocytes of patients with neurodegenerative diseases (Bandopadhyay et al. 2004; Neumann et al. 2004), and obvious oxDJ-1 IR in astrocytes is detected in the striatum, showing IR of the cell bodies and proximal processes of astrocytes (Fig. 10.5a). Astrocytes are neuroprotective, and there is an inverse correlation between astrocyte number and dopaminergic cell loss (Damier et al. 1993; Maragakis and Rothstein 2006). Astrocytes possess high levels of antioxidant enzymes to protect neurons from oxidative stress, and there is high expression of DJ-1, which plays an important role in the function of astrocytes, in particular their neuroprotective effects (Mullett et al. 2009; Larsen et al. 2011; Mullett et al. 2013). Glial cells are considered important contributors in the prevention of oxidative stress in PD, and DJ-1 might play a significant role in the antioxidative function of astrocytes via oxidation of Cys-106.

10.7 Oxidized DJ-1 in the Biological Fluids of Patients with Parkinson's Disease

DJ-1 in biological fluids has attracted interest as a biomarker of PD. Studies of DJ-1 as a biomarker of PD are summarized in Table 10.2. DJ-1 is induced by oxidative stress, and an increase of DJ-1 protein is expected to reflect oxidative stress in patients with PD (Taira et al. 2004; Yanagida et al. 2009). The DJ-1 content of the blood and CSF has been determined (Table 10.2). However, the findings reported are controversial (Waragai et al. 2007; Shi et al. 2010; Hong et al. 2010). DJ-1 content in erythrocytes is remarkably high, and hemolysis and contamination by erythrocytes greatly affect the level of DJ-1 found in plasma and CSF (Hong et al. 2010; Shi et al. 2011). Thus, to determine the DJ-1 content in plasma and CSF, it is necessary to evaluate contamination by hemolysis. Shi et al. reported that when using plasma without hemolysis, the plasma level of DJ-1 was not a useful biomarker for diagnosis of PD or for determining its progression/severity (Shi et al. 2010). By contrast, when CSF without blood cell contaminants is examined, the levels of DJ-1 and α-synuclein are significantly decreased in patients with PD compared with patients with Alzheimer's disease and control subjects (Hong et al. 2010; Shi et al. 2011). This change of DJ-1 in CSF does not differentiate between parkinsonian syndromes, such as dementia with LBs (DLB), PSP, or multiple system atrophy (MSA) (Salvesen et al. 2012). Recently, it has been reported that *DJ-1* expression in peripheral blood mononuclear cells and CSF exosomal RNA is increased in patients with PD (Gui et al. 2015; Yalcinkaya et al. 2016).

In addition to the quantitative levels of DJ-1, its qualitative change, namely, oxidation, is of interest as a candidate biomarker of PD. Using a competitive ELISA for oxDJ-1, an increase in oxDJ-1 levels in erythrocytes was found (Saito et al. 2009; Akazawa et al. 2010; Saito et al. 2016). We have recently reported that the levels of oxDJ-1 in the erythrocytes of unmedicated PD patients (n = 88) are higher than in those in medicated PD patients (treated with L-DOPA and/or a dopamine receptor agonist, n = 62) and healthy control subjects (n = 33) (Fig. 10.7a) (Saito et al. 2016). The term "unmedicated PD patients" refers to patients diagnosed with PD but not yet started on medications such as L-DOPA and/or a dopamine receptor agonist. Thus, unmedicated PD patients are essentially those with PD in its earliest phases. PD was diagnosed on the basis of criteria reported by Calne et al. (1992) and classified into Hoehn-Yahr (H-Y) stages 1–5 (Hoehn and Yahr 1967). Plotting the levels of oxDJ-1 against PD stage (H-Y 1–5) showed that higher levels of oxDJ-1 in erythrocytes were present at early-stage PD, such as H-Y 1 and H-Y 2 (Fig. 10.7b). Furthermore, when the oxDJ-1 levels were plotted against time from onset of PD, higher levels of oxDJ-1 were primarily observed within the first 5 years (Fig. 10.7c). Taken together, these results suggest that the levels of oxDJ-1 in erythrocytes are highest during early-phase PD, particularly in unmedicated PD patients (Saito et al. 2016). In addition to findings in humans, elevation of blood oxDJ-1 levels has been observed in 6-OHDA-treated rats and MPTP-treated monkeys and mice (Fig. 10.7d) (Akazawa et al. 2010; Saito et al. 2016).

Table 10.2 Biomarker study of DJ-1 in PD

Study cited	Disease and number[a]	Material	Method	Outcome reported
Sci Rep 2016;6:30793 (Saito et al. 2016)	PD (unmedicated, 88; medicated, 62), Cont (33)	RBC	ELISA	Oxidized DJ-1 of unmedicated PD patients were higher than medicated PD patients and Cont
Neurosci Lett 2016;615: 72-77 (Yalcinkaya et al. (2016)	PD (postural instability, 11; postural stability, 39), Cont (50)	PBMC	Real-time PCR	DJ-1 expression levels were increased in PD patients compared to control. In PD patients with instability, DJ-1 levels were low
Oncotarget 2015;6;37043-37053 (Gui et al. 2015)	PD (47), AD (28), Cont (27)	CSF (exosomal RNA)	Real-time PCR	DJ-1 expression levels were decreased in PD and AD patients compared to control
Parkin Rel Dis 2015;21:1251-1255	PD (16), Cont (22)	Saliva	Western blot	DJ-1 increased in PD patients compared to control
Front Aging Neurosci 2014; 6:102	PD (285), Cont (91)	Saliva	Luminex assay	DJ-1 levels in H&Y 4 stage of PD were higher than those in H&Y 1–3 stage as well as those in healthy controls
Brain Nerve 2014;66:471-477 (Ogawa et al. 2014)	PD (unmedicated, 13; medicated, 10), Cont (13)	RBC	ELISA	Oxidized DJ-1 of unmedicated PD patients were higher than medicated PD patients and Cont
Neurobiol Aging 2012; 33:836. e5-e7	LRRK2 (26)	CSF	Luminex assay	DJ-1 in LRRK2 CSF do not correlate with striatal dopaminergic function
Parkin Rel Dis 2012;18: 899-901. (Salvesen et al. 2012)	PD (30), DLB (17), PSP (19), MSA (14), CBD (6), unspecified (6)	CSF	ELISA	DJ-1 concentration in CSF does not differentiate among parkinsonian syndromes
Sci Rep 2012;2:954 (Lin et al. 2012)	PD (159), AD (14), Cont (60)	Whole blood	2D-WB	Blood levels of DJ-1 with 4-HNE modifications were altered in late-stage Parkinson's disease
Ann Neurol 2011;69:570–580 (Shi et al. 2011)	PD (126), AD (50), MSA (32), Cont (137)	CSF	Luminex assay	DJ-1 levels were decreased in PD versus Cont or AD
Brain 2011;134:1–5	PD (24), Cont (25)	Saliva	Luminex assay	DJ-1 increased in PD patients compared to control

(continued)

Table 10.2 (continued)

Study cited	Disease and number[a]	Material	Method	Outcome reported
Neurosci Lett 2010;480: 78–82 (Shi et al. 2010)	PD (126), AD (33), Cont (122)	Plasma	Luminex assay	DJ-1 in plasma is not useful as biomarkers for PD diagnosis or progression/severity
Brain 2010;133:713–726 (Hong et al. 2010)	PD (117), AD (50), Cont (132)	CSF	Luminex assay	DJ-1 levels were decreased in PD patients versus Cont or AD patients
Neurosci Lett 2009;465:1-5 (Saito et al. 2009)	PD (unmedicated, 8; medicated, 7), Cont (18)	RBC	ELISA	Oxidized DJ-1 of unmedicated PD patients were higher than medicated PD patients and Cont
Neurosci Lett 2008;431:86-89	PD (95), other disease (30), Cont (24)	Serum	ELISA	There was no significant difference between the levels of serum DJ-1 in PD and Cont

[a]The number of determined samples was shown in parentheses
PD Parkinson's disease, *LRRK2* leucine-rich repeat kinase 2, *DLB* dementia with Lewy body, *PSP* progressive supranuclear palsy, *MSA* multiple system atrophy, *CBD* corticobasal degeneration, *AD* Alzheimer's disease, *RBC* red blood cells, *PBMC* peripheral blood mononuclear cells, *CSF* cerebrospinal fluid, *ELISA* enzyme-linked immunosorbent assay, *PCR* polymerase chain reaction, *2D-PAGE* two-dimensional polyacrylamide gel electrophoresis, *WB* Western blot

The biochemical properties of oxDJ-1 detected in erythrocytes of patients with early-stage PD suggest two forms of oxDJ-1, namely, one at 45 kDa corresponding to a dimer and a second at more than 200 kDa (Fig. 10.8a) (Saito et al. 2016). In the fractionized high-molecular-weight polymer form of oxDJ-1, subunits of a 20S proteasome, which interacts with DJ-1 (Moscovitz et al. 2015), have been detected by LC-MS/MS analysis. Experiments using a 20S proteasome purified from human erythrocytes and recombinant oxDJ-1 suggest an interaction between oxDJ-1 and the 20S proteasome in the erythrocytes of unmedicated PD patients (Fig. 10.8b). DJ-1 has been reported to interact with the 20S proteasome and inhibit its activity (Moscovitz et al. 2015). By contrast, several studies have reported that posttranslationally modified proteins such as α-synuclein oligomer and protein carbonyls accumulate in the erythrocytes of PD patients (Chen et al. 2009; Cristalli et al. 2012; Wang et al. 2015). A change of 4-hydroxy-2-nonenal-modified DJ-1 in the whole blood of PD patients has also been reported (Lin et al. 2012). Together, these results suggest that oxDJ-1 interacts with the 20S proteasome, inhibits its activity, and is related to the accumulation of modified proteins in the erythrocytes of PD patients (Saito et al. 2016). A biochemical alteration in the blood of PD patients could be used as a marker for the early diagnosis of PD.

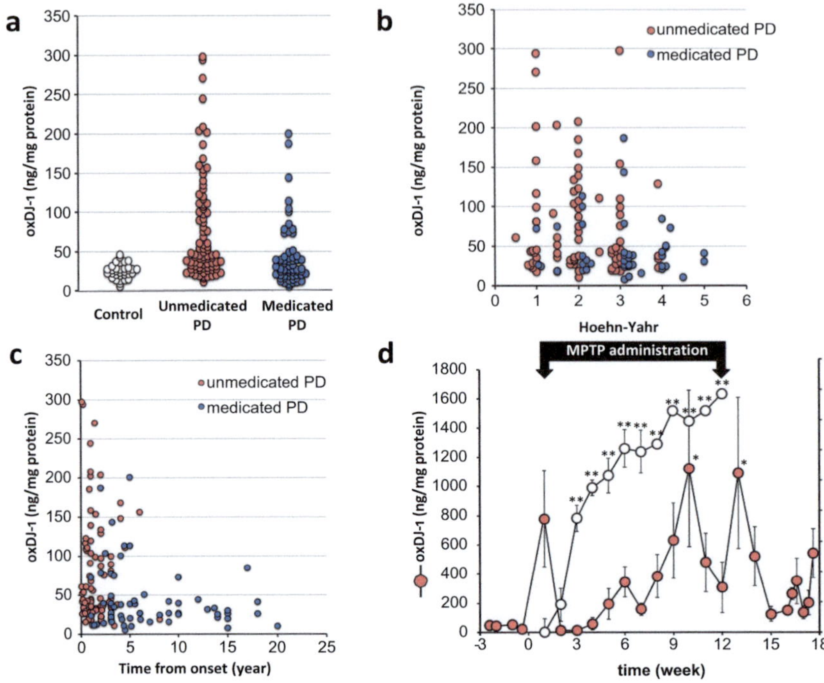

Fig. 10.7 Elevation of oxDJ-1 levels in the erythrocytes of unmedicated patients with Parkinson's disease (PD) and a nonhuman primate model of preclinical PD. (**a–c**) Levels of oxDJ-1 in erythrocytes of healthy controls (n = 33), unmedicated PD patients (n = 88), and medicated PD patients (n = 62). Individual data are shown (**a**). Individual oxDJ-1 levels were plotted against Hoehn-Yahr stage (**b**) and time from onset in years (**c**). (**d**) During MPTP treatment, oxDJ-1 levels in erythrocytes from each monkey were determined using a competitive ELISA. The mean ± SEM of oxDJ-1 levels and neurological score (n = 4) are shown. *$P < 0.05$, **$P < 0.01$, Tukey-Kramer test, ANOVA, when compared with pretreatment (−1 week) (Adapted from Saito et al. 2016 with permission)

10.8 Significance of Oxidized DJ-1 as a Biomarker of Parkinson's Disease

Accumulating evidence indicates that DJ-1 oxidation in erythrocytes and the brain is a common phenomenon in patients with PD and animal models of PD, particularly at early phases (Fig. 10.9) (Saito 2014; Ogawa et al. 2014; Saito et al. 2016). Although it is not clear whether oxDJ-1 in erythrocytes of PD patients is derived from the brain or erythrocytes, the measurement of oxDJ-1 in erythrocytes as a biomarker of PD has several benefits, such as convenience and versatility. At present, the movement dysfunction associated with PD is considered to be "just the tip of the iceberg," and several other changes are known, including the formation of LBs in the peripheral tissues of patients with PD (Braak et al. 2004; Langston 2006). Peripheral nervous system disorders, such as reduced cardiac uptake of radiolabeled

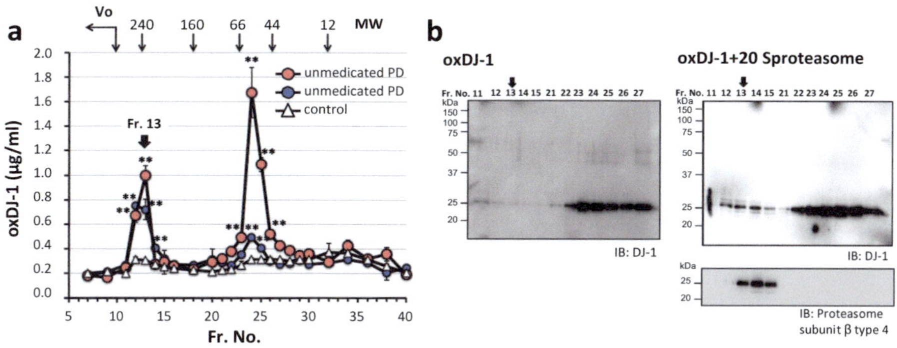

Fig. 10.8 Biochemical properties of the polymer form of oxDJ-1 detected in the erythrocytes of unmedicated PD patients. (**a**) The erythrocyte lysates of unmedicated PD patients and healthy controls (60 mg protein) were separated by gel chromatography, and the oxDJ-1 content of each fraction was determined by using an ELISA (**a**). The mean ± SD (n = 3) is shown. $^{**}P < 0.01$, Tukey-Kramer test, ANOVA, when compared with a buffer control. The major oxDJ-1 peak of the polymer form (fraction 13) is indicated by black arrows. (**b**) 20S proteasome purified from human erythrocytes was incubated for 18 h at 4 °C in either the presence or absence of purified recombinant human oxDJ-1 in 50 mM HEPES buffer (pH 7.4) supplemented with 10% glycerol, 2 mM ATP, and 2 mM DTT. Samples were then separated by gel chromatography, and each fraction was subjected to Western blot analysis. Adapted from Saito et al. (2016) with permission

MIBG, have also been used for the diagnosis of PD. An association between oxDJ-1 levels in erythrocytes and cardiac uptake of MIBG has been recognized (Ogawa et al. 2014). Thus, changes in peripheral tissues could be utilized as biomarkers for the diagnosis of early-phase PD. The substantial relationship between the central nervous system and peripheral tissues in the pathology of PD remains unclear. To elucidate this relationship is not only critical for the understanding of PD pathology but also for the development of a reliable biomarker of PD.

10.9 Conclusion

Several studies indicate a relationship between DJ-1 oxidation, oxidative stress, and the onset and progression of PD. The importance of the prevention of oxidative stress in PD is recognized. The administration of GSH to patients with PD and its beneficial effects have received much attention, and several attempts have been made to improve the dosage and develop a system for measuring the levels of GSH in the brain (Hauser et al. 2009; Holmay et al. 2013; Mischley et al. 2016). Elucidation of the timing and site of oxidative stress in PD might be required to establish this therapy. Achievement of antioxidative therapy for PD using oxDJ-1 levels as a biomarker is highly anticipated (Fig. 10.9).

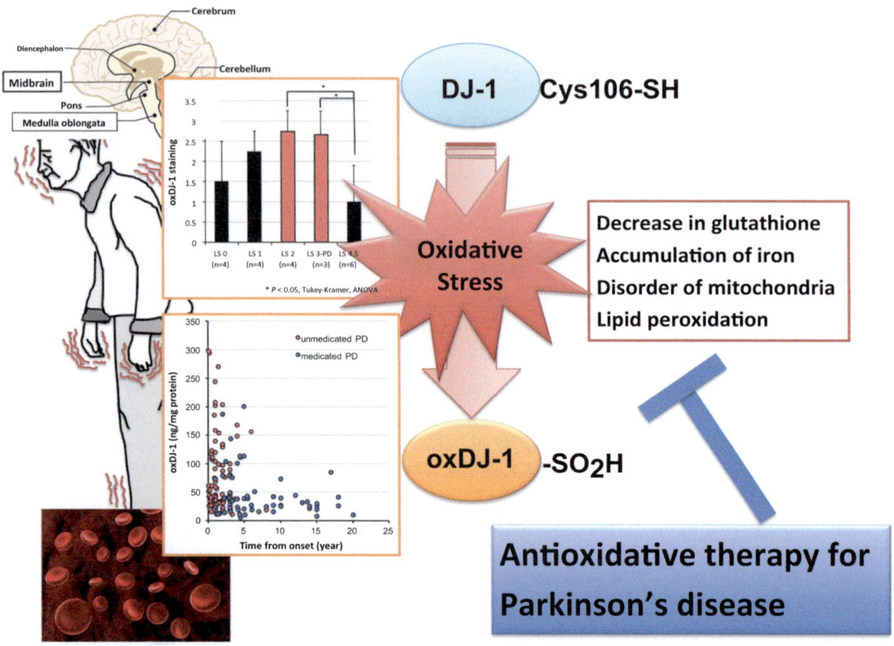

Fig. 10.9 Oxidized DJ-1 as a biomarker of Parkinson's disease

References

Akazawa YO, Saito Y, Hamakubo T, Masuo Y, Yoshida Y, Nishio K et al (2010) Elevation of oxidized DJ-1 in the brain and erythrocytes of Parkinson disease model animals. Neurosci Lett 483:201–205

Andersen JK (2004) Oxidative stress in neurodegeneration: cause or consequence? Nat Med 10(7):S18–S25

Ariga H, Takahashi-Niki K, Kato I, Maita H, Niki T, Iguchi-Ariga SM (2013) Neuroprotective function of DJ-1 in Parkinson's disease. Oxidative Med Cell Longev 2013:683920

Bandopadhyay R, Kingsbury AE, Cookson MR, Reid AR, Evans IM, Hope AD et al (2004) The expression of DJ-1 (PARK7) in normal human CNS and idiopathic Parkinson's disease. Brain 127:420–430

Bellinger FP, Bellinger MT, Seale LA, Takemoto AS, Raman AV, Miki T et al (2011) Glutathione peroxidase 4 is associated with Neuromelanin in Substantia Nigra and dystrophic axons in putamen of Parkinson's brain. Mol Neurodegener 6:8

Blackinton J, Lakshminarasimhan M, Thomas KJ, Ahmad R, Greggio E, Raza AS et al (2009) Formation of a stabilized cysteine sulfinic acid is critical for the mitochondrial function of the parkinsonism protein DJ-1. J Biol Chem 284:6476–6485

Blum D, Torch S, Lambeng N, Nissou M, Benabid AL, Sadoul R et al (2001) Molecular pathways involved in the neurotoxicity of 6-OHDA, dopamine and MPTP: contribution to the apoptotic theory in Parkinson's disease. Prog Neurobiol 65:135–172

Bonifati V (2012) Autosomal recessive parkinsonism. Parkinsonism Relat Disord 18(Suppl 1):S4–S6

Bonifati V, Rizzu P, van Baren MJ, Schaap O, Breedveld GJ, Krieger E et al (2003) Mutations in the DJ-1 gene associated with autosomal recessive early-onset parkinsonism. Science 299:256–259

Booij J, Tissingh G, Winogrodzka A, Boer GJ, Stoof JC, Wolters EC et al (1997) Practical benefit of [123I]FP-CIT SPET in the demonstration of the dopaminergic deficit in Parkinson's disease. Eur J Nucl Med 24:68–71

Braak H, Del Tredici K, Rub U, de Vos RA, Jansen Steur EN, Braak E (2003) Staging of brain pathology related to sporadic Parkinson's disease. Neurobiol Aging 24:197–211

Braak H, Ghebremedhin E, Rub U, Bratzke H, Del Tredici K (2004) Stages in the development of Parkinson's disease-related pathology. Cell Tissue Res 318:121–134

Calne DB, Snow BJ, Lee C (1992) Criteria for diagnosing Parkinson's disease. Ann Neurol 32(S1):S125–S127

Chen CM, Liu JL, Wu YR, Chen YC, Cheng HS, Cheng ML et al (2009) Increased oxidative damage in peripheral blood correlates with severity of Parkinson's disease. Neurobiol Dis 33:429–435

Cookson MR (2010) DJ-1, PINK1, and their effects on mitochondrial pathways. Mov Disord 25(S1):S44–S48

Cristalli DO, Arnal N, Marra FA, de Alaniz MJ, Marra CA (2012) Peripheral markers in neurodegenerative patients and their first-degree relatives. J Neurol Sci 314:48–56

Damier P, Hirsch EC, Zhang P, Agid Y, Javoy-Agid F (1993) Glutathione peroxidase, glial cells and Parkinson's disease. Neuroscience 52:1–6

Dexter DT, Jenner P, Schapira AH, Marsden CD (1992) Alterations in levels of iron, ferritin, and other trace metals in neurodegenerative diseases affecting the basal ganglia. The royal kings and queens Parkinson's disease research group. Ann Neurol 32(Suppl):S94–100

Dias V, Junn E, Mouradian MM (2013) The role of oxidative stress in Parkinson's disease. J Parkinsons Dis 3:461–491

Dickson DW, Fujishiro H, DelleDonne A, Menke J, Ahmed Z, Klos KJ et al (2008) Evidence that incidental Lewy body disease is pre-symptomatic Parkinson's disease. Acta Neuropathol 115:437–444

Dixon SJ, Lemberg KM, Lamprecht MR, Skouta R, Zaitsev EM, Gleason CE et al (2012) Ferroptosis: an iron-dependent form of nonapoptotic cell death. Cell 149:1060–1072

El-Agnaf OM, Salem SA, Paleologou KE, Curran MD, Gibson MJ, Court JA et al (2006) Detection of oligomeric forms of alpha-synuclein protein in human plasma as a potential biomarker for Parkinson's disease. FASEB J 20:419–425

Forno LS (1996) Neuropathology of Parkinson's disease. J Neuropathol Exp Neurol 55:259–272

Gui Y, Liu H, Zhang L, Lv W, Hu X (2015) Altered microRNA profiles in cerebrospinal fluid exosome in Parkinson disease and Alzheimer disease. Oncotarget 6:37043–37053

Guzman JN, Sanchez-Padilla J, Wokosin D, Kondapalli J, Ilijic E, Schumacker PT et al (2010) Oxidant stress evoked by pacemaking in dopaminergic neurons is attenuated by DJ-1. Nature 468:696–700

Hauser RA, Lyons KE, McClain T, Carter S, Perlmutter D (2009) Randomized, double-blind, pilot evaluation of intravenous glutathione in Parkinson's disease. Mov Disord 24:979–983

Heikkila RE, Cohen G (1973) 6-Hydroxydopamine: evidence for superoxide radical as an oxidative intermediate. Science 181:456–457

Hoehn MM, Yahr MD (1967) Parkinsonism: onset, progression and mortality. Neurology 17:427–442

Holmay MJ, Terpstra M, Coles LD, Mishra U, Ahlskog M, Oz G et al (2013) N-acetylcysteine boosts brain and blood glutathione in Gaucher and Parkinson diseases. Clin Neuropharmacol 36:103–106

Hong Z, Shi M, Chung KA, Quinn JF, Peskind ER, Galasko D et al (2010) DJ-1 and alpha-synuclein in human cerebrospinal fluid as biomarkers of Parkinson's disease. Brain 133:713–726

Hughes AJ, Daniel SE, Kilford L, Lees AJ (1992) Accuracy of clinical diagnosis of idiopathic Parkinson's disease: a clinico-pathological study of 100 cases. J Neurol Neurosurg Psychiatry 55:181–184

Inden M, Kitamura Y, Takahashi K, Takata K, Ito N, Niwa R et al (2011) Protection against dopaminergic neurodegeneration in Parkinson's disease-model animals by a modulator of the oxidized form of DJ-1, a wild-type of familial Parkinson's disease-linked PARK7. J Pharmacol Sci 117:189–203

Ishikawa S, Taira T, Niki T, Takahashi-Niki K, Maita C, Maita H et al (2009) Oxidative status of DJ-1-dependent activation of dopamine synthesis through interaction of tyrosine hydroxylase and 4-dihydroxy-L-phenylalanine (L-DOPA) decarboxylase with DJ-1. J Biol Chem 284:28832–28844

Ishiwatari S, Takahashi M, Yasuda C, Nakagawa M, Saito Y, Noguchi N et al (2015) The protective role of DJ-1 in ultraviolet-induced damage of human skin: DJ-1 levels in the stratum corneum as an indicator of antioxidative defense. Arch Dermatol Res 307:925–935

Izumi Y, Sawada H, Sakka N, Yamamoto N, Kume T, Katsuki H et al (2005) P-Quinone mediates 6-hydroxydopamine-induced dopaminergic neuronal death and ferrous iron accelerates the conversion of p-quinone into melanin extracellularly. J Neurosci Res 79:849–860

Jenner P (2003) Oxidative stress in Parkinson's disease. Ann Neurol 53(S3):S26–S38

Jones R (2010) Biomarkers: casting the net wide. Nature 466:S11–S12

Kahle PJ, Waak J, Gasser T (2009) DJ-1 and prevention of oxidative stress in Parkinson's disease and other age-related disorders. Free Radic Biol Med 47:1354–1361

Kato I, Maita H, Takahashi-Niki K, Saito Y, Noguchi N, Iguchi-Ariga SM et al (2013) Oxidized DJ-1 inhibits p53 by sequestering p53 from promoters in a DNA-binding affinity-dependent manner. Mol Cell Biol 33:340–359

Kinumi T, Kimata J, Taira T, Ariga H, Niki E (2004) Cysteine-106 of DJ-1 is the most sensitive cysteine residue to hydrogen peroxide-mediated oxidation in vivo in human umbilical vein endothelial cells. Biochem Biophys Res Commun 317:722–728

Langston JW (2006) The Parkinson's complex: parkinsonism is just the tip of the iceberg. Ann Neurol 59:591–596

Larsen NJ, Ambrosi G, Mullett SJ, Berman SB, Hinkle DA (2011) DJ-1 knock-down impairs astrocyte mitochondrial function. Neuroscience 196:251–264

Lesage S, Brice A (2012) Role of Mendelian genes in "sporadic" Parkinson's disease. Parkinsonism Relat Disord 18(Suppl 1):S66–S70

Lev N, Roncevic D, Ickowicz D, Melamed E, Offen D (2006) Role of DJ-1 in Parkinson's disease. J Mol Neurosci 29:215–226

Lin X, Cook TJ, Zabetian CP, Leverenz JB, Peskind ER, Hu SC et al (2012) DJ-1 isoforms in whole blood as potential biomarkers of Parkinson disease. Sci Rep 2:954

Lotharius J, Brundin P (2002) Pathogenesis of Parkinson's disease: dopamine, vesicles and alpha-synuclein. Nat Rev Neurosci 3:932–942

Lu SC (2000) Regulation of glutathione synthesis. Curr Top Cell Regul 36:95–116

Maker HS, Weiss C, Silides DJ, Cohen G (1981) Coupling of dopamine oxidation (monoamine oxidase activity) to glutathione oxidation via the generation of hydrogen peroxide in rat brain homogenates. J Neurochem 36:589–593

Maragakis NJ, Rothstein JD (2006) Mechanisms of disease: astrocytes in neurodegenerative disease. Nat Clin Pract Neurol 2:679–689

Mischley KL, Conley KE, Shankland EG, Kavanagh TJ, Rosenfeld ME, Duda JE, White CC, Wilbur TK, Torre PU, Padowski JM (2016) Central nervous system uptake of intranasal glutathione in Parkinson's disease. npj Park Dis 2:16002

Mitsumoto A, Nakagawa Y, Takeuchi A, Okawa K, Iwamatsu A, Takanezawa Y (2001) Oxidized forms of peroxiredoxins and DJ-1 on two-dimensional gels increased in response to sublethal levels of paraquat. Free Radic Res 35:301–310

Miyama A, Saito Y, Yamanaka K, Hayashi K, Hamakubo T, Noguchi N (2011) Oxidation of DJ-1 induced by 6-hydroxydopamine decreasing intracellular glutathione. PLoS One 6:e27883

Mollenhauer B, El-Agnaf OM, Marcus K, Trenkwalder C, Schlossmacher MG (2010) Quantification of alpha-synuclein in cerebrospinal fluid as a biomarker candidate: review of the literature and considerations for future studies. Biomark Med 4:683–699

Moscovitz O, Ben-Nissan G, Fainer I, Pollack D, Mizrachi L, Sharon M (2015) The Parkinson's-associated protein DJ-1 regulates the 20S proteasome. Nat Commun 6:6609

Mullett SJ, Hamilton RL, Hinkle DA (2009) DJ-1 immunoreactivity in human brain astrocytes is dependent on infarct presence and infarct age. Neuropathology 29:125–131

Mullett SJ, Di Maio R, Greenamyre JT, Hinkle DA (2013) DJ-1 expression modulates astrocyte-mediated protection against neuronal oxidative stress. J Mol Neurosci 49:507–511

Neumann M, Muller V, Gorner K, Kretzschmar HA, Haass C, Kahle PJ (2004) Pathological properties of the Parkinson's disease-associated protein DJ-1 in alpha-synucleinopathies and tauopathies: relevance for multiple system atrophy and Pick's disease. Acta Neuropathol 107:489–496

Oakley AE, Collingwood JF, Dobson J, Love G, Perrott HR, Edwardson JA et al (2007) Individual dopaminergic neurons show raised iron levels in Parkinson disease. Neurology 68:1820–1825

Ogawa I, Saito Y, Saigoh K, Hosoi Y, Mitsui Y, Noguchi N et al (2014) The significance of oxidized DJ-1 protein (oxDJ-1) as a biomarker for Parkinson's disease. Brain Nerve 66:471–477

Orimo S, Ozawa E, Nakade S, Sugimoto T, Mizusawa H (1999) (123)I-metaiodobenzylguanidine myocardial scintigraphy in Parkinson's disease. J Neurol Neurosurg Psychiatry 67:189–194

Orimo S, Takahashi A, Uchihara T, Mori F, Kakita A, Wakabayashi K et al (2007) Degeneration of cardiac sympathetic nerve begins in the early disease process of Parkinson's disease. Brain Pathol 17:24–30

Parkinson Progression Marker Initiative (2011) The Parkinson progression marker initiative (PPMI). Prog Neurobiol 95:629–635

Parnetti L, Castrioto A, Chiasserini D, Persichetti E, Tambasco N, El-Agnaf O et al (2013) Cerebrospinal fluid biomarkers in Parkinson disease. Nat Rev Neurol 9:131–140

Perry TL, Godin DV, Hansen S (1982) Parkinson's disease: a disorder due to nigral glutathione deficiency? Neurosci Lett 33:305–310

Pezzella A, d'Ischia M, Napolitano A, Misuraca G, Prota G (1997) Iron-mediated generation of the neurotoxin 6-hydroxydopamine quinone by reaction of fatty acid hydroperoxides with dopamine: a possible contributory mechanism for neuronal degeneration in Parkinson's disease. J Med Chem 40:2211–2216

Rybicki BA, Johnson CC, Uman J, Gorell JM (1993) Parkinson's disease mortality and the industrial use of heavy metals in Michigan. Mov Disord 8:87–92

Saito Y (2014) Oxidized DJ-1 as a possible biomarker of Parkinson's disease. J Clin Biochem Nutr 54:138–144

Saito Y, Ruberu NN, Sawabe M, Arai T, Kazama H, Hosoi T et al (2004) Lewy body-related alpha-synucleinopathy in aging. J Neuropathol Exp Neurol 63:742–749

Saito Y, Nishio K, Ogawa Y, Kimata J, Kinumi T, Yoshida Y et al (2006) Turning point in apoptosis/necrosis induced by hydrogen peroxide. Free Radic Res 40:619–630

Saito Y, Nishio K, Ogawa Y, Kinumi T, Yoshida Y, Masuo Y et al (2007a) Molecular mechanisms of 6-hydroxydopamine-induced cytotoxicity in PC12 cells: involvement of hydrogen peroxide-dependent and -independent action. Free Radic Biol Med 42:675–685

Saito Y, Nishio K, Numakawa Y, Ogawa Y, Yoshida Y, Noguchi N et al (2007b) Protective effects of 15-deoxy-Delta12,14-prostaglandin J2 against glutamate-induced cell death in primary cortical neuron cultures: induction of adaptive response and enhancement of cell tolerance primarily through up-regulation of cellular glutathione. J Neurochem 102:1625–1634

Saito Y, Hamakubo T, Yoshida Y, Ogawa Y, Hara Y, Fujimura H et al (2009) Preparation and application of monoclonal antibodies against oxidized DJ-1. Significant elevation of oxidized DJ-1 in erythrocytes of early-stage Parkinson disease patients. Neurosci Lett 465:1–5

Saito Y, Miyasaka T, Hatsuta H, Takahashi-Niki K, Hayashi K, Mita Y et al (2014) Immunostaining of oxidized DJ-1 in human and mouse brains. J Neuropathol Exp Neurol 73:714–728

Saito Y, Akazawa-Ogawa Y, Matsumura A, Saigoh K, Itoh S, Sutou K et al (2016) Oxidation and interaction of DJ-1 with 20S proteasome in the erythrocytes of early stage Parkinson's disease patients. Sci Rep 6:30793

Salvesen L, Bech S, Lokkegaard A, Hjermind LE, Nielsen JE, Pakkenberg B et al (2012) The DJ-1 concentration in cerebrospinal fluid does not differentiate among Parkinsonian syndromes. Parkinsonism Relat Disord 18:899–901

Sayre LM, Perry G, Smith MA (2008) Oxidative stress and neurotoxicity. Chem Res Toxicol 21:172–188

Schapira AH (2013) Recent developments in biomarkers in Parkinson disease. Curr Opin Neurol 26:395–400

Sechi G, Deledda MG, Bua G, Satta WM, Deiana GA, Pes GM et al (1996) Reduced intravenous glutathione in the treatment of early Parkinson's disease. Prog Neuro-Psychopharmacol Biol Psychiatry 20:1159–1170

Semchuk KM, Love EJ, Lee RG (1992) Parkinson's disease and exposure to agricultural work and pesticide chemicals. Neurology 42:1328–1335

Sharma S, Moon CS, Khogali A, Haidous A, Chabenne A, Ojo C et al (2013) Biomarkers in Parkinson's disease (recent update). Neurochem Int 63:201–229

Shi M, Zabetian CP, Hancock AM, Ginghina C, Hong Z, Yearout D et al (2010) Significance and confounders of peripheral DJ-1 and alpha-synuclein in Parkinson's disease. Neurosci Lett 480:78–82

Shi M, Bradner J, Hancock AM, Chung KA, Quinn JF, Peskind ER et al (2011) Cerebrospinal fluid biomarkers for Parkinson disease diagnosis and progression. Ann Neurol 69:570–580

Sian J, Dexter DT, Lees AJ, Daniel S, Agid Y, Javoy-Agid F et al (1994) Alterations in glutathione levels in Parkinson's disease and other neurodegenerative disorders affecting basal ganglia. Ann Neurol 36:348–355

Sies H (1999) Glutathione and its role in cellular functions. Free Radic Biol Med 27:916–921

Taira T, Saito Y, Niki T, Iguchi-Ariga SM, Takahashi K, Ariga H (2004) DJ-1 has a role in antioxidative stress to prevent cell death. EMBO Rep 5:213–218

Takebe G, Yarimizu J, Saito Y, Hayashi T, Nakamura H, Yodoi J et al (2002) A comparative study on the hydroperoxide and thiol specificity of the glutathione peroxidase family and selenoprotein P. J Biol Chem 277:41254–41258

van der Brug MP, Blackinton J, Chandran J, Hao LY, Lal A, Mazan-Mamczarz K et al (2008) RNA binding activity of the recessive parkinsonism protein DJ-1 supports involvement in multiple cellular pathways. Proc Natl Acad Sci U S A 105:10244–10249

Wang X, Michaelis EK (2010) Selective neuronal vulnerability to oxidative stress in the brain. Front Aging Neurosci 2:12

Wang X, Thomas B, Sachdeva R, Arterburn L, Frye L, Hatcher PG et al (2006) Mechanism of arylating quinone toxicity involving Michael adduct formation and induction of endoplasmic reticulum stress. Proc Natl Acad Sci U S A 103:3604–3609

Wang X, Yu S, Li F, Feng T (2015) Detection of alpha-synuclein oligomers in red blood cells as a potential biomarker of Parkinson's disease. Neurosci Lett 599:115–119

Waragai M, Nakai M, Wei J, Fujita M, Mizuno H, Ho G et al (2007) Plasma levels of DJ-1 as a possible marker for progression of sporadic Parkinson's disease. Neurosci Lett 425:18–22

Wilson MA (2011) The role of cysteine oxidation in DJ-1 function and dysfunction. Antioxid Redox Signal 15:111–122

Witt AC, Lakshminarasimhan M, Remington BC, Hasim S, Pozharski E, Wilson MA (2008) Cysteine pKa depression by a protonated glutamic acid in human DJ-1. Biochemistry 47:7430–7440

Yalcinkaya N, Haytural H, Bilgic B, Ozdemir O, Hanagasi H, Kucukali CI et al (2016) Expression changes of genes associated with apoptosis and survival processes in Parkinson's disease. Neurosci Lett 615:72–77

Yanagida T, Tsushima J, Kitamura Y, Yanagisawa D, Takata K, Shibaike T et al (2009) Oxidative stress induction of DJ-1 protein in reactive astrocytes scavenges free radicals and reduces cell injury. Oxidative Med Cell Longev 2:36–42

Yang WS, SriRamaratnam R, Welsch ME, Shimada K, Skouta R, Viswanathan VS et al (2014) Regulation of ferroptotic cancer cell death by GPX4. Cell 156:317–331

Zecca L, Zucca FA, Wilms H, Sulzer D (2003) Neuromelanin of the substantia nigra: a neuronal black hole with protective and toxic characteristics. Trends Neurosci 26:578–580

Zhong N, Kim CY, Rizzu P, Geula C, Porter DR, Pothos EN et al (2006) DJ-1 transcriptionally up-regulates the human tyrosine hydroxylase by inhibiting the sumoylation of pyrimidine tract-binding protein-associated splicing factor. J Biol Chem 281:20940–20948

Zhou W, Freed CR (2005) DJ-1 up-regulates glutathione synthesis during oxidative stress and inhibits A53T alpha-synuclein toxicity. J Biol Chem 280:43150–43158

Zhou W, Zhu M, Wilson MA, Petsko GA, Fink AL (2006) The oxidation state of DJ-1 regulates its chaperone activity toward alpha-synuclein. J Mol Biol 356:1036–1048

Chapter 11
The Role of the Antioxidant Protein DJ-1 in Type 2 Diabetes Mellitus

Daniel Eberhard and Eckhard Lammert

Abstract Type 2 diabetes mellitus (T2DM) is a worldwide escalating health disorder resulting from insulin resistance and functional loss of insulin-producing beta cells that finally cause chronically elevated blood glucose concentrations. Here we review the role of ubiquitously expressed antioxidant protein DJ-1 in the pathogenesis of T2DM. In beta cells, DJ-1 protects against oxidative stress, endoplasmic reticulum stress, and streptozotocin- and cytokine-induced stress and preserves beta cell viability and insulin secretion. In skeletal muscle, DJ-1 controls energy metabolism and efficient fuel utilization, whereas in adipose tissue a role in adipogenesis and obesity-induced inflammation has been reported. This suggests that DJ-1 plays multiple roles in many cell types under metabolically challenging conditions as seen in obesity, insulin resistance, and T2DM.

Keywords DJ-1 • *Park7* • Diabetes mellitus • Obesity • Insulin resistance • Islets of Langerhans • Insulin

---

D. Eberhard (✉)
Institute of Metabolic Physiology, Department of Biology, Heinrich Heine University, D-40225, Düsseldorf, Germany
e-mail: daniel.eberhard@hhu.de

E. Lammert (✉)
Institute of Metabolic Physiology, Department of Biology, Heinrich Heine University, D-40225, Düsseldorf, Germany

Institute for Beta Cell Biology, German Diabetes Center, Leibniz Center for Diabetes, Research at Heinrich Heine University, D-40225 Düsseldorf, Germany

German Center for Diabetes Research (DZD e.V.), D-85764 München-Neuherberg, Germany
e-mail: lammert@hhu.de

H. Ariga, S.M.M. Iguchi-Ariga (eds.), *DJ-1/PARK7 Protein*,
Advances in Experimental Medicine and Biology 1037,
https://doi.org/10.1007/978-981-10-6583-5_11

11.1 Introduction

Diabetes mellitus, a state of chronically increased blood glucose concentrations, is a worldwide escalating health concern affecting around 411 million people worldwide (Welters and Lammert 2014). Type 2 diabetes mellitus (T2DM), the most frequent type of diabetes, results from insulin resistance, i.e., when insulin signaling is blunted in important glucose-metabolizing tissues, and from dysfunction and loss of the insulin-producing beta cells located in the pancreatic islets or islets of Langerhans (Welters and Lammert 2014). Insulin resistance is closely related to obesity, a major risk factor for T2DM (Samuel et al. 2010; Shulman 2014; Sun et al. 2011). In obese individuals, maximally stretched adipocytes insufficiently store lipids, leading to ectopic lipid accumulation in hepatocytes and skeletal muscle (Samuel et al. 2010; Shulman 2014; Sun et al. 2011). It is thought that excess lipid species contribute to impairment of insulin signaling in peripheral tissues, thereby limiting glucose uptake and causing beta cells to produce more insulin to maintain normoglycemia (Shulman 2014; Remedi and Emfinger 2016; Welters and Lammert 2014). However, during this extended period of insulin overproduction, the beta cells are challenged by oxidative and endoplasmic reticulum stress or low-grade inflammation leading to dysfunction and loss of beta cells (Bensellam et al. 2012; Lenzen 2008; Weir and Bonner-Weir 2013). Therefore, agents reducing insulin resistance and agents protecting beta cells from dysfunction and cell death have been in the focus of development of diabetes therapies (Remedi and Emfinger 2016; Welters and Lammert 2014).

The DJ-1 protein is a multifunctional, ubiquitously expressed protein, which was first discovered as an oncogene in conjunction with ras, but later the DJ-1 gene emerged as a causative gene (called *Park7*) of inherited Parkinson's disease (Ariga et al. 2013; Kahle et al. 2009). Among many functions reported, DJ-1 protects from oxidative stress insults, increases cell viability, acts as chaperone and protease, maintains mitochondrial integrity, and contributes to transcriptional regulation, although it does not bind to DNA itself (Ariga et al. 2013; Kahle et al. 2009). The number of proteins, which directly interact or are closely associated with DJ-1, is enormous and includes transcription factors and signal transduction factors.

The activity of DJ-1, which acts as a homodimer, is thought to be modulated by oxidation and posttranslational modifications (sumoylation and S-nitrosylation). DJ-1 contains three cysteine residues which can be oxidized sequentially to SOH, SO_2H, and SO_3H (Ariga et al. 2013). A lower degree of oxidation is thought to correspond to activated DJ-1, whereas complete oxidation renders the protein inactive, possibly even irreversibly (Duan et al. 2008; Choi et al. 2006; Ariga et al. 2013; Kahle et al. 2009). Accumulation of oxidized DJ-1 has been brought in connection with many diseases in which oxidative stress plays a pivotal role. For example, (over-)oxidized DJ-1 forms have been found in the brains of patients with Parkinson's and Alzheimer's disease (Choi et al. 2006).

Oxidative stress, the presence of supraphysiological levels of reactive oxygen species (ROS), which can damage DNA and proteins, has also been linked to the

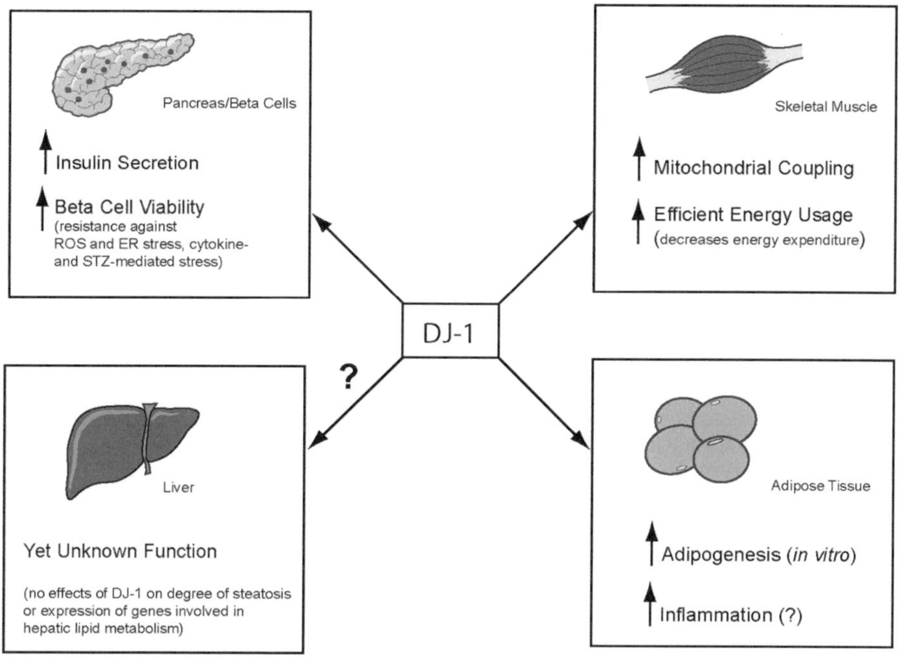

Fig. 11.1 Suggested functions of DJ-1 in tissues relevant for glucose tolerance (Organ icons taken from Welters and Lammert 2014)

pathogenesis of T2DM (Finkel 2011; Houstis et al. 2006). For example, ROS levels rise in response to high glucose exposure in rodent and human islets (Bensellam et al. 2012) and increase in response to a high-fat diet (HFD) or obesity in skeletal muscle (Anderson et al. 2009). However, as ROS are also important signaling molecules, ROS levels may not cause but just correlate with the disease state and may not be harmful under all conditions (Finkel 2011; Tiganis 2011). Here we review the literature about the role of DJ-1 in T2DM in different tissues, including skeletal muscle, adipose tissue, liver, and the islet of Langerhans/pancreatic islets (Fig. 11.1).

11.2 Role of DJ-1 in Glucose Homeostasis In Vivo

Impaired glucose tolerance results from reduced glucose uptake in peripheral tissues and/or insufficient amount of insulin due to islet dysfunction or beta cell loss. This involves many different tissues – e.g., skeletal muscle, liver, adipose tissue, and beta cells – which in concert define the degree of glucose intolerance. The role of DJ-1 in glucose tolerance in vivo has been studied using at least three different DJ-1 deficient mouse models, all harboring a constitutive DJ-1 gene deletion (Shi et al. 2015; Seyfarth et al. 2015; Jain et al. 2015, 2012; Kim et al. 2014) and leading to

Table 11.1 Results of experiments studying glucose tolerance and insulin sensitivity using different DJ-1 KO mouse models

References	Type of DJ-1 KO model (References)	Conditions tested (diet/treatment)	Reported phenotype in DJ-1 KO mice compared to control	Gender differences
Jain et al. (2012)	Deletion of exon 2; Goldberg et al. (2005)	Untreated (chow diet) (aged 8 and 12 weeks)	Increased ROS levels in islets	More pronounced phenotype in male than in female mice
		Short-term 2 week HFD	Decreased insulin secretion Impaired glucose tolerance dependent on age	
Jain et al. (2015)	Deletion of exon 2; Goldberg et al. (2005)	Multiple low-dose streptozotocin (MLDS) treatment	Pronounced diabetic phenotype Deteriorated insulin secretion and glucose tolerance	Only male mice tested
Kim et al. (2014)	Gene trap inserted between exon 6–7; Manning-Bog et al. (2007)	Mice fed with HFD for 12 weeks	Impaired glucose tolerance No changes observed in insulin tolerance test	Only male mice tested
Seyfarth et al. (2015)	Gene trap inserted between exons 6–7; Pham et al. (2010)	Untreated (chow diet)	Transiently increased adiposity in HFD cohort after 14 weeks HFD	Similar results in male and female mice
		HFD for up to 32 weeks	No changes in glucose tolerance observed	
Shi et al. (2015)	Deletion of exons 3–5; Kim et al. (2005)	Untreated (chow diet)	Decreased weight, decreased adiposity, high energy expenditure	More pronounced phenotype in female than in male mice
		HFD for 3 months	Improved glucose tolerance in HFD cohort	

different results (Table 11.1). A first study suggested that DJ-1 deficiency decreases insulin secretion from isolated pancreatic islets (Jain et al. 2012). Consistently, 12-week-old male DJ-1-deficient mice were less glucose tolerant compared to controls, and this was accompanied with elevated ROS levels in DJ-1 KO (knockout) islets, an effect that was age dependent (Jain et al. 2012). The beta cell protective effect of DJ-1 in vivo was confirmed by a follow-up study, showing that injection of DJ-1 KO mice with the beta cell toxin streptozotocin (STZ) led to severe glucose intolerance, dramatically decreased insulin concentrations, as well as decreased beta cell viability and beta cell mass in comparison to STZ-injected control mice (Jain et al. 2015).

A potential role of DJ-1 in diet-induced obesity, i.e., high-fat diet (HFD) feeding, a model to study insulin resistance and glucose tolerance in mice, was analyzed using different experimental strategies (Table 11.1). For example, glucose intolerance was reported in DJ-1 KO mice in combination with a 12-week-long HFD challenge compared to controls (Kim et al. 2014). This phenotype was accompanied with reduced inflammation in adipose tissue (Kim et al. 2014), considered to be beneficial for insulin sensitivity due to reduced levels of circulating pro-inflammatory cytokines. However, as insulin sensitivity was unaltered in DJ-1-deficient mice compared to controls, dysfunctional pancreatic islets may again explain the observed glucose intolerance (Kim et al. 2014).

However, glucose tolerance was not affected in DJ-1 KO mice, fed either with a chow or HFD for 14 weeks (Seyfarth et al. 2015), in another study using mice derived from the same ES cell clone as a previous study with differences in glucose tolerance (Seyfarth et al. 2015; Kim et al. 2014; Manning-Bog et al. 2007; Pham et al. 2010). In contrast to all other studies, a markedly improved glucose tolerance was reported using a third DJ-1 KO mouse model under HFD conditions (Kim et al. 2005; Shi et al. 2015). The effect was explained by higher energy expenditure ("energy wasting") by skeletal muscle of DJ-1 KO mice leading to lower adiposity, improved insulin sensitivity and glucose tolerance (Shi et al. 2015). The adiposity phenotype was in contrast to the phenotype of another DJ-1 model (Kim et al. 2014; Seyfarth et al. 2015), which displayed a transient increase in body fat mass (adjusted to body mass) in male and female DJ-1 KO mice, possibly caused by reduced physical activity of these DJ-1 KO mice (Seyfarth et al. 2015).

The reasons for the discrepancies in glucose homeostasis in the different DJ-1 KO mouse models may be explained by differences in the experimental setup (gender, fasting time, age, length of HFD challenge, or microbiomes). The use of organ- and cell-specific deletions of the DJ-1 gene, as well as mice overexpressing DJ-1, will allow more comprehensive insights in the tissue- and cell-specific role of DJ-1 mice in the future.

11.3 The Role of DJ-1 in the Islet of Langerhans

11.3.1 Protective Effect of DJ-1 in Beta Cells

Dysfunction and progressive loss of the insulin-producing beta cells lead to chronic hyperglycemia (Cnop et al. 2005; Weir and Bonner-Weir 2013; Remedi and Emfinger 2016; Welters and Lammert 2014), which is preceded by a stage of insulin resistance in T2DM. During mild insulin resistance, insulin overproduction by beta cells is still sufficient to maintain normoglycemia. However, even mild hyperglycemia is thought to reduce glucose-stimulated insulin secretion (Weir and Bonner-Weir 2013). With increasing insulin resistance and hyperglycemia, beta cells face additional stresses: Firstly, beta cells are exposed to high levels of ROS (Bensellam et al. 2012) that are predominately produced in mitochondria and quenched by an antioxidant defense system under normal conditions. However, as beta cells only express low levels of antioxidant proteins, the antioxidative capacity is limited in beta cells, thus more easily leading to oxidative stress (Lenzen 2008). Secondly, prolonged periods of insulin overproduction trigger ER stress leading to the recruitment of chaperones that support the folding or degradation of proteins. Excessive and prolonged ER stress, however, also triggers apoptotic pathways causing beta cell death (Weir and Bonner-Weir 2013).

In addition to oxidative stress and ER stress, beta cells in prediabetic subjects and individuals with established T2DM patients encounter low-grade inflammation, i.e., elevated levels of pro-inflammatory cytokines that affect islet function and survival (Abdulreda and Berggren 2013). The triggers for islet inflammation in T2DM still need to be clarified though (Abdulreda and Berggren 2013). Some of the inflammatory pathways may be shared with type 1 diabetes mellitus, in which immune cells infiltrate the islets and secrete cytotoxic pro-inflammatory cytokines promoting beta cell death (Abdulreda and Berggren 2013; Eizirik et al. 2009; Jorns et al. 2014).

A role of DJ-1 in oxidative stress/glucotoxic conditions in the islets of Langerhans was first suggested in a large quantitative proteomic analysis when isolated mouse islets were exposed to high glucose concentrations leading to a twofold upregulation of the DJ-1 protein (Waanders et al. 2009). Thereafter, the beta cell protective role of DJ-1 was shown for many different experimental stress conditions including oxidative stress (Inberg and Linial 2010; Jain et al. 2012; Jo et al. 2016a; Waanders et al. 2009), ER stress (Inberg and Linial 2010) or pro-inflammatory cytokine-induced stress (Jain et al. 2015; Jo et al. 2016b). Moreover, DJ-1 was shown to preserve mitochondrial integrity (Jain et al. 2012, 2015) and insulin secretion in beta cells and experimental beta cell lines (Inberg and Linial 2010; Jain et al. 2012, 2015). In line with increased beta cell viability and function, DJ-1 mRNA and protein are upregulated in beta cell lines or isolated mouse and human islets in response to experimental stress conditions, including treatment with H_2O_2, exposure to high glucose concentrations (Inberg and Linial 2010; Jain et al. 2012; Waanders et al. 2009), or treatment with thapsigargin, an ER stress inducer (Inberg and Linial 2010). Likewise, MIN6 cells with silenced DJ-1 levels are highly sensitive to H_2O_2-mediated oxidative insult compared to cells with normal DJ-1 expression (Inberg and Linial 2010).

Consistent with the DJ-1 loss-of-function experiments described, beta cell survival can be enhanced by DJ-1 overexpression in several beta cell lines. For example, adenoviral overexpression of DJ-1 in mouse insulinoma cells (MIN6) significantly preserved cell viability after H_2O_2 or thapsigargin treatment (Inberg and Linial 2010). Moreover, Jo et al. (2016a, b) demonstrated the protective effect of a cell permeable DJ-1 protein (Tat-DJ-1), which protected rat insulinoma cells

(RINm5F) from H_2O_2 and also from pro-inflammatory cytokine-induced cell stress (Jo et al. 2016a, b).

The protective effects of DJ-1 were also demonstrated in vivo by challenging DJ-1-deficient mice with multiple low doses of STZ (Jain et al. 2015). STZ is a glucose analogue causing DNA alkylation and NAD^+ depletion (due to hyperactivity of poly (ADP-ribose) polymerase, PARP) in beta cells resulting in insulitis and beta cell death. STZ treatment led to a doubled rate of beta cell death, decreased plasma insulin levels, increased fasting blood glucose concentrations, and a dramatically reduced glucose tolerance in DJ-1 KO mice compared to controls. Moreover, the mitochondrial network was reduced and less insulin granules were observed in STZ-treated DJ-1 KO beta cells compared to STZ-treated controls (Jain et al. 2015). Increased cell death rates were also observed ex vivo, when isolated islets of DJ-1 KO mice were treated with either a pro-inflammatory cytokine cocktail (interleukin-1β, tumor necrosis factor α, and interferon γ) or with STZ (Jain et al. 2015).

Most importantly, in human islets, an upregulation of DJ-1 protein was also observed after exposure to high glucose concentrations, indicating a protective role for DJ-1 in human beta cells (Jain et al. 2012). Interestingly, DJ-1 mRNA expression in human islets appears to be age dependent, i.e., DJ-1 expression was increased in islets of elderly humans (with an average age of 74 years) compared to islets of younger humans (with an average age of 44 years) (Jain et al. 2012). Thus, in human beta cells, DJ-1 expression may increase and adjust these cells to an age-related increase in ROS. The finding that DJ-1 expression is significantly reduced in the islets of elderly human subjects with T2DM (Jain et al. 2012) suggests that failure to upregulate DJ-1 weakens the cell stress defense in human islets making them more susceptible to oxidative stress followed by beta cell dysfunction and death.

11.3.2 Molecular Basis of the DJ-1 Protective Effect in Beta Cells

Different scenarios have been proposed to explain the beta cell protective effects of DJ-1 at a molecular level. Firstly, DJ-1 may quench ROS species by oxidation of its cysteine residues (Ariga et al. 2013). However, this capacity is limited (Junn et al. 2005), suggesting that ROS scavenging by DJ-1 alone is insufficient to normalize cellular redox homeostasis. More likely, DJ-1 may act as a stress sensor, which enables transcription factors to translocate to the nucleus, activates survival pathways, or reduces the activity of pro-apoptotic signaling pathways. The ability to serve as a redox sensor is attributed to at least one of its three cysteine residues (C106) thought to activate DJ-1 upon its oxidation (Ariga et al. 2013). In mouse islets and MIN6 cells, DJ-1 isoforms shift from basic to more acidic (oxidized) forms after exposure to H_2O_2 (Inberg and Linial 2010), suggesting that DJ-1 also serves as an oxidative sensor in beta cells as is seen in neurons (Ariga et al. 2013;

Kahle et al. 2009). Consistent with this notion, the beta cell protective effect of DJ-1 is lost when a DJ-1 mutant lacking the oxidative sensitive cysteine residue C106 was used (Jo et al. 2016a).

Several downstream pathways of DJ-1 have been proposed to mediate protection of beta cells. For example, it has been reported that DJ-1 regulates nuclear factor erythroid 2-related factor (Nrf2), a master regulator of cellular oxidative stress defense, in cancer cell lines and mouse fibroblasts (Clements et al. 2006; Ma 2013). Under oxidative stress conditions, DJ-1 sequesters Keap-1 (Kelch-like ECH-associated protein 1), a cytosolic Nrf2-binding protein, which enables free Nrf2 to translocate to the nucleus and induce the expression of antioxidative genes restoring ROS levels to normal (Clements et al. 2006; Ma 2013). Although a potential link between DJ-1 and Nrf2 activation in the beta cell remains to be shown, the Keap1-Nrf2 system was shown to play an important role in beta cell maintenance in response to toxic levels of reactive species (Dinic et al. 2016; Yagishita et al. 2014).

A link between DJ-1 and other pathways affecting beta cell viability, the NF-κB and the mitogen-activated protein kinases (MAPK) pathways, has been suggested under oxidative stress conditions in rat insulinoma cells (Jo et al. 2016a). NF-κB is a transcription factor, which undergoes nuclear translocation, e.g., after oxidative stress or cytokine exposure in many different cell types. In beta cells, a potential activation of NF-κB by ROS or high glucose levels has been debated (Cnop et al. 2005). However, peroxide-challenged RINm5F cells displayed activated NF-κB, and this activation is reduced (less phosphorylation of p65 and IκBα) if the cells are treated with a cell permeable Tat-DJ-1 protein. This effect was abolished if cells were treated with a mutant DJ-1 protein lacking the oxidation-sensitive cysteine C106 (Jo et al. 2016a). Besides its effects on NF-κB, the functional Tat-DJ-1 protein also reduced the phosphorylation of MAP kinases p38, JNK (c-Jun N-terminal kinases), and ERK (extracellular signal-regulated kinases) (Jo et al. 2016a) and attenuated the apoptotic pathway (e.g., resulting in less cleaved caspase-3) after exposure to pro-inflammatory cytokines interleukin-1β, tumor necrosis factor α, and interferon γ (Jo et al. 2016b). As Tat-DJ-1 could reduce both, the peroxide and cytokine-induced high ROS levels, (Jo et al. 2016a, b), the DJ-1 protective effect in beta cells may primarily ground on its antioxidative effect.

Regarding ER stress, another DJ-1 protective mechanism was proposed by Inberg et al. who identified the transcription factor TFII-I as a cytosolic interaction partner of DJ-1 (Inberg and Linial 2010). TFII-I activates the expression of the chaperone BiP, which is part of the unfolded protein response (UPR) in ER stress (Inberg and Linial 2010). Interestingly, under conditions of high DJ-1 expression, BiP expression was less increased after thapsigargin treatment (Inberg and Linial 2010). It was suggested that a high amount of cytosolic DJ-1 reduces TFII-I nuclear translocation, thereby restraining the UPR response (including BiP). However, DJ-1 may also act via an independent mechanism to reduce ER stress.

Finally, DJ-1 protects mitochondrial integrity and function in beta cells via normalizing ROS levels (Jain et al. 2012, 2015). Mitochondria are an important source of ROS, and high ROS levels cause mitochondrial dysfunction and insulin secretion defects in beta cells (Supale et al. 2012). MIN6 cells silenced for DJ-1 display increased mitochondrial ROS levels, which can be restored to normal by transfection with a DJ-1 expression plasmid (Jain et al. 2012). Moreover, significantly more fragmented mitochondria are observed in DJ-1 silenced MIN6 cells and islets of DJ-1 KO mice aged 12–13 weeks (Jain et al. 2012), an effect that could be reversed by the antioxidant N-acetyl-L-cysteine (NAC) (Jain et al. 2012). Moreover, ATP production, an important mediator of insulin secretion in response to high glucose, was decreased in DJ-1-deficient mouse islets compared to control islets, consistent with the mitochondrial phenotype (Jain et al. 2012). The molecular basis of the protective effect of DJ-1 in mitochondria remains to be further explored. In this context, however, it is noteworthy that expression levels of dynamin-like protein (DLP1/DRP1), a regulator of mitochondrial fission, were shown to depend on DJ-1 expression in human neuroblastoma cells (Wang et al. 2012).

In conclusion, DJ-1 protects beta cells from various cell stresses and preserves mitochondrial homeostasis and insulin secretion. However, a common molecular mechanism explaining all cytoprotective effects of DJ-1 has not been proposed so far.

11.4 Adipose Tissue

Adipose tissue plays a central role in the development of insulin resistance (Shulman 2014; Welters and Lammert 2014; McArdle et al. 2013). Adipocytes store excess energy as triglycerides and also contribute to appetite and metabolic control by secreting hormones (McArdle et al. 2013). In obesity, adipose tissue expands due to adipocyte hypertrophy and differentiation of new adipocytes from precursor cells (adipogenesis) (Sun et al. 2011). In severe obesity, hypoxia and a low-grade inflammation introduced by resident and infiltrating immune cells lead to release of free fatty acids and pro-inflammatory cytokines, negatively affecting insulin signaling in other tissues (Sun et al. 2011; Samuel et al. 2010; Shulman 2014; Welters and Lammert 2014; McArdle et al. 2013).

The role of DJ-1 in adipogenesis and inflammation has been studied in vitro and in diet-induced obesity in rodents (Kim et al. 2014; Shi et al. 2015). DJ-1 is upregulated during adipogenic differentiation of fibroblast-like 3T3-L1 cells toward adipocytes in vitro. Interestingly, silencing of DJ-1 in 3T3-L1 reduced adipogenic differentiation and decreased markers of mature adipocytes (PPARγ, LPL, Glut4), showing that DJ-1 is required for adipogenesis in vitro (Kim et al. 2014). However, the role of DJ-1 in adipogenesis in vivo as investigated by Kim et al. (2014) may be different or at least multifaceted, as the expression of adipogenic genes, adipocyte number and size were not different between DJ-1 KO and control mice (Kim et al. 2014). This was partly confirmed for another DJ-1 mouse model that showed no change in adipogenic and lipogenic gene expression in the adipose tissue (Shi et al. 2015). However, the DJ-1 KO mice in this model were leaner and had smaller adipocytes compared to controls (Shi et al. 2015). Interestingly, mononuclear cell infiltration and interleukin-6 serum levels, a measure of adipose tissue inflammation in obesity, were decreased in DJ-1 KO mice suggesting that DJ-1 contributes to inflammation in adipose tissue, which however did not change insulin resistance or glucose tolerance in this mouse (Kim et al. 2014).

11.5 Skeletal Muscle

In humans and rodents, skeletal muscle accounts for most of the postprandial glucose uptake. In response to obesity or HFD, ROS levels rise in skeletal muscle (Anderson et al. 2009; Shi et al. 2015). In contrast to beta cells where extreme ROS levels are deleterious (see section on beta cells), in skeletal muscle, high ROS levels may rather be beneficial, especially as elevated ROS levels are found in conditions associated with increased life span, e.g., during exercise (Tiganis 2011).

In line with this, low DJ-1 expression in skeletal muscle causes a rise in ROS levels, which however did not lead to oxidative stress or mitochondrial disarrangement (Shi et al. 2015). Instead, DJ-1, which is upregulated under HFD conditions, seems to be involved in energy metabolism in the skeletal muscle, and DJ-1 deficiency was reported to be favorable for insulin sensitivity and glucose tolerance in vivo (Shi et al. 2015). It was suggested that elevated mitochondrial ROS levels in DJ-1 KO skeletal muscle induce the expression of uncoupling protein 3 (UCP3), thus facilitating proton leakage (mitochondrial uncoupling), increasing O_2 consumption, but decreasing ATP production. As a consequence of emerging energy depletion, AMPK, a central mediator of cellular energy levels, is activated leading to increased glycolysis, generating a futile cycle thereby "wasting" energy (Shi et al. 2015). In line with this, DJ-1 KO mice consumed more oxygen and energy (even though no change in body temperature could be detected) and displayed a decreased body weight and adiposity. Consistently, the mice were more insulin sensitive and glucose tolerant compared to controls in response to a HFD (Shi et al. 2015). It remains to be resolved if other DJ-1 KO models share this phenotype, as no weight changes (Kim et al. 2014) or even transiently increased adiposity (Seyfarth et al. 2015) was reported for other DJ-1 KO mouse models in response to a HFD.

11.6 Liver

Hepatic insulin resistance leads to an increased glucose output contributing to chronic hyperglycemia (Samuel et al. 2010; Welters and Lammert 2014; Perry et al. 2014). Little is known about a potential function of DJ-1 in hepatic insulin resistance or nonalcoholic fatty liver disease (NAFLD), a condition closely related to hepatic insulin resistance (Samuel et al. 2010; Perry et al. 2014). Even though hepatic ROS levels rise in response to a HFD in mice (Lohr et al. 2016; Shi et al. 2015), DJ-1 mRNA expression is not found to increase (Shi et al. 2015). In addition, ROS levels in the liver of DJ-1 KO mice were not different compared to control mice, and the degree of lipid accumulation as well as expression of genes controlling hepatic lipid metabolism were unchanged, indicating no major role of DJ-1 in hepatic insulin resistance under the experimental conditions studied (Shi et al. 2015).

11.7 Conclusions and Implications for Future Research

DJ-1 is a ubiquitously expressed antioxidative protein with multiple functions. Low levels of DJ-1 are frequently associated with increasing ROS levels, especially under challenging conditions like hyperglycemia. Most likely, elevated DJ-1 levels will help to protect insulin- producing beta cells as well as other cells vulnerable to oxidative stress. Increasing DJ-1 by pharmacological treatment may be one way forward to treat diabetes and its complications, although adverse events of elevated DJ-1 activity have to be carefully considered, since ROS levels may correlate rather than contribute to the pathogenesis of T2DM in some tissues (Finkel 2011). One way forward to increase DJ-1 levels may be the clinically used chemical chaperone 4-phenyl butyric acid (PBA), which increases DJ-1 expression (Zhou et al. 2011) and improves insulin resistance in T2DM rodent mouse model (Ozcan et al. 2006). Alternatively, substances designed to maintain the activity of DJ-1 by preventing its hyperoxidation and inactivation (Inden et al. 2011; Kitamura et al. 2011) could be tested under conditions of T2DM.

References

Abdulreda MH, Berggren PO (2013) Islet inflammation in plain sight. Diabetes Obes Metab 15(Suppl 3):105–116. https://doi.org/10.1111/dom.12160

Anderson EJ, Lustig ME, Boyle KE, Woodlief TL, Kane DA, Lin CT, Price JW 3rd, Kang L, Rabinovitch PS, Szeto HH, Houmard JA, Cortright RN, Wasserman DH, Neufer PD (2009) Mitochondrial H2O2 emission and cellular redox state link excess fat intake to insulin resistance in both rodents and humans. J Clin Invest 119(3):573–581. https://doi.org/10.1172/JCI37048

Ariga H, Takahashi-Niki K, Kato I, Maita H, Niki T, Iguchi-Ariga SM (2013) Neuroprotective function of DJ-1 in Parkinson's disease. Oxidative Med Cell Longev 2013:683920. https://doi.org/10.1155/2013/683920

Bensellam M, Laybutt DR, Jonas JC (2012) The molecular mechanisms of pancreatic beta-cell glucotoxicity: recent findings and future research directions. Mol Cell Endocrinol 364(1–2):1–27. https://doi.org/10.1016/j.mce.2012.08.003

Choi J, Sullards MC, Olzmann JA, Rees HD, Weintraub ST, Bostwick DE, Gearing M, Levey AI, Chin LS, Li L (2006) Oxidative damage of DJ-1 is linked to sporadic Parkinson and Alzheimer diseases. J Biol Chem 281(16):10816–10824. https://doi.org/10.1074/jbc.M509079200

Clements CM, McNally RS, Conti BJ, Mak TW, Ting JP (2006) DJ-1, a cancer- and Parkinson's disease-associated protein, stabilizes the antioxidant transcriptional master regulator Nrf2. Proc Natl Acad Sci U S A 103(41):15091–15096. https://doi.org/10.1073/pnas.0607260103

Cnop M, Welsh N, Jonas JC, Jorns A, Lenzen S, Eizirik DL (2005) Mechanisms of pancreatic beta-cell death in type 1 and type 2 diabetes: many differences, few similarities. Diabetes 54(Suppl 2):S97–107

Dinic S, Grdovic N, Uskokovic A, Dordevic M, Mihailovic M, Jovanovic JA, Poznanovic G, Vidakovic M (2016) CXCL12 protects pancreatic beta-cells from oxidative stress by a Nrf2-induced increase in catalase expression and activity. Proc Jpn Acad Ser B Phys Biol Sci 92(9):436–454. https://doi.org/10.2183/pjab.92.436

Duan X, Kelsen SG, Merali S (2008) Proteomic analysis of oxidative stress-responsive proteins in human pneumocytes: insight into the regulation of DJ-1 expression. J Proteome Res 7(11):4955–4961. https://doi.org/10.1021/pr800295j

Eizirik DL, Colli ML, Ortis F (2009) The role of inflammation in insulitis and beta-cell loss in type 1 diabetes. Nat Rev Endocrinol 5(4):219–226. https://doi.org/10.1038/nrendo.2009.21

Finkel T (2011) Signal transduction by reactive oxygen species. J Cell Biol 194(1):7–15. https://doi.org/10.1083/jcb.201102095

Goldberg MS, Pisani A, Haburcak M, Vortherms TA, Kitada T, Costa C, Tong Y, Martella G, Tscherter A, Martins A, Bernardi G, Roth BL, Pothos EN, Calabresi P, Shen J (2005) Nigrostriatal dopaminergic deficits and hypokinesia caused by inactivation of the familial Parkinsonism-linked gene DJ-1. Neuron 45(4):489–496. https://doi.org/10.1016/j.neuron.2005.01.041

Houstis N, Rosen ED, Lander ES (2006) Reactive oxygen species have a causal role in multiple forms of insulin resistance. Nature 440(7086):944–948. https://doi.org/10.1038/nature04634

Inberg A, Linial M (2010) Protection of pancreatic beta-cells from various stress conditions is mediated by DJ-1. J Biol Chem 285(33):25686–25698. https://doi.org/10.1074/jbc.M110.109751

Inden M, Kitamura Y, Takahashi K, Takata K, Ito N, Niwa R, Funayama R, Nishimura K, Taniguchi T, Honda T, Taira T, Ariga H (2011) Protection against dopaminergic neurodegeneration in Parkinson's disease-model animals by a modulator of the oxidized form of DJ-1, a wild-type of familial Parkinson's disease-linked PARK7. J Pharmacol Sci 117(3):189–203

Jain D, Jain R, Eberhard D, Eglinger J, Bugliani M, Piemonti L, Marchetti P, Lammert E (2012) Age- and diet-dependent requirement of DJ-1 for glucose homeostasis in mice with implications for human type 2 diabetes. J Mol Cell Biol 4(4):221–230. https://doi.org/10.1093/jmcb/mjs025

Jain D, Weber G, Eberhard D, Mehana AE, Eglinger J, Welters A, Bartosinska B, Jeruschke K, Weiss J, Path G, Ariga H, Seufert J, Lammert E (2015) DJ-1 protects pancreatic beta cells from cytokine- and streptozotocin-mediated cell death. PLoS One 10(9):e0138535. https://doi.org/10.1371/journal.pone.0138535

Jo HS, Cha HJ, Kim SJ, Yeo HJ, Cho SB, Park JH, Lee CH, Yeo EJ, Choi YJ, Eum WS, Choi SY (2016a) Tat-DJ-1 inhibits oxidative stress-mediated RINm5F cell death through suppression of NF-kappaB and MAPK activation. Med Chem Res 25(11):2589–2598. https://doi.org/10.1007/s00044-016-1698-4

Jo HS, Yeo HJ, Cha HJ, Kim SJ, Cho SB, Park JH, Lee CH, Yeo EJ, Choi YJ, Eum WS, Choi SY (2016b) Transduced Tat-DJ-1 protein inhibits cytokines-induced pancreatic RINm5F cell death. BMB Rep 49(5):297–302

Jorns A, Arndt T, Meyer zu Vilsendorf A, Klempnauer J, Wedekind D, Hedrich HJ, Marselli L, Marchetti P, Harada N, Nakaya Y, Wang GS, Scott FW, Gysemans C, Mathieu C, Lenzen S (2014) Islet infiltration, cytokine expression and beta cell death in the NOD mouse, BB rat, Komeda rat, LEW.1AR1-iddm rat and humans with type 1 diabetes. Diabetologia 57(3):512–521. https://doi.org/10.1007/s00125-013-3125-4

Junn E, Taniguchi H, Jeong BS, Zhao X, Ichijo H, Mouradian MM (2005) Interaction of DJ-1 with Daxx inhibits apoptosis signal-regulating kinase 1 activity and cell death. Proc Natl Acad Sci U S A 102(27):9691–9696. https://doi.org/10.1073/pnas.0409635102

Kahle PJ, Waak J, Gasser T (2009) DJ-1 and prevention of oxidative stress in Parkinson's disease and other age-related disorders. Free Radic Biol Med 47(10):1354–1361. https://doi.org/10.1016/j.freeradbiomed.2009.08.003

Kim RH, Smith PD, Aleyasin H, Hayley S, Mount MP, Pownall S, Wakeham A, You-Ten AJ, Kalia SK, Horne P, Westaway D, Lozano AM, Anisman H, Park DS, Mak TW (2005) Hypersensitivity of DJ-1-deficient mice to 1-methyl-4-phenyl-1,2,3,6-tetrahydropyrindine (MPTP) and oxidative stress. Proc Natl Acad Sci U S A 102(14):5215–5220. https://doi.org/10.1073/pnas.0501282102

Kim JM, Jang HJ, Choi SY, Park SA, Kim IS, Yang YR, Lee YH, Ryu SH, Suh PG (2014) DJ-1 contributes to adipogenesis and obesity-induced inflammation. Sci Rep 4:4805. https://doi.org/10.1038/srep04805

Kitamura Y, Watanabe S, Taguchi M, Takagi K, Kawata T, Takahashi-Niki K, Yasui H, Maita H, Iguchi-Ariga SM, Ariga H (2011) Neuroprotective effect of a new DJ-1-binding compound against neurodegeneration in Parkinson's disease and stroke model rats. Mol Neurodegener 6(1):48. https://doi.org/10.1186/1750-1326-6-48

Lenzen S (2008) Oxidative stress: the vulnerable beta-cell. Biochem Soc Trans 36(Pt 3):343–347. https://doi.org/10.1042/BST0360343

Lohr K, Pachl F, Moghaddas Gholami A, Geillinger KE, Daniel H, Kuster B, Klingenspor M (2016) Reduced mitochondrial mass and function add to age-related susceptibility toward diet-induced fatty liver in C57BL/6J mice. Physiol Rep 4(19). 10.14814/phy2.12988

Ma Q (2013) Role of nrf2 in oxidative stress and toxicity. Annu Rev Pharmacol Toxicol 53:401–426. https://doi.org/10.1146/annurev-pharmtox-011112-140320

Manning-Bog AB, Caudle WM, Perez XA, Reaney SH, Paletzki R, Isla MZ, Chou VP, McCormack AL, Miller GW, Langston JW, Gerfen CR, Dimonte DA (2007) Increased vulnerability of nigrostriatal terminals in DJ-1-deficient mice is mediated by the dopamine transporter. Neurobiol Dis 27(2):141–150. https://doi.org/10.1016/j.nbd.2007.03.014

McArdle MA, Finucane OM, Connaughton RM, McMorrow AM, Roche HM (2013) Mechanisms of obesity-induced inflammation and insulin resistance: insights into the emerging role of nutritional strategies. Front Endocrinol (Lausanne) 4:52. https://doi.org/10.3389/fendo.2013.00052

Ozcan U, Yilmaz E, Ozcan L, Furuhashi M, Vaillancourt E, Smith RO, Gorgun CZ, Hotamisligil GS (2006) Chemical chaperones reduce ER stress and restore glucose homeostasis in a mouse model of type 2 diabetes. Science 313(5790):1137–1140. https://doi.org/10.1126/science.1128294

Perry RJ, Samuel VT, Petersen KF, Shulman GI (2014) The role of hepatic lipids in hepatic insulin resistance and type 2 diabetes. Nature 510(7503):84–91. https://doi.org/10.1038/nature13478

Pham TT, Giesert F, Rothig A, Floss T, Kallnik M, Weindl K, Holter SM, Ahting U, Prokisch H, Becker L, Klopstock T, Hrabe de Angelis M, Beyer K, Gorner K, Kahle PJ, Vogt Weisenhorn DM, Wurst W (2010) DJ-1-deficient mice show less TH-positive neurons in the ventral tegmental area and exhibit non-motoric behavioural impairments. Genes Brain Behav 9(3):305–317. https://doi.org/10.1111/j.1601-183X.2009.00559.x

Remedi MS, Emfinger C (2016) Pancreatic beta-cell identity in diabetes. Diabetes Obes Metab 18(Suppl 1):110–116. https://doi.org/10.1111/dom.12727

Samuel VT, Petersen KF, Shulman GI (2010) Lipid-induced insulin resistance: unravelling the mechanism. Lancet 375(9733):2267–2277. https://doi.org/10.1016/S0140-6736(10)60408-4

Seyfarth K, Poschmann G, Rozman J, Fromme T, Rink N, Hofmann A, Wurst W, Stuhler K, Klingenspor M (2015) The development of diet-induced obesity and associated metabolic impairments in Dj-1 deficient mice. J Nutr Biochem 26(1):75–81. https://doi.org/10.1016/j.jnutbio.2014.09.002

Shi SY, Lu SY, Sivasubramaniyam T, Revelo XS, Cai EP, Luk CT, Schroer SA, Patel P, Kim RH, Bombardier E, Quadrilatero J, Tupling AR, Mak TW, Winer DA, Woo M (2015) DJ-1 links muscle ROS production with metabolic reprogramming and systemic energy homeostasis in mice. Nat Commun 6:7415. https://doi.org/10.1038/ncomms8415

Shulman GI (2014) Ectopic fat in insulin resistance, dyslipidemia, and cardiometabolic disease. N Engl J Med 371(23):2237–2238. https://doi.org/10.1056/NEJMc1412427

Sun K, Kusminski CM, Scherer PE (2011) Adipose tissue remodeling and obesity. J Clin Invest 121(6):2094–2101. https://doi.org/10.1172/JCI45887

Supale S, Li N, Brun T, Maechler P (2012) Mitochondrial dysfunction in pancreatic beta cells. Trends Endocrinol Metab 23(9):477–487. https://doi.org/10.1016/j.tem.2012.06.002

Tiganis T (2011) Reactive oxygen species and insulin resistance: the good, the bad and the ugly. Trends Pharmacol Sci 32(2):82–89. https://doi.org/10.1016/j.tips.2010.11.006

Waanders LF, Chwalek K, Monetti M, Kumar C, Lammert E, Mann M (2009) Quantitative proteomic analysis of single pancreatic islets. Proc Natl Acad Sci U S A 106(45):18902–18907. https://doi.org/10.1073/pnas.0908351106

Wang X, Petrie TG, Liu Y, Liu J, Fujioka H, Zhu X (2012) Parkinson's disease-associated DJ-1 mutations impair mitochondrial dynamics and cause mitochondrial dysfunction. J Neurochem 121(5):830–839. https://doi.org/10.1111/j.1471-4159.2012.07734.x

Weir GC, Bonner-Weir S (2013) Islet beta cell mass in diabetes and how it relates to function, birth, and death. Ann N Y Acad Sci 1281:92–105. https://doi.org/10.1111/nyas.12031

Welters A, Lammert E (2014) Diabetes Mellitus. In: Lammert E, Zeeb M (eds) Metabolism of human diseases: organ physiology and pathophysiology. Springer, Vienna, pp 163–173

Yagishita Y, Fukutomi T, Sugawara A, Kawamura H, Takahashi T, Pi J, Uruno A, Yamamoto M (2014) Nrf2 protects pancreatic beta-cells from oxidative and nitrosative stress in diabetic model mice. Diabetes 63(2):605–618. https://doi.org/10.2337/db13-0909

Zhou W, Bercury K, Cummiskey J, Luong N, Lebin J, Freed CR (2011) Phenylbutyrate up-regulates the DJ-1 protein and protects neurons in cell culture and in animal models of Parkinson disease. J Biol Chem 286(17):14941–14951. https://doi.org/10.1074/jbc.M110.211029

Chapter 12
Therapeutic Activities of DJ-1 and Its Binding Compounds Against Neurodegenerative Diseases

Masatoshi Inden, Daijiro Yanagisawa, Masanori Hijioka, Hiroyoshi Ariga, and Yoshihisa Kitamura

Abstract Parkinson's disease (PD) is a progressive neurodegenerative disorder that is primarily characterized by the degeneration of dopaminergic neurons in the nigrostriatal pathway. Loss-of-function mutations in the gene encoding PARK7/DJ-1 were identified in familial PD. Wild-type DJ-1 acts as an oxidative stress sensor in neural cells. Previously, we identified binding compounds of DJ-1, including UCP0045037/compound A, UCP0054278/compound B, and compound-23 (comp-23), by in silico virtual screening. These compounds prevented oxidative stress-induced dopaminergic neuronal death and restored locomotion defects in animal models of PD. In addition, these binding partners reduced infarct size in cerebral ischemia in rats. The neuroprotective effects of these compounds are lost in *DJ-1*-knockdown cells and *DJ*-1-knockout animal. These results suggest that these compounds interact with endogenous DJ-1 and then produce antioxidant and neuroprotective responses in both animal models for PD and cerebral ischemia in rats. This raises the possibility that interaction partners of DJ-1, such as UCP0045037, UCP0054278, and comp-23, may represent a novel dopaminergic neuroprotective drug for the treatment of PD.

M. Inden
Laboratory of Medical Therapeutics and Molecular Therapeutics, Gifu Pharmaceutical University, Gifu, Japan

D. Yanagisawa
Molecular Neuroscience Research Center, Shiga University of Medical Science, Otsu, Japan

M. Hijioka • Y. Kitamura (✉)
Laboratory of Pharmacology and Neurobiology, College of Pharmaceutical Sciences, Ritsumeikan University, 1-1-1 Noji-Higashi, Kusatsu, Shiga, Japan
e-mail: yo-kita@fc.ritsumei.ac.jp

H. Ariga
Faculty of Pharmaceutical Sciences, Hokkaido University, Kita 12, Nishi 6, Kita-ku, Sapporo 060-0812, Japan
e-mail: hiro@pharm.hokudai.ac.jp

H. Ariga, S.M.M. Iguchi-Ariga (eds.), *DJ-1/PARK7 Protein*, Advances in Experimental Medicine and Biology 1037,
https://doi.org/10.1007/978-981-10-6583-5_12

Keywords Parkinson's disease • DJ-1 • 6-hydroxydopamine • Rotenone • Neuroprotection

12.1 Introduction

12.1.1 DJ-1

DJ-1 (also known as PARK7) was first discovered as a novel oncogene (Nagakubo et al. 1997) and was later identified as a causative gene of Parkinson's disease (PD) (Bonifati et al. 2003). DJ-1 has multiple functions and plays a role in anti-oxidative stress and transcriptional regulation (Takahashi et al. 2001; Niki et al. 2003; Yokota et al. 2003; Taira et al. 2005; Canet-Aviles et al. 2004; Shendelman et al. 2004; Martinat et al. 2004; Shinbo et al. 2005, 2006; Sekito et al. 2006; Xu et al. 2005; Fan et al. 2008; Ishikawa et al. 2009, 2010). Wild-type DJ-1 plays a key role in antioxidation and neuroprotection in neuronal cells, and mutations in the *DJ-1* gene cause loss of function (Taira et al. 2005; Inden et al. 2006; Yanagisawa et al. 2008; Miyazaki et al. 2008; Yanagida et al. 2009a). We have previously demonstrated that DJ-1 administration reduced dopaminergic neuronal death and restored locomotor function in a rat model of PD (Inden et al., 2006). In addition, we have reported that intrastriatal injection of DJ-1 reduced infarct size in cerebral ischemia in rats (Yanagisawa et al. 2008). These findings suggest that DJ-1 is a pharmaceutical target for PD and cerebral ischemia.

Wild-type DJ-1 has three cysteine residues at amino acid numbers 46, 53, and 106 (C46, C53, and C106) in humans and rats (Kinumi et al. 2004) (Fig. 12.1a). A cysteine residue is oxidized from its reduced form (−SH) for sulfenation (−SOH), sulfination ($-SO_2H$), and sulfonation ($-SO_3H$). Among these three cysteine residues, C106 is the most sensitive to oxidative stress (Kinumi et al. 2004). Previously, we identified UCP0045037/compound A and UCP0054278/compound B as interaction partners of the reduced and the SO_2H-oxidized C106 region of DJ-1, respectively. These compounds were found by screening the university compound library, which contains approximately 30,000 compounds (Miyazaki et al. 2008). Like DJ-1, these compounds prevented oxidative stress-induced dopaminergic neuronal death and restored normal locomotor function in animal models of PD. In addition, they reduced infarct size in cerebral ischemia in rats (Miyazaki et al. 2008; Yamane et al. 2009; Yanagida et al. 2009b; Inden et al. 2011a). We further screened for DJ-1-binding compounds from a Zinc compound library that contains approximately 2,500,000 compounds (Kitamura et al. 2011). This screening identified compound-23 (comp-23), which protected oxidative stress-induced dopaminergic neuronal death in PD and ischemia models (Kitamura et al. 2011; Takahashi-Niki et al. 2015). These results suggest that DJ-1-binding compounds are fundamental drugs for PD therapy.

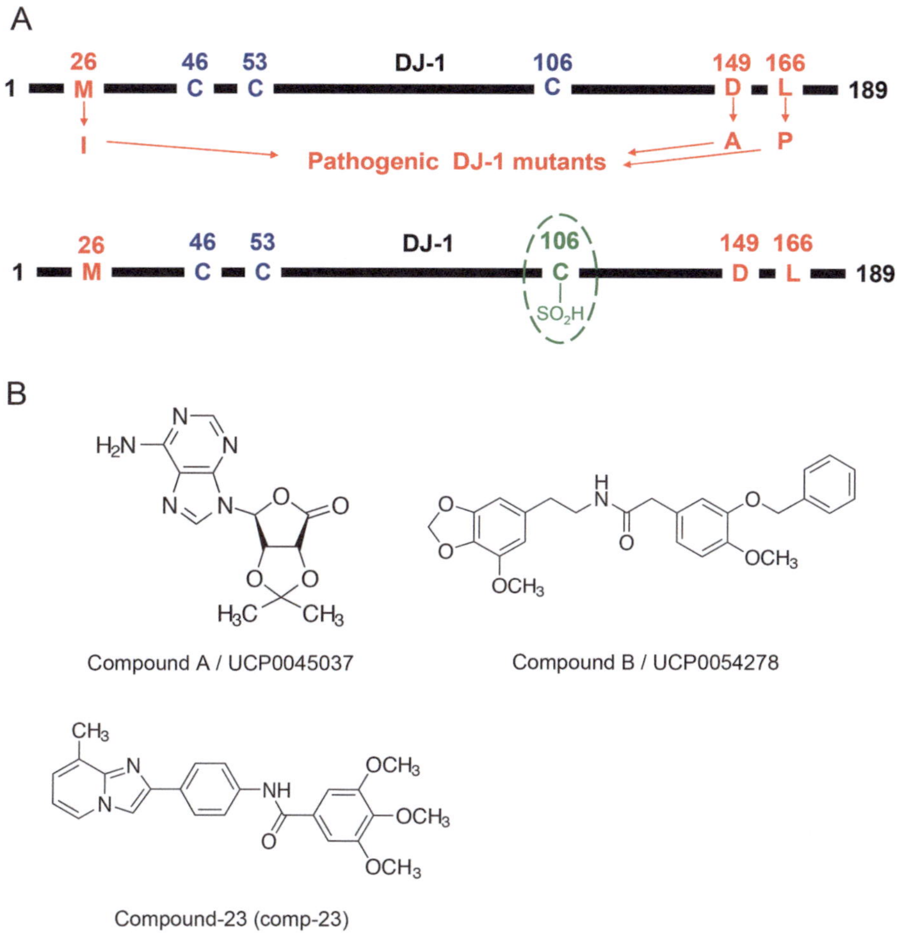

Fig. 12.1 Cysteine oxidation of DJ-1 and DJ-1-binding compounds. (**a**) DJ-1 contains three cysteine residues at amino acid numbers 46, 53, and 106 (C46, C53, and C106). (**b**) Chemical structures of the DJ-1-binding compounds UCP0045037/compound A, UCP0054278/compound B, and compound-23

12.2 DJ-1-Mediated Neuroprotection in PD Models

First, we examined the distribution of endogenous DJ-1 in rat brains treated with 6-hydroxydopamine (6-OHDA). 6-OHDA has long been used as an experimental model to study dopaminergic function in the brain and to evaluate the effects of drugs on central nervous system dopaminergic neurons (Inden et al. 2006). Endogenous expression of DJ-1 is much lower in the dopaminergic neurons of the substantia nigra after 6-OHDA treatment compared with that in glial cells. Therefore, we chose to study the protective effects of exogenous DJ-1 against 6-OHDA-induced parkinsonian in the substantia nigra. DJ-1 was introduced into the brain as

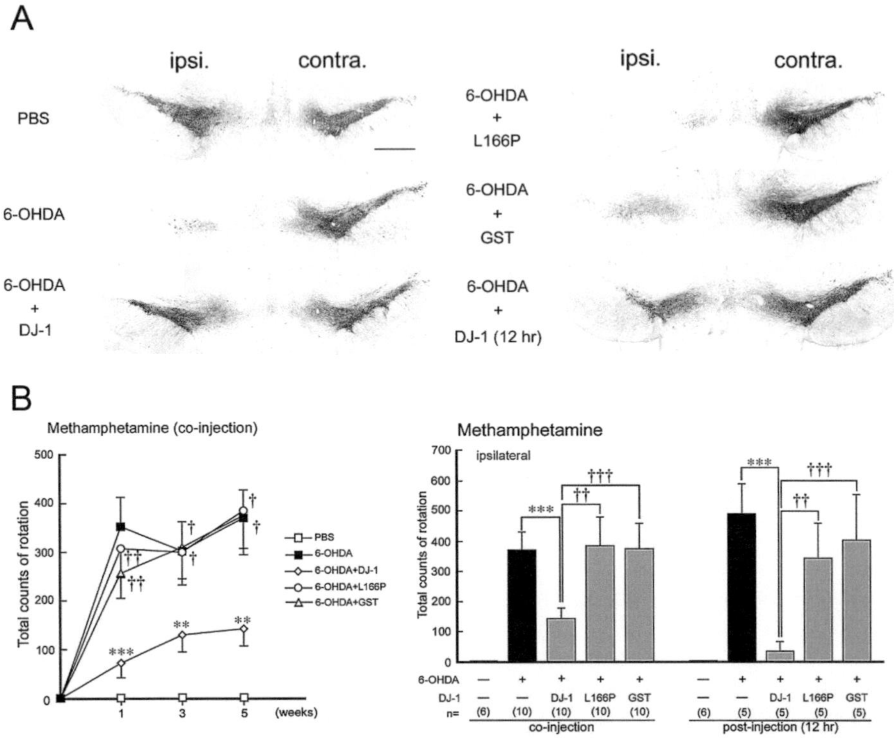

Fig. 12.2 Effect of DJ-1 in 6-OHDA-microinjected rats. (**a**) Changes in the numbers of TH-positive neurons in the substantia nigra. Midbrain slices were stained using an anti-TH antibody. Scale bar, 1 mm. (**b**) Changes in methamphetamine-induced rotational behavior in rats co- or post-treated with DJ-1 (Significance: $^{**}P < 0.01$, $^{***}P < 0.001$ vs. 6-OHDA injection alone. $^{\dagger}P < 0.05$, $^{\dagger\dagger}P < 0.01$, $^{\dagger\dagger\dagger}P < 0.001$ vs. co- or post-administration of DJ-1)

recombinant glutathione S-transferase (GST)-tagged proteins, and GST was injected together with the corresponding vehicle control. Wild-type DJ-1 (DJ-1), GST-tagged mutant DJ-1 (L166P), and vehicle controls were microinjected into the substantia nigra together with 6-OHDA. Dopaminergic neurons were visualized by tyrosine hydroxylase (TH) staining (Fig. 12.2a). Quantification of TH staining revealed that 6-OHDA caused a significant loss of dopaminergic neurons in the substantia nigra and these neurons were preserved in DJ-1-treated rats. However, loss of dopaminergic neurons was not inhibited by the co-administration of L166P and GST. We also investigated whether DJ-1 could protect against 6-OHDA-induced neuronal loss 12 h after 6-OHDA injection. Administration of DJ-1 but not of L166P and GST at 12 h after 6-OHDA injection protected against 6-OHDA-induced loss of dopaminergic neurons in the substantia nigra.

Injection of dopamine-releasing methamphetamine into animals with 6-OHDA-induced unilateral lesions induced movement ipsilateral to the injection site. This rotational behavior was significantly reduced by co-administration of DJ-1 but not of L166P or GST at 1, 3, and 5 weeks after injection (Fig. 12.2b). Further, we examined the effect of post-administration of DJ-1 in 6-OHDA-lesioned hemiparkinsonian rats. Methamphetamine-induced rotational behavior was significantly reduced by the administration of DJ-1 12 h after 6-OHDA injection.

12.3 Neuroprotective Effect of DJ-1 in Ischemia Models

Cerebral ischemia occurs as a result of a local reduction or arrest of blood supply and leads to neuronal cell death in the ischemic region. The pathophysiological mechanisms of ischemia/reperfusion injury are complex. Neurons and glial cells are lethally damaged by a number of events. These include the production of reactive oxygen species (ROS), which occurs immediately after ischemia/reperfusion, followed by more delayed postischemic inflammation and apoptosis. In addition, peri-infarct depolarization occurs within hours of ischemia, and together, these events contribute to the progression and expansion of brain injury. Controlling ROS formation is important for neuroprotection against ischemia/reperfusion injury (Yanagisawa et al. 2008). Three days after middle cerebral artery occlusion (MCAO), a regional loss of 2,3,5-triphenyltetrazolium chloride (TTC) staining was observed in the ipsilateral cerebral cortex and the striatum. This effect was reduced by the injection of DJ-1 110 min after the onset of MCAO (10 min before the reperfusion) (Fig. 12.3a) (Yanagisawa et al. 2008). Quantitative analysis showed that the infarct area was reduced in a dose-dependent manner by administration of DJ-1, and compared with the PBS- and GST-injected groups, the total infarct volume was significantly reduced by DJ-1 injection in a dose-dependent manner (Fig. 12.3b). Delayed administration of DJ-1 at 60 and 110 min after the onset of MCAO significantly reduced the infarct volume. However, the protective effect was lost when the injection was delayed for 180 min (Fig. 12.3c). After 180 min, the total infarct volume did not significantly change. To evaluate the effect of DJ-1 on behavioral function, MCAO or sham operations were performed in animals followed by injection with PBS, GST, or DJ-1 at 110 min after the onset of MCAO (Fig. 12.3d) (Yanagisawa et al. 2008). Neurological tests revealed improved functional outcomes in DJ-1-injected rats 1 day after MCAO. Improved functional outcomes continued in DJ-1-injected rats for 3, 4, 5, and 6 days after MCAO, compared with the PBS-injected group, and for 5 and 6 days after MCAO, compared with the GST-injected group. There were no differences in rotarod test performance between PBS-, GST-, and DJ-1-injected groups 1 day before MCAO (pre-performance). Two and 5 days after MCAO, the PBS- and GST-injected groups showed a marked reduction in rotarod performance, compared with pre-performance. These reductions were inhibited by DJ-1 administration, and there were significant improvements 2 and 5 days after MCAO, compared with PBS-injected and GST-injected animals (Fig. 12.3e).

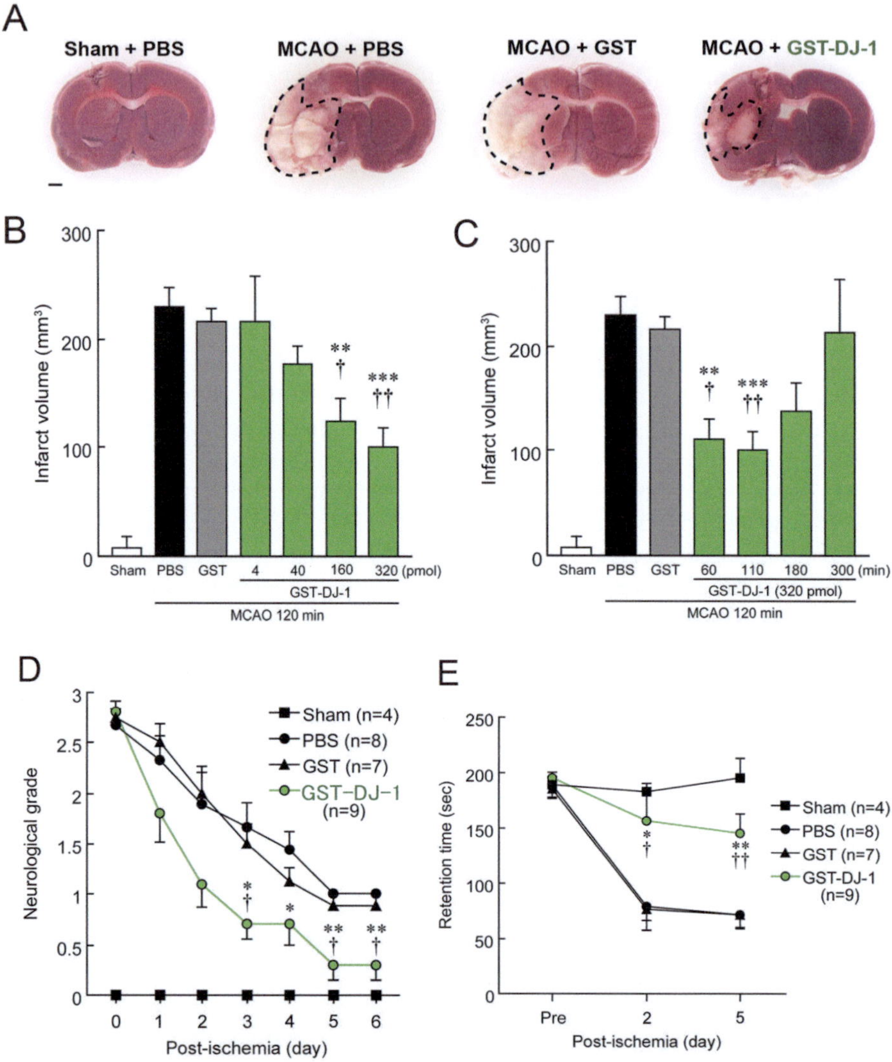

Fig. 12.3 Effect of DJ-1 in ischemia models. (**a**) Representative images of coronal brain sections at +1 mm anterior–posterior from the bregma with TTC staining 3 days after MCAO. (**b**) Dose-dependent protection by DJ-1. (**c**) Time-window of DJ-1-mediated neuroprotection. (**d**) Neurological evaluation was performed daily after reperfusion. (**e**) Spontaneous motor activity was evaluated by the rotarod test before MCAO (pre) and at 2 and 5 days after MCAO (Significance: $*P < 0.05$, $**P < 0.01$, $***P < 0.001$ vs. PBS-injected rats. $\dagger P < 0.05$, $\dagger\dagger P < 0.01$, $\dagger\dagger\dagger P < 0.001$ vs. GST-injected rats. Scale bar: 1 mm (**a**))

12.4 DJ-1 Reduces ROS Production and Protects Against H_2O_2-Induced Cell Death

To investigate whether DJ-1 or L166P affected ROS production, we performed confocal microscopy analysis with a ROS-specific fluorogen (C-DCF) (Inden et al. 2006; Yanagisawa et al. 2008). H_2O_2-induced intracellular C-DCF fluorescence markedly increased in normal human SH-SY5Y cells treated with 100-μmol/L H_2O_2 for 1 h (Fig. 12.4). After 1 h, H_2O_2-induced intracellular ROS production was significantly inhibited by 1-μmol/L DJ-1 but not by GST or L166P (Fig. 12.4). This was also true 24 h after the H_2O_2 treatment.

12.5 Identification of DJ-1-Binding Compounds and Their Neuroprotective Effects in PD and Ischemia Models

C106 is the most sensitive cysteine residue in DJ-1 to oxidative stress. It has been reported that a reduced form and/or partial oxidation of C106 activates DJ-1, which may exert an anti-oxidative response. In contrast, excessive oxidation of DJ-1 may lead to a loss of function (Fig. 12.1a). In addition, an excessively oxidized and inactive form of DJ-1 has been observed in brains of patients with PD, Alzheimer's disease, and Huntington's disease (Bandopadhyay et al. 2004; Choi et al. 2006; Sajjad et al. 2014). Therefore, drugs that inhibit excessive oxidation of C106 may be useful for PD therapy. To search for modulators of DJ-1 binding, we performed virtual screening (in silico) using the X-ray crystal structure of the reduced and

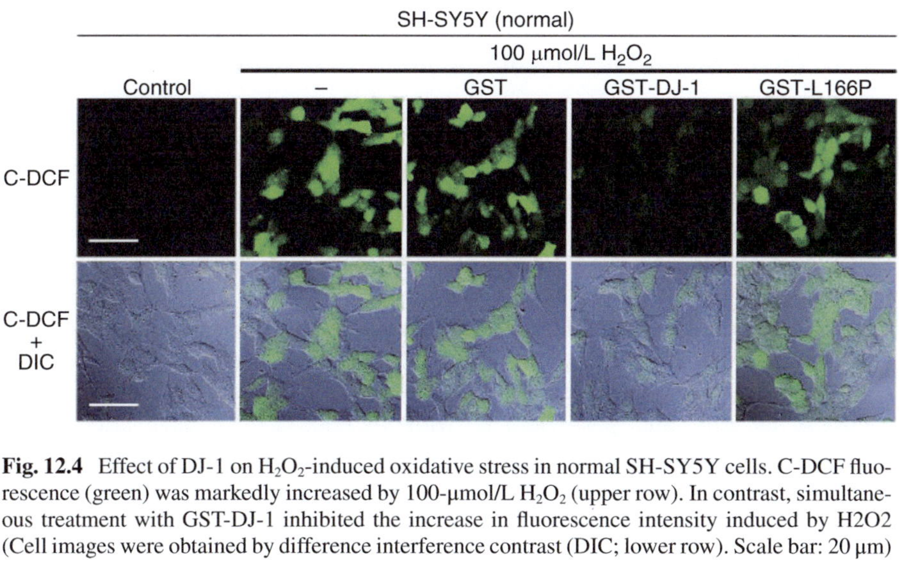

Fig. 12.4 Effect of DJ-1 on H_2O_2-induced oxidative stress in normal SH-SY5Y cells. C-DCF fluorescence (green) was markedly increased by 100-μmol/L H_2O_2 (upper row). In contrast, simultaneous treatment with GST-DJ-1 inhibited the increase in fluorescence intensity induced by H2O2 (Cell images were obtained by difference interference contrast (DIC; lower row). Scale bar: 20 μm)

SO_2H-oxidized C106 region of DJ-1 and the three-dimensional coordinate data of about 30,000 chemical compounds in the University Compound Project at the Foundation for Education of Science and Technology in Japan. Among the DJ-1-binding compounds identified in silico, UCP0045037/compound A and UCP0054278/compound B had the highest binding constant (docking score) for the reduced and SO_2H-oxidized C106 region of DJ-1, respectively (Miyazaki et al. 2008) (Fig. 12.1b).

To examine the effect of UCP0045037 and UCP0054278 on PD phenotypes in vivo, we used the 6-OHDA PD rat model. 6-OHDA was stereotaxically microinjected into the unilateral mesencephalon, and intranigral co-injection of UCP0045037 and UCP0054278 restored the 6-OHDA-induced locomotor defect and prevented 6-OHDA-induced dopaminergic neural death (Miyazaki et al. 2008). In addition, methamphetamine-induced rotational behavior was significantly reduced by co-administration of UCP0045037 and UCP0054278 7 days after injection. TH-positive neurons were preserved in the ipsilateral substantia nigra of UCP0045037- and UCP0054278-treated animals, compared with animals injected with 6-OHDA alone 10 days post-lesion.

In addition, the intrastriatal pre-injection of UCP0045037 and UCP0054278 inhibited neurodegeneration induced by MCAO and reperfusion in rats (Yamane et al. 2009; Yanagida et al. 2009b). TTC staining was markedly decreased 24 h after MCAO in the ipsilateral cerebral cortex and striatum of vehicle-injected rats, and this effect was reduced by the microinjection of UCP0045037 and UCP0054278. Each infarct area was quantitatively smaller, and the total infarct volume was significantly reduced by UCP0045037 and UCP0054278 administration, compared with vehicle-injected rats.

12.6 Effect of UCP0054278 on 6-OHDA-Induced Locomotor Defects and Dopaminergic Neuronal Death

We examined the neuroprotective effect of UCP0054278 on 6-OHDA-induced dopaminergic neuronal death in an in vivo PD model. Pretreatment or posttreatment of UCP0054278 showed a tendency to reduce the number of methamphetamine-induced rotations compared with untreated rats, but this difference was not significant. However, pre- and posttreatment with UCP0054278 significantly ameliorated the methamphetamine-induced behavioral impairment (Fig. 12.5a). To examine the effect of UCP0054278 on 6-OHDA-induced dopaminergic neuronal death, 1-mg/kg UCP0054278 was intraperitoneally injected before treatment, after treatment, or before and after treatment. TH-positive neurons were protected from the effects of 6-OHDA by pre- and posttreatment with UCP0054278 7 days after lesion (Fig. 12.5b). Stereological analysis of nigral TH-positive neurons showed that loss of dopaminergic neurons in the substantia nigra was significantly inhibited by both

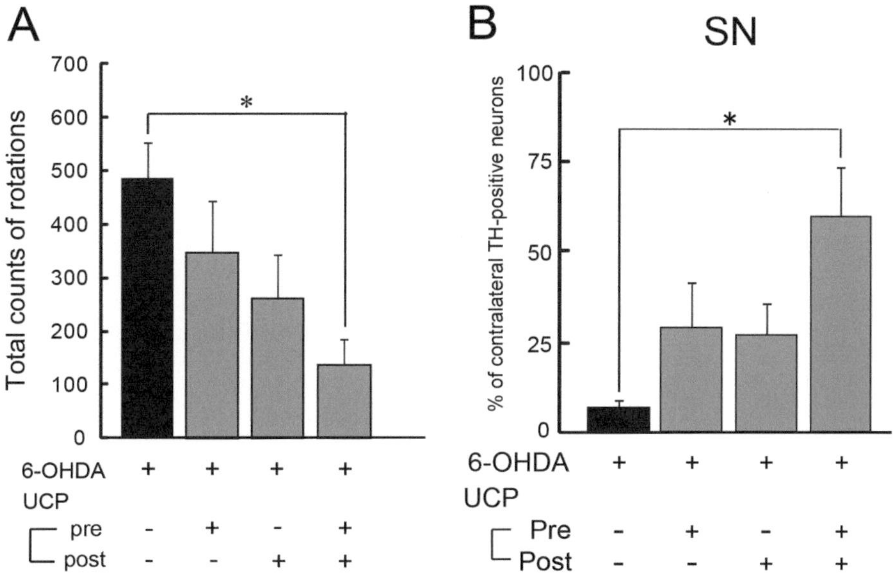

Fig. 12.5 Effect of UCP0054278 on methamphetamine-induced rotation behavior and 6-OHDA-induced dopaminergic neurodegeneration in rats with intranigral 6-OHDA lesions. (**a**) 6-OHDA-microinjected rats were intraperitoneally injected with 1 mg/kg of vehicle or UCP0054278 at 24 h and 30 min before microinjection (pre-treatment), 24, 48 and 72 h after microinjection (posttreatment) or pre- and posttreatment. Seven days later, methamphetamine-induced rotation tests were performed. (**b**) Stereological analysis of TH-positive neurons in the substantia nigra (SN) (Significance: *$P < 0.05$ vs. 6-OHDA alone)

pre- and posttreatment with UCP0054278. However, the loss of dopaminergic neurons was not inhibited by either pre- or posttreatment with UCP0054278.

12.7 Effect of UCP0054278 on Rotenone-Induced Locomotor Defects and Dopaminergic Neuronal Death

We also investigated the neuroprotective effects of UCP0054278 in the rotenone-induced PD model (Inden et al. 2011a). To identify deficits in motor coordination, rotenone-treated mice were tested weekly on the accelerating rotarod (Inden et al. 2011b). Rotenone-treated mice showed marked reductions in endurance time and ability to remain on the rotarod. Treatment with 1- and 3-mg/kg UCP0054278 significantly rescued the retention time on the rotarod (Fig. 12.6a).

To investigate whether UCP0054278 protects dopaminergic neurons in the substantia nigra from rotenone-induced damage, we treated animals with either 1- or 3-mg/kg UCP0054278 for 30 min before the oral administration of rotenone. The

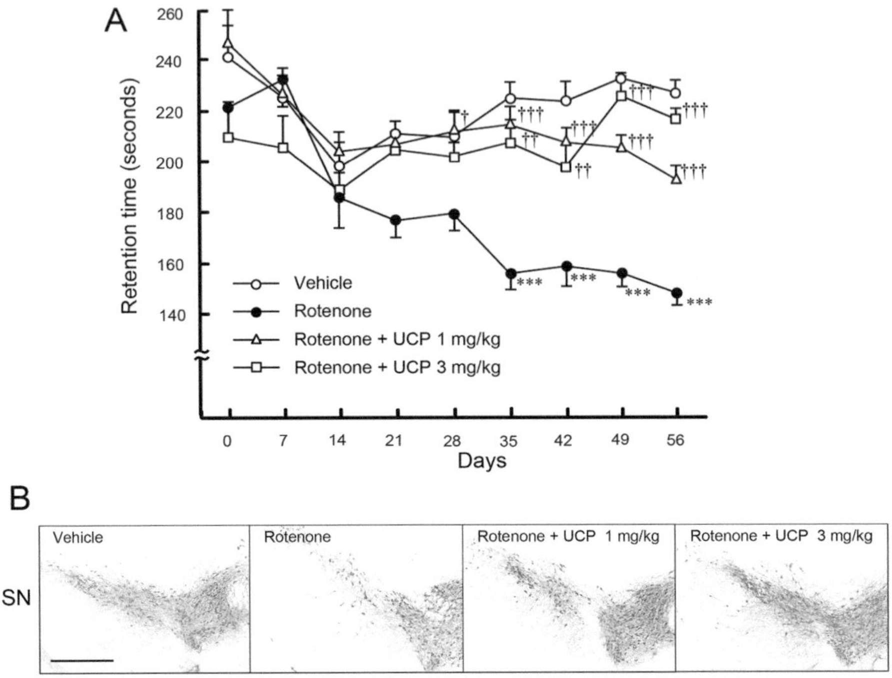

Fig. 12.6 Suppression of rotenone-induced behavioral dysfunction and dopaminergic neurodegeneration by UCP0054278. Rotenone was orally administered to C57BL/6 mice at a dose of 30-mg/kg per day for 56 days. Mice were injected with 1- or 3-mg/kg UCP0054278 once daily for 56 days, 30 min before the oral administration of rotenone. (**a**) The rotarod test was performed once every week (Significance: ***$P < 0.001$ vs. vehicle. †, $P < 0.05$, ††, $P < 0.01$, †††, $P < 0.001$ vs. rotenone alone. (**b**) Representative images of TH immunoreactivity in the substantia nigra (SN). Scale bar: 500 μm)

rotenone-induced loss of TH-positive neurons in the substantia nigra was significantly inhibited by the injection of both 1- and 3-mg/kg UCP0054278 (Fig. 12.6b).

12.8 Effect of UCP0054278 on Intracellular α-Synuclein Expression

A previous study showed that the oral administration of 30-mg/kg rotenone for 56 days produced some TH-positive neurons with a high level of cytoplasmic α-synuclein immunoreactivity in the substantia nigra (Inden et al. 2007, 2011a). To examine the effect of UCP0054278 on intracellular α-synuclein expression in the substantia nigra, we performed confocal microscopic analysis (Fig. 12.7). TH-positive neurons were clearly detected in the substantia nigra of vehicle-treated mice; however, these TH-positive neurons did not show α-synuclein immunoreactivity. Rotenone treatment reduced the number of TH-positive neurons in the

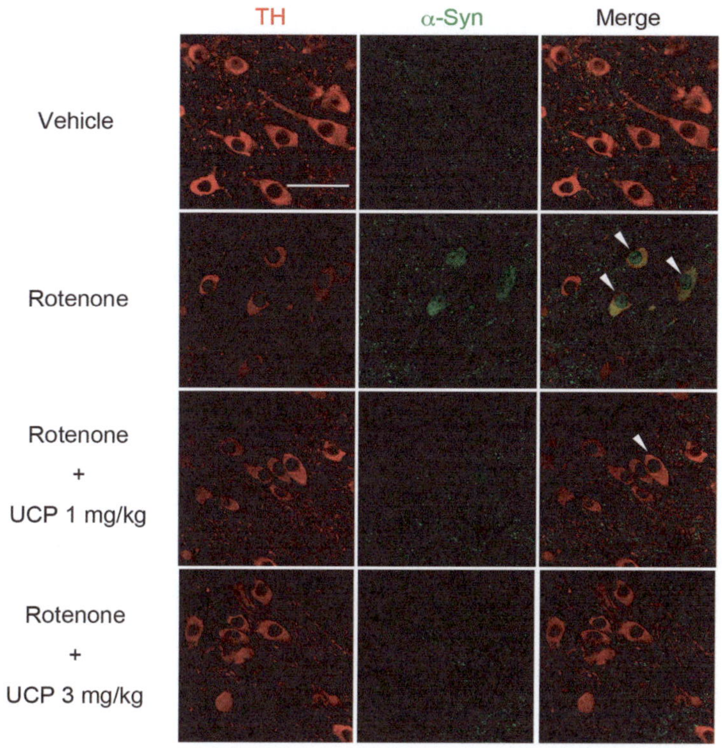

Fig. 12.7 Intracellular expression of α-synuclein in the substantia nigra. Nigral slices from mice treated with 30-mg/kg vehicle or rotenone in the absence or presence of 1- or 3-mg/kg UCP0054278 were labeled with antibodies against TH (red) and α-synuclein (green), and images were captured by laser scanning confocal microscopy. Scale bar: 50 μm

substantia nigra, but α-synuclein immunoreactivity was detected in some of the surviving TH-positive neurons. The number of neurons expressing both α-synuclein and TH was significantly increased in the substantia nigra of rotenone-lesioned mice. Interestingly, α-synuclein expression was significantly decreased in the surviving TH-positive neurons of UCP0054278-treated rotenone mice. In addition, the number of neurons expressing both α-synuclein and TH was significantly decreased by UCP0054278 treatment.

12.9 Effect of UCP0054278 on In Vitro Cultures

We previously established *DJ-1*-knockdown SH-SY5Y cells, in which the endogenous DJ-1 expression was reduced by approximately 76%. These cells serve as an in vitro model of PD (Yanagisawa et al. 2008), and we used this model to examine the effect of UCP0054278 on 6-OHDA-mediated cell death (Yanagisawa et al. 2008).

In normal human SH-SY5Y cells, 6-OHDA caused cell death after 24 h in a concentration-dependent manner (Fig. 12.8a). Cell death was significantly inhibited by UCP0054278 treatment in a concentration-dependent manner. In *DJ-1*-knockdown SH-SY5Y cells, 6-OHDA-induced apoptosis also occurred (Fig. 12.8b), but this was not rescued by UCP0054278 treatment. Similarly, in normal human SH-SY5Y cells, rotenone caused cell death after 48 h in a concentration-dependent manner. Cell death was significantly inhibited by UCP0054278 treatment in a concentration-dependent manner. However, in *DJ-1*-knockdown cells, UCP0054278 did not protect against rotenone-induced cell death. These findings suggest that UCP0054278 exerts its neuroprotective effects against 6-OHDA and rotenone through interactions with DJ-1.

In normal SH-SY5Y cells, incubation with 50-μM 6-OHDA for 1 h induced marked intracellular ROS production (Fig. 12.9a), while 25-μM 6-OHDA had only a slight effect. In *DJ-1*-knockdown cells, 25-μM 6-OHDA was sufficient to significantly enhance ROS production after 1 h (Fig. 12.9b). Simultaneous treatment with 10-μM UCP0054278 significantly inhibited 6-OHDA-induced ROS production in SH-SY5Y cells (Fig. 12.9c). In *DJ-1*-knockdown cells, the inhibitory effect of UCP0054278 on 6-OHDA-induced ROS production was lost (Fig. 12.9d). This suggests that inhibition of ROS production by UCP0054278 depends upon interaction with DJ-1.

12.10 Compound 23

We performed further screening for DJ-1-binding compounds from the Zinc compound library containing approximately 2,500,000 compounds (Fig. 12.1b). We identified compound-23 (comp-23), which prevented oxidative stress-induced death of SH-SY5Y cells and primary neuronal cells of the ventral mesencephalon, but not that of *DJ-1*-knockdown SH-SY5Y cells (Kitamura et al. 2011). Comp-23 inhibited ROS production induced by oxidative stress and prevented excess oxidation of DJ-1. In addition, comp-23 prevented dopaminergic neuronal death and restored normal locomotor function in 6-OHDA-injected rats and rotenone-treated mice. Comp-23 also reduced the infarct size of cerebral ischemia induced by MCAO in rats. The protective activity of comp-23 was stronger than that of UCP0054278. Furthermore, comp-23 inhibited MPTP-induced reduction of the rotarod retention time, dopaminergic neuronal death in the substantia nigra, and striatal dopamine levels in wild-type mice but not in *DJ-1*-knockout mice. These results indicate that comp-23 can cross the blood–brain barrier and that its effects in the brain are dependent upon interaction with DJ-1 (Takahashi-Niki, et al. 2015).

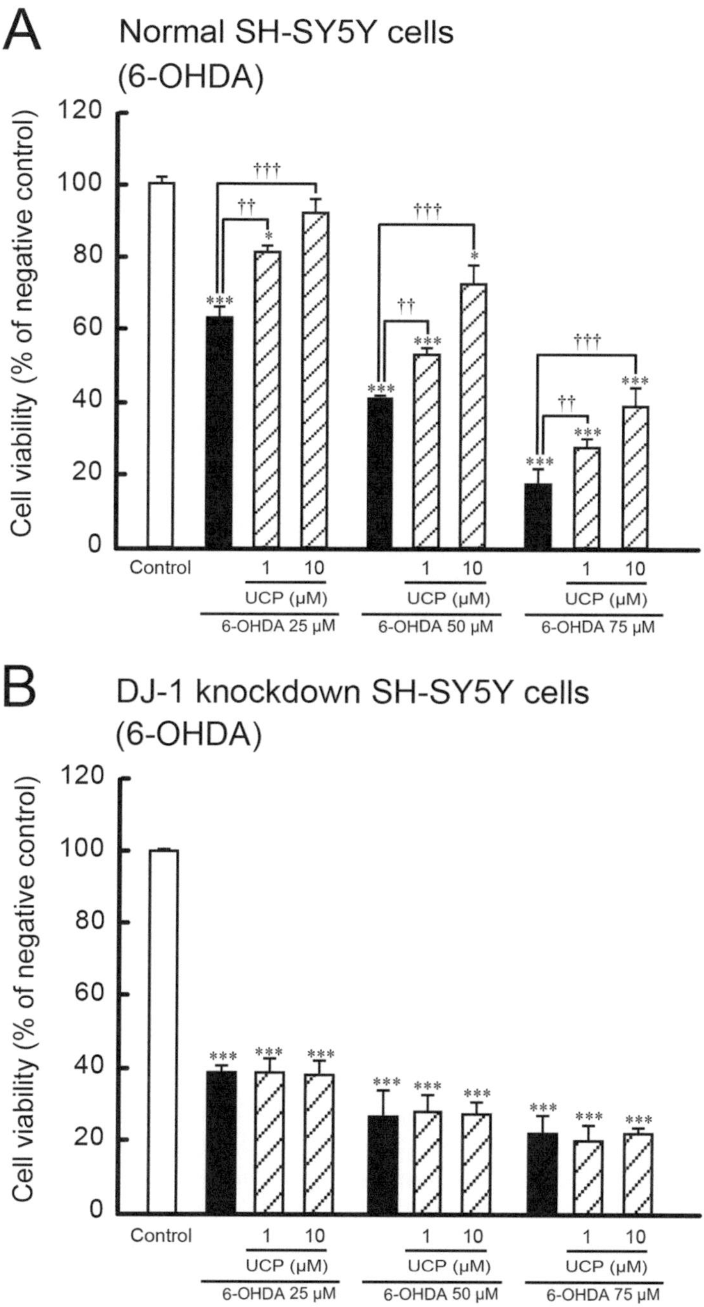

Fig. 12.8 Effect of UCP0054278 on 6-OHDA-induced oxidative stress in human SH-SY5Y cells (**a**, normal) and DJ-1-knockdown cells (**b**, DJ-1 knockdown) (Significance: *, $P < 0.05$, **, $P < 0.01$, ***, $P < 0.001$ vs. treatment with control. ††, $P < 0.01$, †††, $P < 0.001$ vs. treatment with 6-OHDA alone)

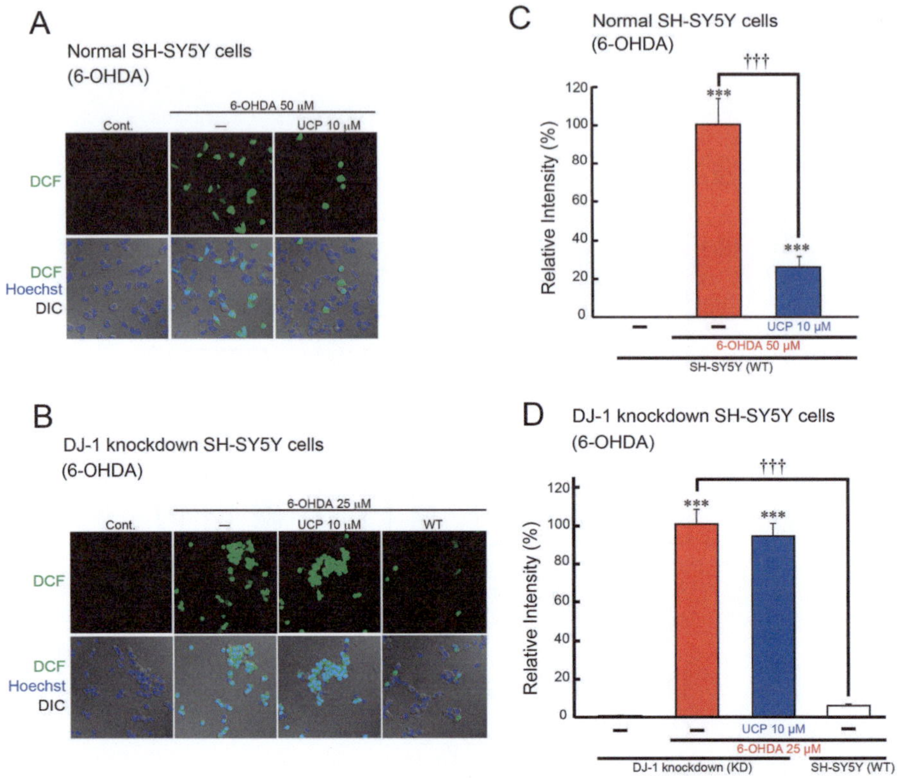

Fig. 12.9 Effect of UCP0054278 on 6-OHDA-induced ROS production. Normal and DJ-1-knockdown SH-SY5Y cells were treated with 6-OHDA in the presence of vehicle (0.01% DMSO) or 10-μM UCP0054278 for 2.5 h. Subsequently, CM-H_2DCFDA was added, and the fluorescence intensity of oxidized DCF (green) was examined by confocal microscopy. Nuclei were stained with Hoechst 33,258 (blue), and cell images were obtained by difference interference contrast (DIC) (Significance: ***$P < 0.001$ vs. vehicle. †††$P < 0.001$ vs. 6-OHDA alone. Scale bar: 50 μm)

12.11 Conclusion

DJ-1-binding compounds, including UCP0045037/compound A, UCP0054278/compound B, and compound-23, were identified by in silico screening using the university compound library and Zinc compound library. These compounds inhibited excessive oxidation of the C106 residue of DJ-1 and maintained the reduced form of DJ-1. This inhibited oxidative stress-induced neuronal cell death and restored locomotor function in animal models of PD and cerebral ischemia. These compounds were able to cross the blood–brain barrier and exerted their protective effects by interacting with DJ-1; protective functions were lost in *DJ-1*-knockdown cells and *DJ-1*-knockout animals. These DJ-1-binding compounds represent promising therapeutic targets for a wide range of neurodegenerative diseases.

References

Bandopadhyay R, Kingsbury AE, Cookson MR, Reid AR, Evans IM, Hope AD, Pittman AM, Lashley T, Canet-Aviles R, Miller DW, McLendon C, Strand C, Leonard AJ, Abou-Sleiman PM, Healy DG, Ariga H, Wood NW, de Silva R, Revesz T, Hardy JA, Lees AJ (2004) The expression of DJ-1 (PARK7) in normal human CNS and idiopathic Parkinson's disease. Brain 127:420–430

Bonifati V, Rizzu P, van Baren MJ, Schaap O, Breedveld GJ, Krieger E, Dekker MC, Squitieri F, Ibanez P, Joosse M, van Dongen JW, Vanacore N, van Swieten JC, Brice A, Meco G, van Duijn CM, Oostra BA, Heutink P (2003) Mutations in the DJ-1 gene associated with autosomal recessive early-onset parkinsonism. Science 299:256–259

Canet-Aviles RM, Wilson MA, Miller DW, Ahmad R, McLendon C, Bandyopadhyay S, Baptista MJ, Ringe D, Petsko GA, Cookson MR (2004) The Parkinson's disease protein DJ-1 is neuroprotective due to cysteine-sulfinic acid-driven mitochondrial localization. Proc Natl Acad Sci U S A 101:9103–9108

Choi J, Sullards MC, Olzmann JA, Rees HD, Weintraub ST, Bostwick DE, Gearing M, Levey AI, Chin LS, Li L (2006) Oxidative damage of DJ-1 is linked to sporadic Parkinson and Alzheimer diseases. J Biol Chem 281:10816–10824

Fan J, Ren H, Ji N, Fei E, Zhou T, Jiang P, Wu M, Wang G (2008) DJ-1 decreases bax expression through repressing p53 transcriptional activity. J Biol Chem 283:4022–4030

Inden M, Taira T, Kitamura Y, Yanagida T, Tsuchiya D, Takata K, Yanagisawa D, Nishimura K, Taniguchi T, Kiso Y, Yoshimoto K, Agatsuma T, Koide-Yoshida S, Iguchi-Ariga SMM, Shimohama S, Ariga H (2006) PARK7 DJ-1 protects against degeneration of nigral dopaminergic neurons in Parkinson's disease rat model. Neurobiol Dis 24:144–158

Inden M, Kitamura Y, Takeuchi H, Yanagida T, Takata K, Kobayashi Y, Taniguchi T, Yoshimoto K, Kaneko M, Okuma Y, Taira T, Ariga H, Shimohama S (2007) Neurodegeneration of mouse nigrostriatal dopaminergic system induced by repeated oral administration of rotenone is prevented by 4-phenylbutyrate, a chemical chaperone. J Neurochem 101:1491–1504

Inden M, Kitamura Y, Takahashi K, Takata K, Ito N, Niwa R, Funayama R, Nishimura K, Taniguchi T, Honda T, Taira T, Ariga H (2011a) Protection against dopaminergic neurodegeneration in Parkinson's disease-model animals by a modulator of the oxidized form of DJ-1, a wild-type of familial Parkinson's disease-linked PARK7. J Pharmacol Sci 117:189–203

Inden M, Kitamura Y, Abe M, Tamaki A, Takata K, Taniguchi T (2011b) Parkinsonian rotenone mouse model: reevaluation of long-term administration of rotenone in C57BL/6 mice. Biol Pharm Bull 34:92–96

Ishikawa S, Taira T, Niki T, Takahashi-Niki K, Maita C, Maita H, Ariga H, Iguchi-Ariga SM (2009) Oxidative status of DJ-1-dependent activation of dopamine synthesis through interaction of tyrosine hydroxylase and 4-dihydroxy-L-phenylalamine (L-DOPA) decarboxylase with DJ-1. J Biol Chem 284:28832–28844

Ishikawa S, Taira T, Niki T, Takahashi-Niki K, Ariga H, Iguchi-Ariga SM (2010) Human DJ-1-specific transcriptional activation of tyrosine hydroxylase gene. J Biol Chem 285:39718–39731

Kinumi T, Kimata J, Taira T, Ariga H, Niki E (2004) Cysteine-106 of DJ-1 is the most sensitive cysteine residue to hydrogen peroxide-mediated oxidation in vivo in human umbilical vein endothelial cells. Biochem Biophys Res Commun 317:722–728

Kitamura Y, Watanabe S, Taguchi M, Takagi K, Kawata T, Takahashi-Niki K, Yasui H, Maita H, Iguchi-Ariga SM, Ariga H (2011) Neuroprotective effect of a new DJ-1-binding compound against neurodegeneration in Parkinson's disease and stroke model rats. Mol Neurodegener 6:48

Martinat C, Shendelman S, Jonason A, Leete T, Beal MF, Yang L, Floss T, Abeliovich A (2004) Sensitivity to oxidative stress in DJ-1-deficient dopamine neurons: an ES- derived cell model of primary parkinsonism. PLoS Biol 2:e327

Miyazaki S, Yanagida T, Nunome K, Ishikawa S, Inden M, Kitamura Y, Nakagawa S, Taira T, Hirota K, Niwa M, Iguchi-Ariga SM, Ariga H (2008) DJ-1-binding compounds prevent oxidative stress-induced cell death and movement defect in Parkinson's disease model rats. J Neurochem 105:2418–2434

Nagakubo D, Taira T, Kitaura H, Ikeda M, Tamai K, Iguchi-Ariga SMM, Ariga H (1997) DJ-1, a novel oncogene which transforms mouse NIH3T3 cells in cooperation with ras. Biochem Biophys Res Commum 31:509–513

Niki T, Takahashi-Niki K, Taira T, Iguchi-Ariga SMM, Ariga H (2003) DJBP: a novel DJ-1-binding protein, negatively regulates the androgen receptor by recruiting histone deacetylase complex, and DJ-1 antagonizes this inhibition by abrogation of this complex. Mol Cancer Res 1:247–261

Sajjad MU, Green EW, Miller-Fleming L, Hands S, Herrera F, Campesan S, Khoshnan A, Outeiro TF, Giorgini F, Wyttenbach A (2014) DJ-1 modulates aggregation and pathogenesis in models of Huntington's disease. Hum Mol Genet 23:755–766

Sekito A, Koide-Yoshida S, Niki T, Taira T, Iguchi-Ariga SM, Ariga H (2006) DJ-1 interacts with HIPK1 and affects H_2O_2-induced cell death. Free Radic Res 40:155–165

Shendelman S, Jonason A, Martinat C, Leete T, Abeliovich A (2004) DJ-1 is a redox-dependent molecular chaperone that inhibits alpha-synuclein aggregate formation. PLoS Biol 2:e362

Shinbo Y, Taira T, Niki T, Iguchi-Ariga SMM, Ariga H (2005) DJ-1 restores p53 transcription activity inhibited by Topors/p53BP3. Int J Oncol 26:641–648

Shinbo Y, Niki T, Taira T, Ooe H, Takahashi-Niki K, Maita C, Seino C, Iguchi-Ariga SMM, Ariga H (2006) Proper SUMO-1 conjugation is essential to DJ-1 to exert its full activities. Cell Death Diff 13:96–108

Taira T, Saito Y, Niki T, Iguchi-Ariga SM, Takahashi K, Ariga H (2005) DJ-1 has a role in antioxidative stress to prevent cell death. EMBO Rep 5:213–218

Takahashi K, Taira T, Niki T, Seino C, Iguchi-Ariga SMM, Ariga H (2001) DJ-1 positively regulates the androgen receptor by impairing the binding of PIASx alpha to the receptor. J Biol Chem 276:37556–37563

Takahashi-Niki K, Inafune A, Michitani N, Hatakeyama Y, Suzuki K, Sasaki M, Kitamura Y, Niki T, Iguchi-Ariga SM, Ariga H (2015) DJ-1-dependent protective activity of DJ-1-binding compound no. 23 against neuronal cell death in MPTP-treated mouse model of Parkinson's disease. J Pharmacol Sci 127:305–310

Xu J, Zhong N, Wang H, Elias JE, Kim CY, Woldman I, Pifl C, Gygi SP, Geula C, Yankner BA (2005) The Parkinson's disease-associated DJ-1 protein is a transcriptional co-activator that protects against neuronal apoptosis. Hum Mol Genet 14:1231–1241

Yamane K, Kitamura Y, Yanagida T, Takata K, Yanagisawa D, Taniguchi T, Taira T, Ariga H (2009) Oxidative neurodegeneration is prevented by UCP0045037, an allosteric modulator for the reduced form of DJ-1, a wild-type of familial Parkinson's disease-linked PARK7. Int J Mol Sci 10:4789–4804

Yanagida T, Tsushima J, Kitamura Y, Yanagisawa D, Takata K, Shibaike T, Yamamoto A, Taniguchi T, Yasui H, Taira T, Morikawa S, Inubushi T, Tooyama I, Ariga H (2009a) Oxidative stress induction of DJ-1 protein in reactive astrocytes scavenges free radicals and reduces cell injury. Oxidative Med Cell Longev 2:36–42

Yanagida T, Kitamura Y, Yamane K, Takahashi K, Takata K, Yanagisawa D, Yasui H, Taniguchi T, Taira T, Honda T, Ariga H (2009b) Protection against oxidative stress-induced neurodegeneration by a modulator for DJ-1, the wild-type of familial Parkinson's disease-linked PARK7. J Pharmacol Sci 109:463–468

Yanagisawa D, Kitamura Y, Inden M, Takata T, Taniguchi T, Morikawa S, Morira M, Inubushi T, Tooyama I, Taira T, Iguchi-Ariga SMM, Akaike A, Ariga H (2008) DJ-1 protects against neurodegeneration caused by focal cerebral ischemia and reperfusion in rats. J Cereb Blood Flow Metab 28:563–578

Yokota T, Sugawara K, Ito K, Takahashi R, Ariga H, Mizusawa H (2003) Down-regulation of DJ-1 enhances cell death by oxidative stress, ER stress, and proteasome inhibition. Biochem Biophys Res Commun 312:1342–1348

Chapter 13
DJ-1 as a Therapeutic Target Against Cancer

Ji Cao, Xiaobing Chen, Meidan Ying, Qiaojun He, and Bo Yang

Abstract DJ-1 is a gene involved in various cellular processes, including transcriptional regulation, oxidative stress response, fertilization, mitochondrial regulation, inflammatory and fibrogenic niche formation, and glycation damage prevention. Although a disease-associated genetic study within the past decade has demonstrated that the mutation of DJ-1 is associated with autosomal early-onset Parkinson's disease, increasing evidence suggests that DJ-1 also plays a critical role in tumor development and progression. In this review, we provide an overview of current knowledge concerning the role and the mechanism of DJ-1 in cancer and also discuss the possibility of DJ-1 as a therapeutic target against cancer.

Keywords DJ-1 • Oncogene • Cancer therapeutics • Target • Mechanism

13.1 Introduction

At the genomic level, the human DJ-1 (RS/PARK7/CAP1) gene comprises seven exons spanning 24 kb and is mapped at chromosome 1p36.2–1p36.3, which is a region that has been shown to be a hot spot of chromosomal abnormalities in several tumor cell types, including non-Hodgkin's lymphoma, leiomyoma, acute myeloid leukemia, astrocytoma, neuroblastoma, adenoma, and adenocarcinoma (Taira et al. 2001). DJ-1 encodes a 189-amino-acid protein that is highly conserved across diverse species and ubiquitously expressed in more than 22 human tissues (Hod et al. 1999). The X-ray crystal structure of DJ-1 shows that it contains an α/β-fold, which is conserved among members of the DJ-1/ThiJ/PfpI superfamily (Tao and Tong 2003). Interestingly, DJ-1 also contains an extra helix at the C terminus, which mediates a novel mode of dimerization of DJ-1 proteins (Tao and Tong 2003; Wilson et al. 2003). Studies with the disease-causing L166P mutant suggest that this

J. Cao • X. Chen • M. Ying • Q. He • B. Yang (✉)
Zhejiang Province Key Laboratory of Anti-Cancer Drug Research, College of Pharmaceutical Sciences, Zhejiang University, Hangzhou, China
e-mail: yang924@zju.edu.cn

H. Ariga, S.M.M. Iguchi-Ariga (eds.), *DJ-1/PARK7 Protein*, Advances in Experimental Medicine and Biology 1037,
https://doi.org/10.1007/978-981-10-6583-5_13

mutation, which is associated with autosomal recessive early-onset Parkinson's disease (PD) (Bonifati et al. 2003), disrupts the C-terminal region and the dimerization of the protein, leading to DJ-1 degradation through the ubiquitin-proteasome system and a low level of DJ-1 expression (Miller et al. 2003).

DJ-1 was originally identified as a novel oncogene and exhibits significant transforming activity in cooperation with c-Myc or H-Ras (Nagakubo et al. 1997). Accumulating studies reveal that DJ-1 has multiple functions, such as the regulatory subunit of RNA-binding protein (Hod et al. 1999), a redox-regulated chaperone (Shendelman et al. 2004), cysteine protease (Olzmann et al. 2004), transcriptional coactivator (Xu et al. 2005), and protein deglycase (Richarme et al. 2015). Therefore, DJ-1 is involved in various biological processes, including transcriptional regulation (Fan et al. 2008a, b; Kato et al. 2013), oxidative stress response (Takahashi-Niki et al. 2004; Clements et al. 2006), fertilization (Okada et al. 2002; Yoshida et al. 2003), mitochondrial regulation (Mccoy and Cookson 2011; Shimizu et al. 2016), inflammatory and fibrogenic niche formation (Chen et al. 2016), and glycation damage prevention (Advedissian et al. 2016). In this section, we summarize the current knowledge concerning the role of DJ-1 in tumor development and progression as well as describe the mechanism-based approaches to therapeutically target oncogenic DJ-1 signaling in cancer.

13.2 DJ-1 Status in Human Cancers

13.2.1 DJ-1 Is Overexpressed in Human Cancers and Related to Poor Prognosis

An increasing number of studies have found that the amount of DJ-1 is overexpressed in most known human cancers, which is consistent with its role as an oncogene (Table 13.1). Compared with normal adjacent control tissue, overexpressed DJ-1 can be detected in neoplastic thyroid tissues (Srisomsap et al. 2002), prostate cancer (Hod 2004), renal carcinoma (Sitaram et al. 2009), hepatocellular carcinoma (Liu et al. 2010), ovarian carcinoma (Davidson et al. 2008), uveal melanoma (Pardo et al. 2006), non-small cell lung carcinoma (NSCLC) (MacKeigan et al. 2003), breast cancer (Le Naour et al. 2001), acute leukemia (Liu et al. 2008), cervical cancer (Arnouk et al. 2009), papillary thyroid cancer (Giusti et al. 2008), pancreatic ductal adenocarcinoma (PDAC) (Tian et al. 2008), laryngeal squamous cell carcinoma (Shen et al. 2011), oral squamous cell carcinoma (Xu et al. 2016), and esophageal squamous cell carcinoma (ESCC) (Yuen et al. 2008). Interestingly, primary human malignant NSCLC and ESCC tumor samples showed DJ-1 overexpression at both the mRNA and protein levels, suggesting that DJ-1 overexpression is largely restricted to tumor cells and increases with cellular transformation (MacKeigan et al. 2003; Yuen et al. 2008).

Given that DJ-1 is frequently overexpressed in many cancers, its physiological role in cancer progression has also been studied. Small interfering RNA (siRNA)-mediated DJ-1 knockdown could induce the inhibition of cell proliferation and enhance the sensitivity of tumor cells to chemotherapeutic drugs, such as

Table 13.1 Overview of DJ-1 overexpression investigated in human tumors and cancer cell lines

Tumor type – human	Clinical impact	Refs.
Thyroid cancer	Protects cancer cells from apoptosis induced by TRAIL	Srisomsap et al. 2002
Prostate cancer	Resistance to cytotoxic agents	Hod 2004
Renal carcinoma	None	Sitaram et al. 2009
Hepatocellular carcinoma	Preoperative AFP, liver cirrhosis, vein invasion, differentiation, and overall survival	Liu et al. 2010
Ovarian carcinoma	Overall survival	Davidson et al. 2008
Uveal melanoma	Serum biomarker	Pardo et al. 2006
	Tumor invasion and poor prognosis	Chen et al. 2015
Non-small cell lung carcinoma (NSCLC)	Cell survival	MacKeigan et al. 2003
	Cisplatin resistance and poor prognosis	Zeng et al. 2011
	Relapse incidence	Kim et al. 2005
Breast cancer	Circulating antigen	Le Naour et al. 2001
	Predictor of CR after neoadjuvant chemotherapy	Kawate et al. 2013
	Diagnostic biomarker	Oda et al. 2012/Kawate et al. 2015
Acute leukemia	Leukemogenesis and/or disease progression	Liu et al. 2008
Cervical cancer	Biomarker of disease progression	Arnouk et al. 2009
Papillary thyroid cancer	Diagnostic biomarker	Giusti et al. 2008
Pancreatic ductal adenocarcinoma (PDAC)	Tumor differentiation	Tian et al. 2008
	Tumor invasion and poor prognosis	He et al. 2012
Pancreatic cancer	Tumor differentiation and overall survival	Chen et al. 2012
Laryngeal squamous cell carcinoma	Tumor stage, cell differentiation, overall survival	Shen et al. 2011
Oral squamous cell carcinoma	Proliferation and invasion	Xu et al. 2016
Esophageal squamous cell carcinoma (ESCC)	Overall survival	Yuen et al. 2008
Glottic squamous cell carcinoma	Overall survival	Zhu et al. 2010

TRAIL (Hod 2004), gemcitabine (Chen et al. 2012), 2′-benzoyloxycinnamaldehyde (Ismail et al. 2012), dihydroartemisinin (Zhu et al. 2014), etoposide (Liu et al. 2008), Taxol, cisplatin (Zeng et al. 2011), and fenretinide(N-(4-hydroxyphenyl) retinamide, 4-HPR) (Cao et al. 2014). Moreover, evidence from both in vitro and in vivo studies imply that elevated DJ-1 expression is associated with metastasis and differentiation of ESCC (Yuen et al. 2008), PDAC (He et al. 2012), and pancreatic cancer (Chen et al. 2012). Biopsy analysis has proven that DJ-1 expression is positively related to tumor progression and decreased survival in patients with pancreatic cancer (He et al. 2012), cervical cancer (Arnouk et al. 2009), ESCC (Yuen et al. 2008), NSCLC (Kim et al. 2005), and glottic squamous cell carcinoma (Zhu et al.

2010). Interestingly, DJ-1 has been further related to tumor recurrence in clinical studies. The 3-year recurrence of patients with lung cancer expressing high levels of DJ-1 mRNA was increased compared with disease-matched patients who had a low level of DJ-1 mRNA (MacKeigan et al. 2003; Kim et al. 2005), suggesting that DJ-1 expression may be useful for identifying patients who are more likely to relapse. These findings demonstrate that DJ-1 is functionally related to poor prognosis in a majority of tumor types.

13.2.2 DJ-1 Is Secreted by Human Cancers and Can Serve as a Biomarker

Additionally, DJ-1 could be secreted into the peripheral circulation. Accumulating evidence has shown that DJ-1 might be a promising serum marker for cancer diagnosis, monitoring, and prognosis. The level of DJ-1 in pancreatic juice is increased in pancreatic cancer patients compared with healthy controls (Tian et al. 2008); this finding raises the intriguing possibility of extracellular secretion of DJ-1 (Melle et al. 2007). Several studies indicate that serum (Le Naour et al. 2001) and nipple fluid (Oda et al. 2012) obtained from a large fraction of newly diagnosed breast cancer patients had a high concentration of DJ-1, and serum DJ-1 levels were significantly higher in patients with metastatic uveal melanoma compared with patients who were disease free for at least 10 years or with controls (Chen et al. 2015). Interestingly, high levels of DJ-1, likely an isoform at pI 6.3, were examined in the serum of breast cancer patients, indicating that an increase in the isoform with basic pI can distinguish breast cancer patients from those with Parkinson's disease and other neurodegenerative disorders (Kawate et al. 2015). Low DJ-1 protein expression was detected in breast cancer patients with pathological complete remission after neoadjuvant chemotherapy (Kawate et al. 2013). Altogether, these findings indicate that DJ-1 overexpression and secretion is a frequent event in cancer cells and further emphasize its potential prognostic value to predict survival in patients with various tumor types.

13.2.3 DJ-1 Is Involved in Cancer Inflammatory Niche Formation

Tumors are composed of an array of cell types, including not only cancer cells but also non-cancer cells. The tumor microenvironment (niche) is critical for tumor initiation and progression (Borovski et al. 2011; Quail and Joyce 2013). Most recently, evidence indicates that DJ-1 is also involved in creating an inflammatory tumor microenvironment (Chien et al. 2015; Chen et al. 2016). Chien et al. used WT and DJ-1 KO mice and found that IL-1β levels were elevated both in DJ-1 KO mice and in cultured macrophage cells with DJ-1 knockdown (Chien et al. 2015). Elevated IL-1β levels correlated with the formation of melanoma nodules in the lungs of DJ-1 KO mice, and these levels are decreased by treating mice with IL-1β-neutralizing antibodies. Similarly, Chen et al. also observed that DJ-1 deficiency negatively regulates liver progenitor cell (LPC) proliferation by impairing the formation of LPC-associated fibrosis and inflammatory niches (Chen et al. 2016); this finding is consistent with their previous result that DJ-1 has a critical role in initiating an inflammatory response. Therefore, these results introduce the concept that the inflammatory microenvironment generated by dysregulated DJ-1 affects cancer progression.

13.3 Mechanisms of DJ-1 Functions in Tumor Promotion

13.3.1 Oxidative Stress Response

As a redox-sensitive chaperone and an intracellular sensor for oxidative stress (Shendelman et al. 2004), DJ-1 protects tumor cells from various oxidative agents, including UV irradiation (Mo et al. 2008), H_2O_2 (Taira et al. 2004), and anticancer drugs (Zeng et al. 2011; Chen et al. 2012), and determines the sensitivity of cancer cells to chemotherapy-induced apoptosis (Fig. 13.1). Under oxidative stress, DJ-1

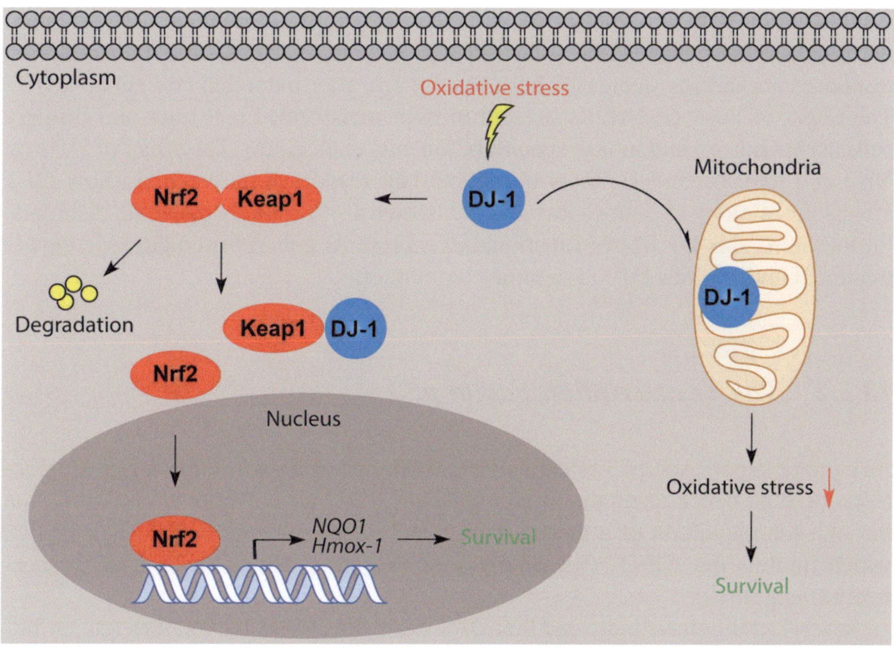

Fig. 13.1 Schematic representation of DJ-1 antioxidative mechanism of action. Under oxidative stress, DJ-1 translocates from the cytoplasm to the mitochondria and stabilizes mitochondrial permeability, providing DJ-1 with stronger cytoprotective activity and reducing oxidative stress. On the other hand, DJ-1 stabilizes the antioxidant transcriptional master regulator Nrf2, which regulates the expression of several genes, including the prototypic Nrf2-regulated antioxidant enzyme NQO1 and Hmox-1

translocates from the cytoplasm to the mitochondria and stabilizes mitochondrial permeability, providing DJ-1 with stronger cytoprotective activity and reducing oxidative stress (Canet-Aviles et al. 2004; Junn et al. 2009). DJ-1 deficiency or suppression of its translocation to the mitochondria sensitizes cells to death caused by oxidative and endoplasmic reticulum stresses and proteasome inhibitors (Yokota et al. 2003; Taira et al. 2004; Canet-Aviles et al. 2004; Zeng et al. 2011; Chen et al. 2012). Furthermore, recent evidence also shows that DJ-1 could regulate mitochondrial fission and mitophagy to protect cells from death (Zhou et al. 2015; Shimizu et al. 2016). Thus, DJ-1, activated by oxidative stress, is critical to maintain the antioxidative activity of the mitochondria and survival of tumor cells under physiological conditions.

In addition to regulating mitochondrial function, some studies further suggest that DJ-1 stabilizes the antioxidant transcriptional master regulator Nrf2, which orchestrates the expression of genes coding for the stress response and antioxidant proteins (Clements et al. 2006). DJ-1 is required for the expression of several genes, including the prototypic Nrf2-regulated antioxidant enzyme NQO1[NAD(P)H quinoneoxidoreductase 1] and Hmox-1, and enzymes that generate antioxidant molecules, such as glutathione. Accumulating evidence has implicated the effects of DJ-1 on Nrf2 in the development of Parkinson's disease and cancer and present potential therapeutic targets (Kahle et al. 2009). DJ-1 stabilizes Nrf2 by preventing its association with its inhibitor protein, Keap1, and the subsequent ubiquitination of Nrf2. Without intact DJ-1, the Nrf2 protein is unstable, and transcriptional responses are thereby decreased both basally and after induction (Gan et al. 2010). The effect of DJ-1 on Nrf2 is present in both transformed cell lines and primary cells across human and mouse species (Clements et al. 2006). The effect of DJ-1 on Nrf2 and its subsequent effects on antioxidant responses may explain how DJ-1 affects the etiology of both cancer and PD, which are seemingly disparate disorders. Furthermore, this DJ-1/Nrf2 functional axis presents a therapeutic target in cancer treatment and justifies DJ-1 as a tumor biomarker.

13.3.2 The Transcription Factor p53

As a tumor suppressor, p53 selectively regulates cell cycle arrest, DNA repair, apoptosis, or senescence to maintain genomic integrity and prevents tumor formation through the regulation of a group of its target genes, such as IGF-BP3 or PTEN, which inhibits the IGF-1/AKT pathway or decreases activation of AKT, respectively (Feng 2010).

Several studies have indicated that DJ-1 exerts its cytoprotection through inhibiting the p53-Bax-caspase pathway (Fan et al. 2008a, b) given the existing link between p53 and DJ-1 (Fig. 13.2). DJ-1 represses p53 transcriptional activity via binding to the DNA-binding region of p53 in a manner that is dependent on the p53 DNA-binding affinity and oxidation of C106 (Kato et al. 2013), thereby interfering with the binding of p53 to the DNA promoter. Overexpression of DJ-1 decreases the

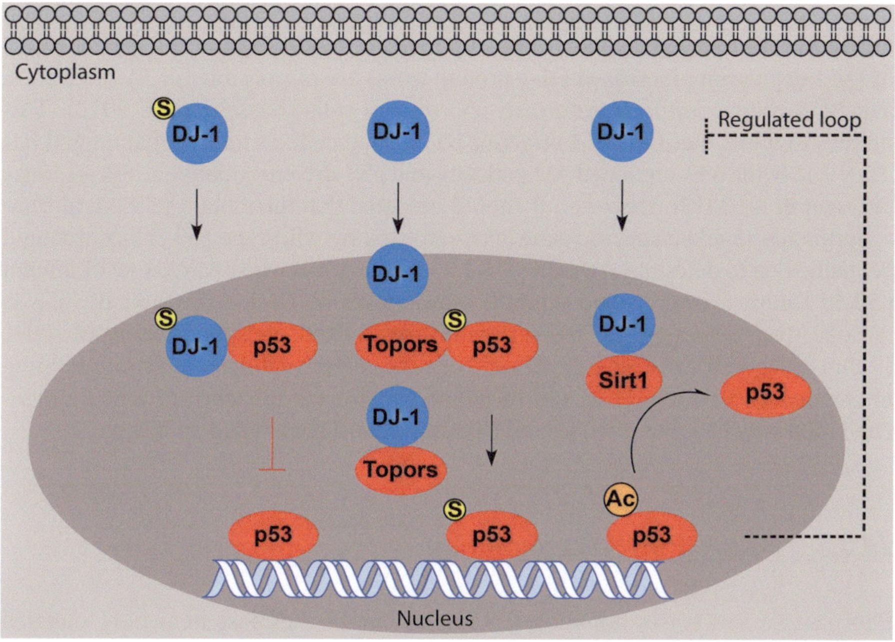

Fig. 13.2 Schematic representation of the effect of DJ-1 on p53 transcription factor. DJ-1 represses p53 transcriptional activity via binding to the DNA-binding region of p53 in a manner that is dependent on the sumoylation of DJ-1. However, other studies have suggested that DJ-1 binds to Topors/p53BP3 in vitro and in vivo, thereby releasing the sumoylated form of p53, which contributes to restoring p53 transcriptional activity. Additionally, DJ-1 could directly bind to Sirt1 and stimulate the deacetylase activity of SIRT1. Subsequently, SIRT1-suppressed transcriptional activity of SIRT1-target p53 was further decreased by DJ-1. Moreover, p53 prevents the accumulation of the DJ-1 protein, whereas loss of p53 leads to stabilization and enhancement of DJ-1 expression. S, sumoylation; Ac, acetylation

expression of Bax and inhibits caspase activation, whereas knockdown of DJ-1 increases Bax protein levels and accelerates caspase-3 activation and cell death induced by UV exposure. In addition, sumoylation is critical for DJ-1 to repress p53 transcriptional activity, whereas DJ-1(K130R), a non-sumoylatable mutant form of DJ-1, shifts from the nucleus to cytoplasm, fails to repress p53 transcriptional activity, and loses its protective function against ultraviolet-induced cell death (Fan et al. 2008a, b). However, other studies have suggested that DJ-1 positively regulates p53 through Topors-mediated sumoylation (Shinbo et al. 2005). DJ-1 binds to Topors/p53BP3 in vitro and in vivo, thereby releasing the sumoylated form of p53, which contributes to restoring p53 transcriptional activity. Finally, new evidence suggests that DJ-1 could directly bind to Sirt1 and stimulate the deacetylase activity of SIRT1. Subsequently, SIRT1-suppressed transcriptional activity of SIRT1-target p53 was further decreased by DJ-1, suggesting a novel method by which DJ-1 regulates p53 activity (Takahashi-Niki et al. 2016).

Moreover, DJ-1 and p53 seem to be tightly "linked"; p53 prevents the accumulation of the DJ-1 protein, whereas loss of p53 leads to stabilization and enhancement of DJ-1 expression. Increased DJ-1 protein levels are responsible for Akt activation and ROS production in transformed p53-mutant cells (Vasseur et al. 2012). This finding suggests that DJ-1 is a target of p53 during cellular transformation and has a key role in the p53-regulated Akt pathway and p53-driven oxidative stress response (Vasseur et al. 2012). However, it should be noted that the mutant p53 might transcriptionally regulate specific genes not activated by wild-type p53. Thus, it would be interesting to determine whether DJ-1 is the specific downstream target of mutant p53 in future research. One possible mechanism of DJ-1 repression by p53 is through the induction of transcription-independent phosphorylation of DJ-1 (Rahman-Roblick et al. 2008). Taken together, these findings confirm the strong interconnection between p53 and DJ-1 and suggest the existence of a finely regulated loop between these two proteins during tumorigenesis and apoptosis.

13.3.3 PTEN/PI3K/Akt Signaling

Phosphatase and tensin homolog (PTEN) is one of the most frequently mutated tumor suppressor genes in human cancers. Using a genetic screen of *Drosophila* gain-of-function mutants, Kim et al. first identified DJ-1 as a suppressor of PTEN function (Kim et al. 2005). These researchers demonstrated that reduced and increased expression levels of DJ-1 lead to hypo- and hyperphosphorylation of PKB/AKT, resulting in the activation and inactivation of survival signals, respectively. Using primary breast cancer samples, they showed that DJ-1 expression correlates negatively with PTEN immunoreactivity and positively with PKB/Akt hyperphosphorylation (Fig. 13.3). Of note, increasing evidence supports that DJ-1 promotes cell survival through PTEN/PI3K/Akt signaling in laryngeal squamous cell carcinoma (Wang et al. 2014). Though mechanistic details delineating how DJ-1 regulates PTEN activity remain unknown, some clues suggest that DJ-1 inhibits PTEN phosphatase activity through direct interaction of DJ-1 with PTEN in a manner dependent on the oxidative status of C106 (Kim et al. 2009). Moreover, Choi et al. proposed that PTEN phosphatase activity is inhibited by DJ-1 via a transnitrosylation reaction. Specifically, DJ-1 is S-nitrosylated; subsequently, the NO group is transferred from DJ-1 to PTEN by transnitrosylation (Choi et al. 2014). Taken together, these studies show that DJ-1 suppresses PTEN-dependent cell death by negatively regulating PTEN phosphatase activity.

PTEN is regulated by DJ-1, and the downstream component of this signaling pathway is also controlled by DJ-1. SG2NA was initially characterized as a nuclear localized tumor antigen whose expression is augmented during the S and G2 phases of cell cycle proteins. Tanti et al. reported that SG2NA protects cells from oxidative stress by recruiting DJ-1 and Akt to the mitochondria (Tanti and Goswami 2014). Most recently, their data also show that SG2NA protects DJ-1 from proteasomal degradation in cancer cells and that ROS enhances the formation of SG2NA, DJ-1,

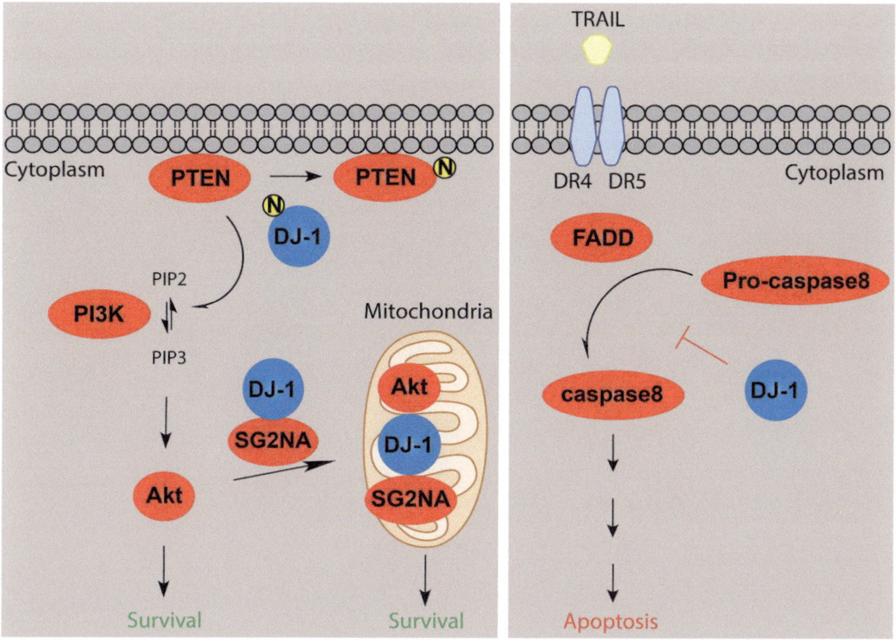

Fig. 13.3 Schematic representation of the effect of DJ-1 on PTEN-PI3K-Akt signaling and death-inducing signaling complex. (Left panel) DJ-1 is a suppressor of PTEN function, and increased expression levels of DJ-1 lead to hyperphosphorylation of PKB/AKT, resulting in the activation of survival signals. Mechanistically, DJ-1 is S-nitrosylated; subsequently, the NO group is transferred from DJ-1 to PTEN by transnitrosylation. Moreover, SG2NA protects cells from oxidative stress by recruiting DJ-1 and Akt to the mitochondria and enhances the formation of SG2NA, DJ-1, and Akt trimerization. N, nitrosylation. (Right panel) DJ-1 inhibits TRAIL-induced apoptosis by competing with pro-caspase-8 to bind to FADD at the death effector domain, thereby repressing the recruitment and activation of pro-caspase-8 to the active form of caspase-8

and Akt trimerization (Tanti et al. 2015). These data support the novel idea that DJ-1 inhibition might be an effective strategy to modulate PTEN/PI3K/Akt signaling in cancer.

13.3.4 MAPK Signaling

The mitogen-activated protein kinase (MAPK) signaling pathways are involved in diverse physiological processes and are critical for the induction of oxidative stress responses. Several components of this signaling, particularly apoptosis signal-regulating kinase 1 (ASK1), c-Jun N-terminal kinase (JNK), p38, and extracellular signal-regulated kinase (ERK), have been directly implicated in oncogenesis.

ASK1, which is finely and positively regulated by the death protein Daxx, is a stress-responsive MAP kinase kinase kinase (MAP3K). Daxx is activated by ROS,

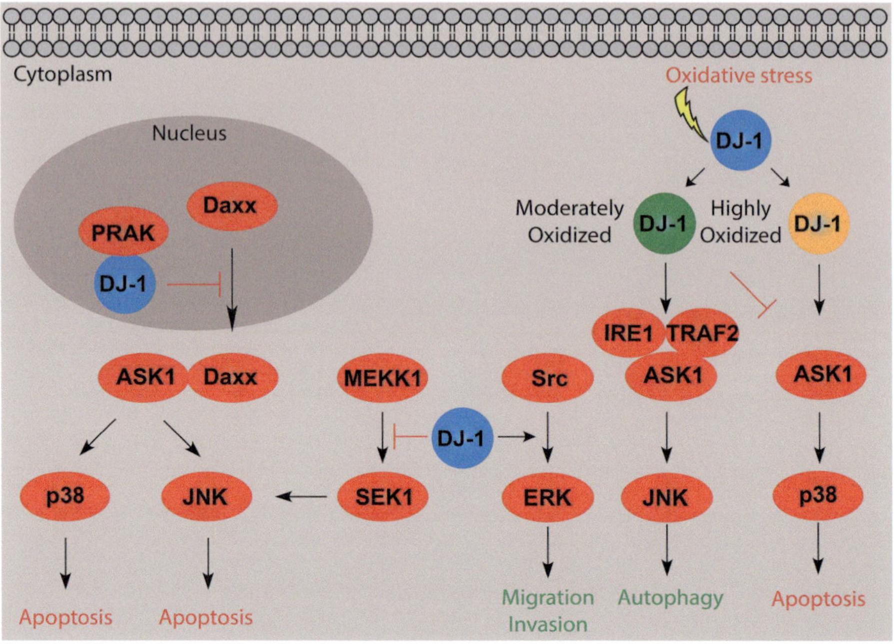

Fig. 13.4 Schematic representation of the effect of DJ-1 on MAPK signaling. DJ-1 sequesters Daxx in the nucleus, preventing it from gaining access to the cytoplasm, from binding to and activating its effector kinase ASK1. On the other hand, DJ-1 physically binds to MEKK1 and sequesters MEKK1 within the cytoplasm, and suppresses the downstream activation of SEK1 and JNK1. Moreover, we found that under mild oxidative stress, moderate oxidation of DJ-1 is recruited to inhibit the activity of ASK1 and maintains cell viability by activating autophagy. Under a lethal level of oxidative stress, excessive oxidized DJ-1 dissociates from ASK1 and activates it, thereby initiating p38 activation and apoptosis. In addition to regulating cell survival, DJ-1 is associated with cancer cell migration and invasion via activating the ERK/SRC phosphorylation cascade

increases the interaction between Daxx and ASK1, promotes ASK1 oligomerization, and subsequently mediates downstream activation of JNK and p38, leading to apoptosis (Runchel et al. 2011). A yeast two-hybrid screen found that the death protein Daxx is a DJ-1-interacting partner (Junn et al. 2005). Further, mechanistic studies have shown that wild-type DJ-1 sequesters Daxx in the nucleus, preventing it from gaining access to the cytoplasm, from binding to and activating its effector kinase ASK1, and therefore from triggering the ensuing death pathway (Fig. 13.4). All these steps are impaired by the disease-causing L166P mutant isoform of DJ-1 (Junn et al. 2005; Waak et al. 2009). Moreover, a recent study has shown that p38 regulated/activated kinase (PRAK), a core member of the p38 MAPK signal transduction pathway, was also directly regulated by DJ-1. DJ-1 colocalized with PRAK in the nucleus. Under stress conditions, the majority of endogenous DJ-1 in PRAK(+/+) cells still remained in the nucleus, whereas most DJ-1 in PRAK(−/−) cells translocated from the nucleus into the cytoplasm, indicating that PRAK is essential for DJ-1 to localize in the nucleus. Interestingly, PRAK-associated phosphorylation of DJ-1 was observed in vitro and in vivo in H_2O_2-challenged PRAK(+/+) cells. Cytoplasmic translocation of DJ-1 to the nucleus is lost in in H_2O_2-treated PRAK(−/−) cells, and as a result, Daxx is localized to the cytoplasm and triggers cell death (Tang et al. 2014). Overall, these findings highlight that DJ-1 prevents cell death by sequestering Daxx in the nucleus, which is facilitated by PRAK in response to oxidative stress.

Furthermore, there is evidence to suggest that DJ-1 protects against UV-induced cell death through the mitogen-activated protein kinase/extracellular signal-regulated kinase kinase kinase 1 (MEKK1)/SEK1/JNK1 signaling cascade (Mo et al. 2008). Specifically, DJ-1 physically binds to MEKK1 and sequesters MEKK1 within the cytoplasm, thus blocking UV-induced translocation of MEKK1 into the nucleus and suppressing the downstream activation of SEK1 and JNK1. The L166P mutant of DJ-1 leads to an impaired physical association with MEKK1 and facilitates its translocation to the nucleus, whereas both L166P mutant and DJ-1 knockdown by siRNA render the cells highly susceptible to UV-induced activation of the MEKK1/SEK1/JNK1 signaling pathway and cell death (Mo et al. 2008). These findings suggest that several components of MAPK signaling are tightly regulated by DJ-1, and this axis is a critical mechanism by which DJ-1 exerts its cytoprotective function.

We have found that MAPK signaling is critical for the biological function of DJ-1 in cell survival, and we have linked the different branches of MAPK signaling regulated by DJ-1 to two types of programed cell death (autophagy and apoptosis) (Cao et al. 2014). Specifically, when we treated cancer cells with 4-HPR, a chemopreventive reagent tested in clinical trials that triggers ROS in tumor cells, we found that a modest level of ROS promotes tumor survival, whereas an excessive level paradoxically directs cells toward apoptosis. A mechanistic study revealed that under mild oxidative stress induced by low concentrations of 4-HPR, moderate oxidation of DJ-1 is recruited to inhibit the activity of ASK1 and maintain cell viability by activating autophagy. Under a lethal level of oxidative stress, excessive oxidized DJ-1 dissociates from ASK1 and activates it, thereby initiating p38 activation and apoptosis (Cao et al. 2014). Therefore, our results reveal that the different oxidation states of DJ-1 function as a cellular redox sensor of ROS caused by 4-HPR chemotherapy and determine the cancer cell fate of autophagy or apoptosis. In conclusion, DJ-1 partially affects cell death decisions via modulating autophagy and determines the progression and treatment of cancer.

In addition to regulating cell survival, DJ-1 is associated with cancer cell migration and invasion via activating the ERK/SRC phosphorylation cascade. Knockdown of DJ-1 expression resulted in decreased ERK1/2 and SRC phosphorylation and reduced invasion and the cell migration potential in pancreatic cancer cells (He et al. 2012). These findings are consistent with another report that shows the importance of the ERK1/2 pathway during pancreatic cancer metastasis (Tan et al. 2005). Overall, these reports show that DJ-1 prevents cell death and promotes invasion/migration by negative or positive regulation of the MAPK pathway.

13.3.5 Transcription Factor HIF-1

The microenvironment surrounding a tumor is characterized by regions of fluctuating or chronic hypoxia and nutrient depletion. Hypoxia-inducible factor-1α (HIF-1α), a master transcription factor that is stabilized in hypoxic conditions, regulates the expression of multiple target genes to help tumor cells adapt to hypoxia. Given that the expression of HIF-1α subunits during hypoxia is at least partly dependent on the PI3K/Akt/mTOR pathway-controlled translation step (Zhong et al. 2000), it is expected that DJ-1 is involved in regulating HIF-1α transcriptional activity in hypoxic conditions. Loss of DJ-1 decreases HIF-1α expression in neuroblastoma cells and the transcription of a variety of HIF-1-responsive genes during hypoxia in human osteosarcoma cells and transformed mouse fibroblasts (Vasseur et al. 2009). Another report identifies the Von Hippel-Lindau (VHL) protein as a critical DJ-1-interacting protein (Lonser et al. 2003). DJ-1 negatively regulates VHL ubiquitination activity of the α-subunit of hypoxia-inducible factor-1 (HIF-1α) by inhibiting the HIF–VHL interaction. Consistent with this observation, DJ-1 deficiency reduces HIF-1α levels in models of both hypoxia and oxidative stress, which are two stresses known to stabilize HIF-1α (Parsanejad et al. 2014). These results suggest that both the translation and degradation of HIF-1α are regulated by DJ-1, which might be related to its role in tumorigenesis.

13.3.6 Death-Inducing Signaling Complex (DISC)

The death-inducing signaling complex (DISC) is a multi-protein complex formed by members of the "death receptor" family of apoptosis-inducing cellular receptors. The DISC is composed of a death receptor, Fas-associated protein death domain (FADD), and caspase-8. Formation of DISC triggers auto-processing and activation of caspase-8 that subsequently leads to apoptosis (Muzio et al. 1997). Forced exogenous expression of DJ-1 significantly suppressed cell death induced by TRAIL (tumor necrosis factor-related apoptosis-inducing ligand) in human neoplastic thyroid cells (Zhang et al. 2008) (Fig. 13.3). Fu et al. revealed that DJ-1 inhibits TRAIL-induced apoptosis by competing with pro-caspase-8 to bind to FADD at the death effector domain, thereby repressing the recruitment and activation of pro-caspase-8 to the active form of caspase-8 (Fu et al. 2012). Thus, these studies suggest that DJ-1 protects against TRAIL-induced apoptosis through the regulation of DISC formation.

13.3.7 20S Proteasome

In general, both unstructured proteins and relatively unfolded regions of proteins are substrates of the 20S proteasome in cells. This class of proteins can be degraded by two proteasomal degradation mechanisms: ubiquitin-dependent degradation by the 26S proteasome and ubiquitin-independent degradation by the 20S proteasome (Coux et al. 1996). Unlike degradation by the 26S proteasome, degradation by the 20S proteasome is a passive process that relies on the inherent disorder of the protein being degraded. Moscovitz et al. shows that DJ-1 physically binds the 20S proteasome and inhibits its activity, rescuing partially unfolded proteins from degradation, such as α-synuclein and p53 (Moscovitz et al. 2015). These results suggest a regulatory circuit in which DJ-1, under conditions of oxidative stress, both upregulates and inhibits the 20S proteasome, providing a rigorous control mechanism at a time when the 20S proteasome becomes the major proteolytic machinery.

13.4 DJ-1 Could Be a Promising Therapeutic Target for Cancer

Although it has been accepted that DJ-1 can act as a molecular biomarker for cancer diagnosis, it is unknown whether DJ-1 also could serve as a target for cancer treatment. Some studies have shown that modulating DJ-1 protein levels by chemical or biological methods affects the outcome of chemotherapy, suggesting that suppression of its tumor protection function may inhibit tumor growth, serving as a therapeutic target in cancer. In principle, DJ-1 interventions could be afforded by (i) modulating transcription/translation process, for instance, through an antisense approach, (ii) inhibiting protein complex formation, and (iii) interfering with protein function. We will comment on those avenues where progress can be made in DJ-1 intervention.

13.4.1 Blocking DJ-1 Protein Expression by RNA Interference

RNA interference (RNAi) is a ubiquitous pathway that regulates gene expression by precisely inhibiting the expression of targeted mRNAs, accelerating their degradation and suppressing their translation into protein. More than 20 RNAi-based therapeutics are currently in clinical trials. Previous reports found that DJ-1 knockdown by small interfering RNA (siRNA) leads to impaired cell growth and enhanced sensitivity of tumor cells to chemotherapeutic drugs, such as TRAIL (Hod 2004), gemcitabine (Chen et al. 2012), 2′-benzoyloxycinnamaldehyde (MacKeigan et al. 2003), dihydroartemisinin (Zhu et al. 2014), etoposide (Liu et al. 2008), Taxol, and cisplatin (Zeng et al. 2011). Our lab has also confirmed that depletion of DJ-1 renders tumor cells susceptible to the cancer chemopreventive and therapeutic agent 4-HPR both in vitro and in vivo (Cao et al. 2014). Moreover, DJ-1 is related to tumor metastasis and differentiation of ESCC (Yuen et al. 2008), PDAC (He et al. 2012), and pancreatic cancer (Chen et al. 2012), whereas knockdown of DJ-1 expression in PDAC cell lines reduces cell migration and invasion potential in vitro and inhibits metastasis in vivo (He et al. 2012). DJ-1 is a potential target for drug development for cancer, and its downregulation potentiates the antitumor effect of chemotherapeutic and chemosensitizing agents. Remarkably, siRNA-mediated silencing of DJ-1 expression substantially enhances the anticancer activity in several types of cancers. However, despite the success in preclinical reports, initial difficulties remain to be solved in achieving efficacious results with RNAi-based DJ-1 intervention therapeutics without toxic side effects, advances in delivery, and improved chemistry.

In addition, there may be more compounds that target DJ-1 and inhibit its expression directly to inhibit tumor progression. For example, in a preclinical model of ovarian cancer (OVCA), DJ-1 expression increased with OVCA progression, and Ashwagandha (ASH, a natural product) reduced the expression of DJ-1 and OVCA metastasis (Gupta et al. 2014). ASH may be used to enhance cytotoxicity of conventional chemotherapeutics by preventing DJ-1 expression.

13.4.2 Disrupting DJ-1-Associated Protein Complex Formation

DJ-1 is a dimeric cytoprotective protein that defends against oxidative stress and preserves mitochondrial function. Dimerization of DJ-1 is thought to be essential for this function, as some disease-associated mutations, such as L166P, cause poor folding and disrupt the DJ-1 dimer (Moore et al. 2003), thereby promoting its degradation by the ubiquitin-proteasome system (Miller et al. 2003). One hot spot identified on the surface of DJ-1 provides new hypotheses for binding sites of key players in the dimer interface, where a pharmacological chaperone could bind to increase or decrease the stability of the dimeric structure (Landon et al. 2009). Therefore, a small molecule disrupting DJ-1 dimerization may contribute to anticancer activity in tumors with DJ-1 overexpression.

Moreover, increasing evidence suggests that several proteins, such as ASK1 (Cao et al. 2014), Trx (Im et al. 2010), Daxx (Junn et al. 2005), PTEN (Kim et al. 2005), and MEKK1 (Mo et al. 2008), which are involved in either MAPK signaling or PI3K/Akt signaling, can directly bind to DJ-1. Whether these binding partners bind to the same domain of DJ-1 is largely unknown, yet research on DJ-1 still provides the new opportunity to modulate these key signaling pathways in cancer cells by developing an inhibitor targeting the DJ-1-associated protein complex formation.

13.4.3 Interference with Protein Function

A conserved cysteine residue in DJ-1 (Cys106) is both functionally essential and subject to oxidation to cysteine sulfinate and cysteine sulfonate (Wilson 2011). Accumulating evidence has shown that the oxidative modification of Cys106 allows DJ-1 to act as a sensor of cellular redox homeostasis and to participate in cytoprotective signaling pathways in the cell by effecting DJ-1 localization to the mitochondria (Blackinton et al. 2009) and p53-binding affinity and p53-dependent repression of gene transcription (Kato et al. 2013). In addition, oxidation-induced mitochondrial relocalization of DJ-1 and protection against cell death were abrogated in C106A (Canet-Aviles et al. 2004) and C106S (Takahashi-Niki et al. 2004).

Considering the clinical limitations of drug delivery of siRNA therapeutics and the identification of a ligand-binding hot spot for DJ-1 in the region containing C106 (Landon et al. 2009), the design of small molecular compounds targeting the cytoprotective function of DJ-1 may be an alternative method for inactivating DJ-1 to improve the efficacy of chemotherapies. A recent study has identified a DJ-1 modulator that may bind to the SO_2H-oxidized C106 region in the endogenous DJ-1 protein (Yanagida et al. 2009). Using the X-ray crystal structure and virtual screening (*in silico*) of oxidized DJ-1 at C106 with the SO_2H form and the three-dimensional coordinate data of approximately 30,000 chemical compounds in the University Compound Project (UCP) at the Foundation for Education of Science and Technology, 2-[3-(benzyloxy)-4-methoxyphenyl]-N-[2-(7-methoxy-1,3-benzodioxol-5-yl)ethyl]acetamide (UCP0054278) was identified as a stimulatory modulator with the highest binding constant (docking score) toward the pocket of the SO_2H-oxidized C106 region. UCP0054278 significantly inhibited H_2O_2-induced cell death and the production of ROS in normal neuroblastoma SH-SY5Y cells but not in DJ-1-knockdown cells (Yanagida et al. 2009). Overall, these results provide compelling evidence that the design of small molecular compounds that selectively bind to DJ-1 to prevent oxidization at its C106 residue is an effective strategy to effectively improve the outcome of chemotherapies.

13.5 Conclusions

In summary, DJ-1, which is overexpressed in various human malignances, is closely related to the proliferation, metastasis, occurrence, and prognosis of malignant tumors. DJ-1 exerts its cytoprotective and tumorigenic functions through negative regulation of tumor suppression genes and the careful modulation of autophagy and apoptosis. Our current knowledge suggests that DJ-1 serum levels may be a potential biomarker for several types of human tumors and that DJ-1 plays critical roles in tumor chemoresistance, supporting the development of chemotherapeutic approaches targeting this oncogene.

References

Advedissian T, Deshayes F, Poirier F, Viguier M, Richarme G (2016) The Parkinsonism-associated protein DJ-1/Park7 prevents glycation damage in human keratinocyte. Biochem Biophys Res Comm 473:87–91

Arnouk H, Merkley MA, Podolsky RH, Stoppler H, Santos C, Alvarez M, Mariategui J, Ferris D, Lee JR, Dynan WS (2009) Characterization of molecular markers indicative of cervical cancer progression. Proteomics Clin Appl 3:516–527

Blackinton J, Lakshminarasimhan M, Thomas KJ, Ahmad R, Greggio E, Raza AS, Cookson MR, Wilson MA (2009) Formation of a stabilized cysteine sulfinic acid is critical for the mitochondrial function of the Parkinsonism protein DJ-1. J Biol Chem 284:6476–6485

Bonifati V, Rizzu P, van Baren MJ, Schaap O, Breedveld GJ, Krieger E, Dekker MC, Squitieri F, Ibanez P, Joosse M, van Dongen JW, Vanacore N, van Swieten JC, Brice A, Meco G, van Duijn CM, Oostra BA, Heutink P (2003) Mutations in the DJ-1 gene associated with autosomal recessive early-onset Parkinsonism. Science 299:256–259

Borovski T, De Sousa EMF, Vermeulen L, Medema JP (2011) Cancer stem cell niche: the place to be. Cancer Res 71:634–639

Canet-Aviles RM, Wilson MA, Miller DW, Ahmad R, McLendon C, Bandyopadhyay S, Baptista MJ, Ringe D, Petsko GA, Cookson MR (2004) The Parkinson's disease protein DJ-1 is neuroprotective due to cysteine-sulfinic acid-driven mitochondrial localization. Proc Natl Acad Sci U S A 101:9103–9108

Cao J, Ying M, Xie N, Lin G, Dong R, Zhang J, Yan H, Yang X, He Q, Yang B (2014) The oxidation states of DJ-1 dictate the cell fate in response to oxidative stress triggered by 4-hpr: autophagy or apoptosis? Antioxid Redox Signal 21:1443–1459

Chen Y, Kang M, Lu W, Guo Q, Zhang B, Xie Q, Wu Y (2012) DJ-1, a novel biomarker and a selected target gene for overcoming chemoresistance in pancreatic cancer. J Cancer Res Clin Oncol 138:1463–1474

Chen LL, Tian JJ, Su L, Jing Y, Zhang SC, Zhang HX, Wang XQ, Zhu CB (2015) DJ-1: a promising marker in metastatic uveal melanoma. J Cancer Res Clin Oncol 141:315–321

Chen L, Luo M, Sun X, Qin J, Yu C, Wen Y, Zhang Q, Gu J, Xia Q, Kong X (2016) DJ-1 deficiency attenuates expansion of liver progenitor cells through modulating the inflammatory and fibrogenic niches. Cell Death Dis 7:e2257

Chien CH, Lee MJ, Liou HC, Liou HH, Fu WM (2015) Local immunosuppressive microenvironment enhances migration of melanoma cells to lungs in DJ-1 knockout mice. PloS one 10(2) e0115827

Choi MS, Nakamura T, Cho SJ, Han XM, Holland EA, Qu J, Petsko GA, Yates JR, Liddington RC, Lipton SA (2014) Transnitrosylation from DJ-1 to PTEN attenuates neuronal cell death in Parkinson's disease models. J Neurosci 34:15123–15131

Clements CM, McNally RS, Conti BJ, Mak TW, Ting JPY (2006) DJ-1, a cancer- and Parkinson's disease-associated protein, stabilizes the antioxidant transcriptional master regulator Nrf2. Proc Natl Acad Sci U S A 103(41):15091–15096. https://doi.org/10.1073/pnas.0607260103

Coux O, Tanaka K, Goldberg AL (1996) Structure and functions of the 20S and 26S proteasomes. Annu Rev Biochem 65:801–847

Davidson B, Hadar R, Schlossberg A, Sternlicht T, Slipicevic A, Skrede M, Risberg B, Florenes VA, Kopolovic J, Reich R (2008) Expression and clinical role of DJ-1, a negative regulator of PTEN, in ovarian carcinoma. Hum Pathol 39:87–95

Fan J, Ren HG, Fei EK, Jia NL, Ying Z, Jiang P, MA W, Wang GH (2008a) Sumoylation is critical for DJ-1 to repress p53 transcriptional activity. FEBS Lett 582:1151–1156

Fan J, Ren HG, Jia NL, Fei E, Zhou T, Jiang P, Wu M, Wang GH (2008b) DJ-1 decreases Bax expression through repressing p53 transcriptional activity. J Biol Chem 283:4022–4030

Feng ZH (2010) p53 regulation of the IGF-1/AKT/mTOR pathways and the endosomal compartment. Cold Spring Harb Perspect Biol 2:a001057

Fu K, Ren H, Wang Y, Fei E, Wang H, Wang G (2012) DJ-1 inhibits TRAIL-induced apoptosis by blocking pro-caspase-8 recruitment to FADD. Oncogene 31:1311–1322

Gan L, Johnson DA, Johnson JA (2010) Keap1-Nrf2 activation in the presence and absence of DJ-1. Eur J Neurosci 31:967–977

Giusti L, Iacconi P, Ciregia F, Giannaccini G, Donatini GL, Basolo F, Miccoli P, Pinchera A, Lucacchini A (2008) Fine-needle aspiration of thyroid nodules: proteomic analysis to identify cancer biomarkers. J Proteome Res 7:4079–4088

Gupta N, Bahr JM, Sharma S, Bitterman P, Guirguis AS, Basu S, Barua A (2014) Inhibition of ovarian tumor-associated DJ-1 expression and tumor progression. J Clin Oncol 32:e16527

He X, Zheng Z, Li J, Ben Q, Liu J, Zhang J, Ji J, Yu B, Chen X, Su L, Zhou L, Liu B, Yuan Y (2012) DJ-1 promotes invasion and metastasis of pancreatic cancer cells by activating SRC/ERK/uPA. Carcinogenesis 33:555–562

Hod Y (2004) Differential control of apoptosis by DJ-1 in prostate benign and cancer cells. J Cell Biochem 92:1221–1233

Hod Y, Pentyala SN, Whyard TC, El-Maghrabi MR (1999) Identification and characterization of a novel protein that regulates RNA-protein interaction. J Cell Biochem 72:435–444

Im JY, Lee KW, Junn E, Mouradian MM (2010) DJ-1 protects against oxidative damage by regulating the thioredoxin/ASK1 complex. Neurosci Res 67:203–208

Ismail IA, Kang HS, Lee HJ, Kwon BM, Hong SH (2012) 2′-Benzoyloxycinnamaldehyde-mediated DJ-1 upregulation protects MCF-7 cells from mitochondrial damage. Biol Pharm Bull 35:895–902

Junn E, Taniguchi H, Jeong BS, Zhao X, Ichijo H, Mouradian MM (2005) Interaction of DJ-1 with Daxx inhibits apoptosis signal-regulating kinase 1 activity and cell death. Proc Natl Acad Sci U S A 102:9691–9696

Junn E, Jang WH, Zhao X, Jeong BS, Mouradian MM (2009) Mitochondrial localization of DJ-1 leads to enhanced Neuroprotection. J Neurosci Res 87:123–129

Kahle PJ, Waak J, Gasser T (2009) DJ-1 and prevention of oxidative stress in Parkinson's disease and other age-related disorders. Free Radic Biol Med 47:1354–1361

Kato I, Maita H, Takahashi-Niki K, Saito Y, Noguchi N, Iguchi-Ariga SMM, Ariga H (2013) Oxidized DJ-1 inhibits p53 by sequestering p53 from promoters in a DNA-binding affinity-dependent manner. Mol Cell Biol 33:340–359

Kawate T, Iwaya K, Kikuchi R, Kaise H, Oda M, Sato E, Hiroi S, Matsubara O, Kohno N (2013) DJ-1 protein expression as a predictor of pathological complete remission after neoadjuvant chemotherapy in breast cancer patients. Breast Cancer Res Treat 139:51–59

Kawate T, Iwaya K, Koshikawa K, Moriya T, Yamasaki T, Hasegawa S, Kaise H, Fujita T, Matsuo H, Nakamura T, Ishikawa T, Hiroi S, Iguchi-Ariga SMM, Ariga H, Murota K, Fujimori M, Yamamoto J, Matsubara O, Kohno N (2015) High levels of DJ-1 protein and isoelectric point 6.3 Isoform in sera of breast cancer patients. Cancer Sci 106:938–943

Kim RH, Peters M, Jang Y, Shi W, Pintilie M, Fletcher GC, DeLuca C, Liepa J, Zhou L, Snow B, Binari RC, Manoukian AS, Bray MR, Liu FF, Tsao MS, Mak TW (2005) DJ-1, a novel regulator of the tumor suppressor PTEN. Cancer Cell 7:263–273

Kim YC, Kitaura H, Taira T, Iguchi-Ariga SMM, Ariga H (2009) Oxidation of DJ-1-dependent cell transformation through direct binding of DJ-1 to PTEN. Int J Oncol 35:1331–1341

Landon MR, Lieberman RL, Hoang QQ, SL J, Caaveiro JMM, Orwig SD, Kozakov D, Brenke R, Chuang GY, Beglov D, Vajda S, Petsko GA, Ringe D (2009) Detection of ligand binding hot spots on protein surfaces via fragment-based methods: application to DJ-1 and glucocerebrosidase. J Comput Aided Mol Des 23:491–500

Le Naour F, Misek DE, Krause MC, Deneux L, Giordano TJ, Scholl S, Hanash SM (2001) Proteomics-based identification of RS/DJ-1 as a novel circulating tumor antigen in breast cancer. Clin Cancer Res 7:3328–3335

Liu H, Wang M, Li M, Wang D, Rao Q, Wang Y, Xu Z, Wang J (2008) Expression and role of DJ-1 in leukemia. Biochem Biophys Res Comm 375:477–483

Liu S, Yang Z, Wei H, Shen W, Liu J, Yin Q, Li X, Yi J (2010) Increased DJ-1 and its prognostic significance in hepatocellular carcinoma. Hepato-Gastroenterology 57:1247–1256

Lonser RR, Glenn GM, Walther M, Chew EY, Libutti SK, Linehan WM, Oldfield EH (2003) von Hippel-Lindau disease. Lancet 361:2059–2067

MacKeigan JP, Clements CM, Lich JD, Pope RM, Hod Y, Ting JP (2003) Proteomic profiling drug-induced apoptosis in non-small cell lung carcinoma: identification of RS/DJ-1 and RhoGDIalpha. Cancer Res 63:6928–6934

Mccoy MK, Cookson MR (2011) DJ-1 regulation of mitochondrial function and autophagy through oxidative stress. Autophagy 7:531–532

Melle C, Ernst G, Escher N, Hartmann D, Schimmel B, Bleul A, Thieme H, Kaufmann R, Felix K, Friess HM, Settmacher U, Hommann M, Richter KK, Daffner W, Taubig H, Manger T,

Claussen U, von Eggeling F (2007) Protein profiling of microdissected pancreas carcinoma and identification of HSP27 as a potential serum marker. Clin Chem 53:629–635
Miller DW, Ahmad R, Hague S, Baptista MJ, Canet-Aviles R, McLendon C, Carter DM, Zhu PP, Stadler J, Chandran J, Klinefelter GR, Blackstone C, Cookson MR (2003) L166P mutant DJ-1, causative for recessive Parkinson's disease, is degraded through the ubiquitin-proteasome system. J Biol Chem 278:36588–36595
Mo JS, Kim MY, Ann EJ, Hong JA, Park HS (2008) DJ-1 modulates UV-induced oxidative stress signaling through the suppression of MEKK1 and cell death. Cell Death Diff 15:1030–1041
Moore DJ, Zhang L, Dawson TM, Dawson VL (2003) A missense mutation (L166P) in DJ-1, linked to familial Parkinson's disease, confers reduced protein stability and impairs homo-oligomerization. J Neurochem 87:1558–1567
Moscovitz O, Ben-Nissan G, Fainer I, Pollack D, Mizrachi L, Sharon M (2015) The Parkinson's-associated protein DJ-1 regulates the 20S proteasome. Nat Commun 6:6609
Muzio M, Salvesen GS, Dixit VM (1997) FLICE induced apoptosis in a cell-free system – cleavage of caspase zymogens. J Biol Chem 272:2952–2956
Nagakubo D, Taira T, Kitaura H, Ikeda M, Tamai K, Iguchi-Ariga SMM, Ariga H (1997) DJ-1, a novel oncogene which transforms mouse NIH3T3 cells in cooperation with ras. Biochem Biophys Res Comm 23:509–513
Oda M, Makita M, Iwaya K, Akiyama F, Kohno N, Tsuchiya B, Iwase T, Matsubara O (2012) High levels of DJ-1 protein in nipple fluid of patients with breast cancer. Cancer Sci 103:1172–1176
Okada M, Matsumoto K, Niki T, Taira T, Iguchi-Ariga SMM, Ariga H (2002) DJ-1, a target protein for an endocrine disrupter, participates in the fertilization in mice. Biol Pharm Bull 25:853–856
Olzmann JA, Brown K, Wilkinson KD, Rees HD, Huai Q, Ke H, Levey AI, Li L, Chin LS (2004) Familial Parkinson's disease-associated L166P mutation disrupts DJ-1 protein folding and function. J Biol Chem 279:8506–8515
Pardo M, Garcia A, Thomas B, Pineiro A, Akoulitchev A, Dwek RA, Zitzmann N (2006) The characterization of the invasion phenotype of uveal melanoma tumour cells shows the presence of MUC18 and HMG-1 metastasis markers and leads to the identification of DJ-1 as a potential serum biomarker. Int J Cancer 119:1014–1022
Parsanejad M, Zhang Y, DB Q, Irrcher I, Rousseaux MWC, Aleyasin H, Kamkar F, Callaghan S, Slack RS, Mak TW, Lee S, Figeys D, Park DS (2014) Regulation of the VHL/HIF-1 pathway by DJ-1. J Neurosci 34:8043–8050
Quail DF, Joyce JA (2013) Microenvironmental regulation of tumor progression and metastasis. Nat Med 19:1423–1437
Rahman-Roblick R, Hellman U, Becker S, Bader FG, Auer G, Wiman KG, Roblick UJ (2008) Proteomic identification of p53-dependent protein phosphorylation. Oncogene 27:4854–4859
Richarme G, Mihoub M, Dairou J, Bui LC, Leger T, Lamouri A (2015) Parkinsonism-associated protein DJ-1/Park7 is a major protein deglycase that repairs methylglyoxal- and glyoxal-glycated cysteine, arginine, and lysine residues. J Biol Chem 290:1885–1897
Runchel C, Matsuzawa A, Ichijo H (2011) Mitogen-activated protein kinases in mammalian oxidative stress responses. Antioxid Redox Signal 15:205–218
Shen Z, Ren Y, Ye D, Guo J, Kang C, Ding H (2011) Significance and relationship between DJ-1 gene and surviving gene expression in laryngeal carcinoma. Eur J Histochem 55:e9
Shendelman S, Jonason A, Martinat C, Leete T, Abeliovich A (2004) DJ-1 is a redox-dependent molecular chaperone that inhibits alpha-synuclein aggregate formation. PLoS Biol 2:1764–1773
Shimizu Y, Lambert JP, Nicholson CK, Kim JJ, Wolfson DW, Cho HC, Husain A, Naqvi N, Chin LS, Li L, Calvert JW (2016) DJ-1 protects the heart against ischemia-reperfusion injury by regulating mitochondrial fission. J Mol Cell Cardiol 97:56–66
Shinbo Y, Taira T, Niki T, Iguchi-Ariga SMM, Ariga H (2005) DJ-1 restores p53 transcription activity inhibited by Topors/p53BP3. Int J Oncol 26:641–648

Sitaram RT, Cairney CJ, Grabowski P, Keith WN, Hallberg B, Ljungberg B, Roos G (2009) The PTEN regulator DJ-1 is associated with hTERT expression in clear cell renal cell carcinoma. Int J Cancer 125:783–790

Srisomsap C, Subhasitanont P, Otto A, Mueller EC, Punyarit P, Wittmann-Liebold B, Svasti J (2002) Detection of cathepsin B up-regulation in neoplastic thyroid tissues by proteomic analysis. Proteomics 2:706–712

Taira T, Takahashi K, Kitagawa R, Iguchi-Ariga SM, Ariga H (2001) Molecular cloning of human and mouse DJ-1 genes and identification of Sp1-dependent activation of the human DJ-1 promoter. Gene 263:285–292

Taira T, Saito Y, Niki T, Iguchi-Ariga SMM, Takahashi K, Ariga H (2004) DJ-1 has a role in antioxidative stress to prevent cell death. EMBO Rep 5:213–218

Takahashi-Niki K, Niki T, Taira T, Iguchi-Ariga SMM, Ariga H (2004) Reduced anti-oxidative stress activities of DJ-1 mutants found in Parkinson's disease patients. Biochem Biophys Res Comm 320:389–397

Takahashi-Niki K, Ganaha Y, Niki T, Nakagawa S, Kato-Ose I, Iguchi-Ariga SMM, Ariga H (2016) DJ-1 activates SIRT1 through its direct binding to SIRT1. Biochem Biophys Res Comm 474:131–136

Tan X, Egami H, Abe M, Nozawa F, Hirota M, Ogawa M (2005) Involvement of MMP-7 in invasion of pancreatic cancer cells through activation of the EGFR mediated MEK-ERK signal transduction pathway. J Clin Pathol 58:1242–1248

Tang J, Liu J, Li X, Zhong Y, Zhong T, Liu Y, Wang JH, Jiang Y (2014) PRAK interacts with DJ-1 and prevents oxidative stress-induced cell death. Oxid Med Cell Long 2014:735618

Tanti GK, Goswami SK (2014) SG2NA recruits DJ-1 and Akt into the mitochondria and membrane to protect cells from oxidative damage. Free Radic Biol Med 75:1–13

Tanti GK, Pandey S, Goswami SK (2015) SG2NA enhances cancer cell survival by stabilizing DJ-1 and thus activating Akt. Biochem Biophys Res Comm 463:524–531

Tao X, Tong L (2003) Crystal structure of human DJ-1, a protein associated with early onset Parkinson's disease. J Biol Chem 278:31372–31379

Tian M, Cui YZ, Song GH, Zong MJ, Zhou XY, Chen Y, Han JX (2008) Proteomic analysis identifies MMP-9, DJ-1 and A1BG as overexpressed proteins in pancreatic juice from pancreatic ductal adenocarcinoma patients. BMC Cancer 8:241

Vasseur S, Afzal S, Tardivel-Lacombe J, Park DS, Iovanna JL, Mak TW (2009) DJ-1/PARK7 is an important mediator of hypoxia-induced cellular responses. Proc Natl Acad Sci U S A 106:1111–1116

Vasseur S, Afzal S, Tomasini R, Guillaumond F, Tardivel-Lacombe J, Mak TW, Iovanna JL (2012) Consequences of DJ-1 upregulation following p53 loss and cell transformation. Oncogene 31:664–670

Waak J, Weber SS, Gorner K, Schall C, Ichijo H, Stehle T, Kahle PJ (2009) Oxidizable residues mediating protein stability and cytoprotective interaction of DJ-1 with apoptosis signal-regulating kinase 1. J Biol Chem 284:14245–14257

Wang B, Qin H, Wang Y, Chen W, Luo J, Zhu X, Wen W, Lei W (2014) Effect of DJ-1 overexpression on the proliferation, apoptosis, invasion and migration of laryngeal squamous cell carcinoma SNU-46 cells through PI3K/AKT/mTOR. Oncol Rep 32:1108–1116

Wilson MA (2011) The role of cysteine oxidation in DJ-1 function and dysfunction. Antioxid Redox Signal 15:111–122

Wilson MA, Collins JL, Hod Y, Ringe D, Petsko GA (2003) The 1.1-A resolution crystal structure of DJ-1, the protein mutated in autosomal recessive early onset Parkinson's disease. Proc Natl Acad Sci U S A 100:9256–9261

Xu J, Zhong N, Wang H, Elias JE, Kim CY, Woldman I, Pifl C, Gygi SP, Geula C, Yankner BA (2005) The Parkinson's disease-associated DJ-1 protein is a transcriptional co-activator that protects against neuronal apoptosis. Hum Mol Genet 14:1231–1241

Xu S, Ma D, Zhuang R, Sun W, Liu Y, Wen J, Cui L (2016) DJ-1 is upregulated in oral squamous cell carcinoma and promotes oral cancer cell proliferation and invasion. J Cancer 7:1020–1028

Yanagida T, Kitamura Y, Yamane K, Takahashi K, Takata K, Yanagisawa D, Yasui H, Taniguchi T, Taira T, Honda T, Ariga H (2009) Protection against oxidative stress-induced neurodegeneration by a modulator for DJ-1, the wild-type of familial Parkinson's disease-linked PARK7. J Pharmacol Sci 109:463–468

Yokota T, Sugawara K, Ito K, Takahashi R, Ariga H, Mizusawa H (2003) Down regulation of DJ-1 enhances cell death by oxidative stress, ER stress, and proteasome inhibition. Biochem Biophys Res Comm 312:1342–1348

Yoshida K, Sato Y, Yoshiike M, Nozawa S, Ariga H, Iwamoto T (2003) Immunocytochemical localization of DJ-1 in human male reproductive tissue. Mol Reprod Dev 66:391–397

Yuen HF, Chan YP, Law S, Srivastava G, El-Tanani M, Mak TW, Chan KW (2008) DJ-1 could predict worse prognosis in esophageal squamous cell carcinoma. Cancer Epidemiol Biomark Prev 17:3593–3602

Zeng HZ, YQ Q, Zhang WJ, Xiu B, Deng AM, Liang AB (2011) Proteomic analysis identified DJ-1 as a cisplatin resistant marker in non-small cell lung cancer. Int J Mol Sci 12:3489–3499

Zhang HY, Wang HQ, Liu HM, Guan Y, ZX D (2008) Regulation of tumor necrosis factor-related apoptosis-inducing ligand-induced apoptosis by DJ-1 in thyroid cancer cells. Endocr Relat Cancer 15:535–544

Zhong H, Chiles K, Feldser D, Laughner E, Hanrahan C, Georgescu MM, Simons JW, Semenza GL (2000) Modulation of hypoxia-inducible factor 1alpha expression by the epidermal growth factor/phosphatidylinositol 3-kinase/PTEN/AKT/FRAP pathway in human prostate cancer cells: implications for tumor angiogenesis and therapeutics. Cancer Res 60:1541–1545

Zhou M, Xia ZY, Lei SQ, Leng Y, Xue R (2015) Role of mitophagy regulated by Parkin/DJ-1 in remote ischemic postconditioning-induced mitigation of focal cerebral ischemia-reperfusion injury in rats. Eur Rev Med Pharmacol Sci 19:4866–4871

Zhu XL, Wang ZF, Lei WB, Zhuang HW, Jiang HY, Wen WP (2010) DJ-1: a novel independent prognostic marker for survival in glottic squamous cell carcinoma. Cancer Sci 101:1320–1325

Zhu H, Liao SD, Shi JJ, Chang LL, Tong YG, Cao J, YY F, Chen XP, Ying MD, Yang B, He QJ, JJ L (2014) DJ-1 mediates the resistance of cancer cells to dihydroartemisinin through reactive oxygen species removal. Free Radic Biol Med 71:121–132

MIX
Papier aus verantwortungsvollen Quellen
Paper from responsible sources
FSC® C105338

If you have any concerns about our products,
you can contact us on
ProductSafety@springernature.com

In case Publisher is established outside the EU,
the EU authorized representative is:
Springer Nature Customer Service Center GmbH
Europaplatz 3, 69115 Heidelberg, Germany

Printed by Libri Plureos GmbH
in Hamburg, Germany